R P Gilbert and R J Weinacht (Editors)

University of Delaware

Function theoretic methods in differential equations

Pitman Publishing

LONDON · SAN FRANCISCO · MELBOURNE

AMS Subject Classifications: 30A92, 30A97, 45E05, 35C15

PITMAN PUBLISHING LTD
Pitman House, 39 Parker Street, London WC2B 5PB, UK

PITMAN PUBLISHING CORPORATION
6 Davis Drive, Belmont, California 94002, USA

PITMAN PUBLISHING PTY LTD
Pitman House, 158 Bouverie Street, Carlton, Victoria 3053, Australia

PITMAN PUBLISHING
COPP CLARK PUBLISHING
517 Wellington Street West, Toronto M5V 1G1, Canada

SIR ISAAC PITMAN AND SONS LTD
Banda Street, PO Box 46038, Nairobi, Kenya

PITMAN PUBLISHING CO SA (PTY) LTD
Craighall Mews, Jan Smuts Avenue, Craighall Park,
Johannesburg 2001, South Africa

Library of Congress Cataloging in Publication Data
Main entry under title:

Function theoretic methods in differential equations.

(Research notes in mathematics; 8)
Papers contributed in honor of the 85th birthday of N. I. Muskhelishvili.
CONTENTS: Bednar, J.B. and English, C. On the convolution product in discrete function theory.—Begehr, H. Das Schwarzsche Lemma and verwandte Sätze für pseudoanalytische Funktionen. (etc.)
1. Boundary value problems—Addresses, essays, lectures. 2. Analytic functions—Addresses, essays, lectures. 3. Integral operators—Addresses, essays, lectures. I. Gilbert, Robert P., 1932- II. Weinacht, R. J., 1931- III. Muskhelishvili, Nikolaĭ Ivanovich, 1891- IV. Series.
QA379.F86 515′.35 76—22502
ISBN 0—273—00306—2

ISBN 0 273 00306 2

Reproduced and printed by photolithography and bound in Great Britain at The Pitman Press, Bath

3509:97

Preface

This collection of papers arose out of an issue of Applicable Analysis which was planned for the 85th birthday of Academician N.I. Muskhelishvili.* The response to our request for papers was enormous and more than could be handled by the normal publication schedule of the journal. The idea occurred that, in view of Academician Muskhelishvili's many admirers, it would be more appropriate to organize a volume around the central theme of Function Theoretic Methods in Differential Equations.

The authors of the manuscripts greeted this new concept wholeheartedly, and Pitman Publishing encouraged us further in supporting this project.

Since this volume is dedicated to Academician Muskhelishvili, we are happy to be able to include several recent works by members of the Georgian Academy of Science. These manuscripts were communicated to us by the current President of the Georgian Academy, Academician Ilya N. Vekua. One of the Editors wishes to express his gratitude at this time for the hospitality shown him at the Georgian Academy where in June 1974, after scientific discussions with the members of the Tbilisi State University Institutes of Mathematics and the Georgian Academy of Science, the idea of this volume was born.

The manuscripts are divided into three broad categories:

(i) Generalizations of analytic function theory ;

(ii) Integral operators ;

(iii) Boundary value problems.

There have been many extensions of the concept of analyticity, such as the generalized analytic functions of Vekua [5], [10], the pseudoanalytic functions of Bers [1], the hyperanalytic and monogenic functions [3-6], and generalized hyperanalytic functions [6], [8]. The principal idea here is to see what basic properties associated with analyticity are preserved when a related family of functions is considered. For

---

* Just before this volume went to press, the Editors received the sad news of the death of Academician Muskhelishvili.

example, in the theory of generalized analytic functions one studies weak solutions of the equation

$$\frac{\partial w}{\partial \bar{z}} + a(z)w + b(z)\bar{w} = 0, \qquad w = u + iv. \tag{1}$$

Here a, b are in an L_p-space, $p > 2$. It is well known that a similarity principle holds for these solutions, which indicates that the polar and zero behaviour of these functions are the same as those of analytic functions. Furthermore, it can be shown that a Cauchy integral formula, generalized Taylor series, maximum principle, and many other 'analytic' properties carry over to this class of functions.

If one seeks bounded solutions of (1) in the plane, then upon suitable conditions on the coefficients, a pair of functions can be found $\{w^{(1)}(z), w^{(2)}(z)\}$ such that $w^{(1)}(\infty) = -i$, $w^{(2)}(\infty) = 1$. Such a pair Bers has called a 'generating pair' and used it to define a particular type of derivative which is the basis of the function theory he developed. It is interesting that his theory, though built on another basis, yields many of the results of the Vekua [10] theory.

These ideas have recently been extended to elliptic systems of 2n-equations in the plane [6], [8] and to elliptic systems in $\mathbb{R}^n$ [8].

The study of singular integral operators is closely tied to investigations of boundary value problems for elliptic systems. In particular, as applied to the boundary value problems occurring in the theory of functions of a complex variable, potential theory, the theory of elasticity and the theory of fluid mechanics, this subject was to a large extent developed by N.I. Muskhelishvili in collaboration with I.N. Vekua and the Tbilisi Mathematical Institute. We are very pleased to have several contributions from this Institute which deal with their latest findings. Furthermore, it was of exceptional courtesy that the authors of these manuscripts were willing to present their work to the Western mathematical community in English. These articles can be found in Section II, along with related papers on singular and other integral operators.

The final section deals with boundary value problems. Several of these papers make use of integral operator techniques to reduce the boundary value problem to a functional equation. The influence of the Muskhelishvili School can be easily recognized in this collection of works.

Robert P. Gilbert
Richard J. Weinacht
Department of Mathematics and
Institute for Mathematical Sciences
University of Delaware
Newark, Delaware 19711
USA

References

1 L. Bers, Theory of Pseudo-analytic Functions. Lecture Notes, New York University (1953).

2 B.V. Bojarskii, Theory of generalized analytic vectors. Annales Polon. Math. 17 (1966) 281-320.

3 R. Delanghe, On regular-analytic functions with values in a Clifford algebra. Math. Ann. 185 (1970) 91-111.

4 A. Douglis, A function-theoretic approach to elliptic systems of equations in two variables. Comm. Pure Appl. Math. 6 (1953) 259-289.

5 R.P. Gilbert, Constructive Methods for Elliptic Partial Differential Equations. Lecture Notes in Mathematics, Springer, Berlin, (1974) (# 365).

6 R.P. Gilbert and G.N. Hile, Generalized hyperanalytic function theory. Trans. Amer. Math. Soc. 78 (1974) 998-1001.

7 W. Haack and W. Wendland, Lectures on Partial and Pfaffian Differential Equations. Pergamon,London (1971).

8 G.N. Hile, Hypercomplex Function Theory Applied to Partial Differential Equations. Ph.D. Dissertation, Indiana University (1972).

9 N.I. Muskhelishvili, Singular Integral Equations. Noordhoff, Groningen (1953).

10 I.N. Vekua, Generalized Analytic Functions. Pergamon, London (1962).

Contents

J B BEDNAR and C ENGLISH

1 On the convolution product in discrete function theory

1. INTRODUCTION

In attempting to extend the associativity of the convolution products [2] of discrete analytic functions to convolution regions more general than rectangular ones, it quickly became apparent to the authors that the chief problem facing them was a lack of intuition about what types of regions in the discrete complex plane constituted convolution regions. To facilitate the checking of various regions to determine whether or not they were convolution regions, a computer program was written to check a given region and simultaneously compute the convolution product of three selected discrete analytic functions to see if associativity is valid for the given region and selected functions. An option was provided in the program to bypass the convolution check and do only the convolution product check.

Throughout the paper the notation, terminology, and fundamental properties discussed and developed in [1] and [2] are used without further mention. Indeed, the discrete complex plane Z_d is the set of all complex numbers $x+iy$ where x and y are integers and a <u>region</u> in Z_d is any finite union of sets of the form $\{z_0, z_0+1, z_0+(1+i), z_0+i\}$ where $z_0 \in Z_d$. A <u>chain</u> of points in Z_d is a finite set of points $z_0, z_1, ---, z_n$ such that $|z_i - z_{i-1}| = 1$ for $1 \leq 1 \leq n$. A <u>closed chain</u> is a chain for which $z_n = z_0$. Given any chain $C = \{z_0 = 0, z_1, ---, z_n\}$ from zero to z_n the <u>counter chain</u> to C is the chain $C' = \{z_n - z_0 = z_n, z_n - z_1, ---, z_n - z_n = 0\}$ from z_n to 0. For a diagram of a chain versus counter chain, see [1].

Also of particular interest is the double dot integral of Duffin [1], defined for two discrete functions f and g to be

$$\int_{z_0}^{z_n} f(z) : g(z)\, \delta z = \sum_{i=1}^{n} \frac{1}{4} [f(z_i) + f(z_{i-1})][g(z_i) + g(z_{i-1})] \times (z_i - z_{i-1}) . \tag{1.1}$$

where, $z_0 = x_0 + iy_0$ and $z_n = x_n + iy_n$ are two points in the discrete complex plane and $\{z_0, z_1, \ldots, z_n\}$ is a chain of points joining z_0 to z_n.

Section 2 below is devoted to a fairly detailed discussion of what the aforementioned program does, as well as to the presentation of several examples of convolution and nonconvolution regions. Section 3 discusses semi-convex convolution regions. Basically, these are rather general convolution regions for which the associativity of the convolution product is valid. Finally, Section 4 provides an example of a non-semi-convex convolution region for which associativity remains valid.

2. THE CHAINING PROGRAM

Recall that a convolution region is defined to be a region R containing the origin and having the property that for every point $(z \in R)$ there exists at least one chain $0 = z_0, z_1, \ldots, z_m = z$ in R such that the counter chain $z - z_0, z - z_1, \ldots, z - z_m$ lies in R. After only a few frustrated attempts at constructing complicated examples of convolution regions, we realized that some method had to be found to simplify the checking of various discrete regions to facilitate determining if they were indeed convolution regions. Accordingly, we designed a FORTRAN program to:

1. Determine if a given region R containing the origin is a convolution region, and
2. Compute the sums.

$$h*(f*g)(\omega) = \int_0^{\omega} h(\omega - \xi) : \int_0^{\xi} f(\xi - t) : g(t)\, \delta t\, \delta\xi \qquad (2.1)$$

and

$$(h*f)*g(\omega) = \int_0^{\omega}\int_0^{\omega-\xi} h(\omega-\xi-t) : f(t)\, \delta t : g(\xi)\, \delta\xi \qquad (2.2)$$

for selected discrete analytic functions h, f and g on R.

We define an associative region as a convolution region R for which

$$h*(f*g) = (h*f)*g. \qquad (2.3)$$

Since the integrals on the right hand sides of (2.1) and (2.2) are easily calculated, once proper chains and counter chains are determined, it is clear that the only real difficulty as far as programming is concerned is the construction of general enough chaining methods. Two chaining methods were employed and are exemplified by Fig. 1 below. Given a point $z = x + iy$ in Z_d, the first method, begins at the origin and proceeds along the x-axis until the value x is reached (left or right according to the sign of x) or the end of the convolution region occurs. At this point the chain is extended within the region either following the boundary of just straight to the point. The second method employs the same technique except that the program chains along the y-axis is to y and then over to $x + iy$.

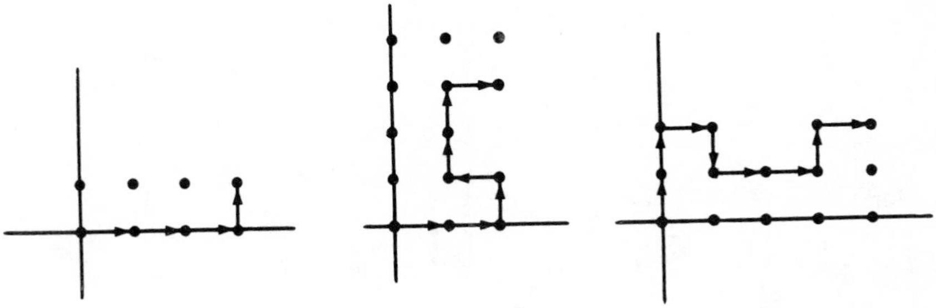

Fig. 1.

Examples of very complex regions which proved to be convolution regions are given by Figs. 2, 4 and 5, while Fig. 6 is an example of a region which is not a convolution region. These particular regions were chosen for two reasons. Firstly, they exemplify regions generally larger than those one would construct by hand, and secondly, they display all of the features of what will be called semi-convex convolution regions in the subsequent section. For examples of simpler convolution regions, see [1] or construct a rectangle containing the origin.

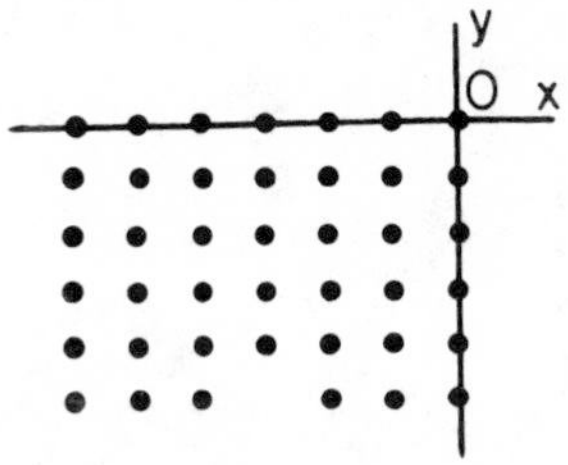

Fig. 2.

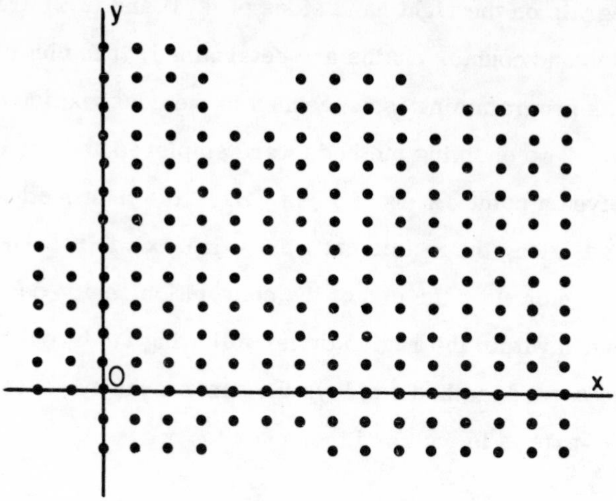

Fig. 3

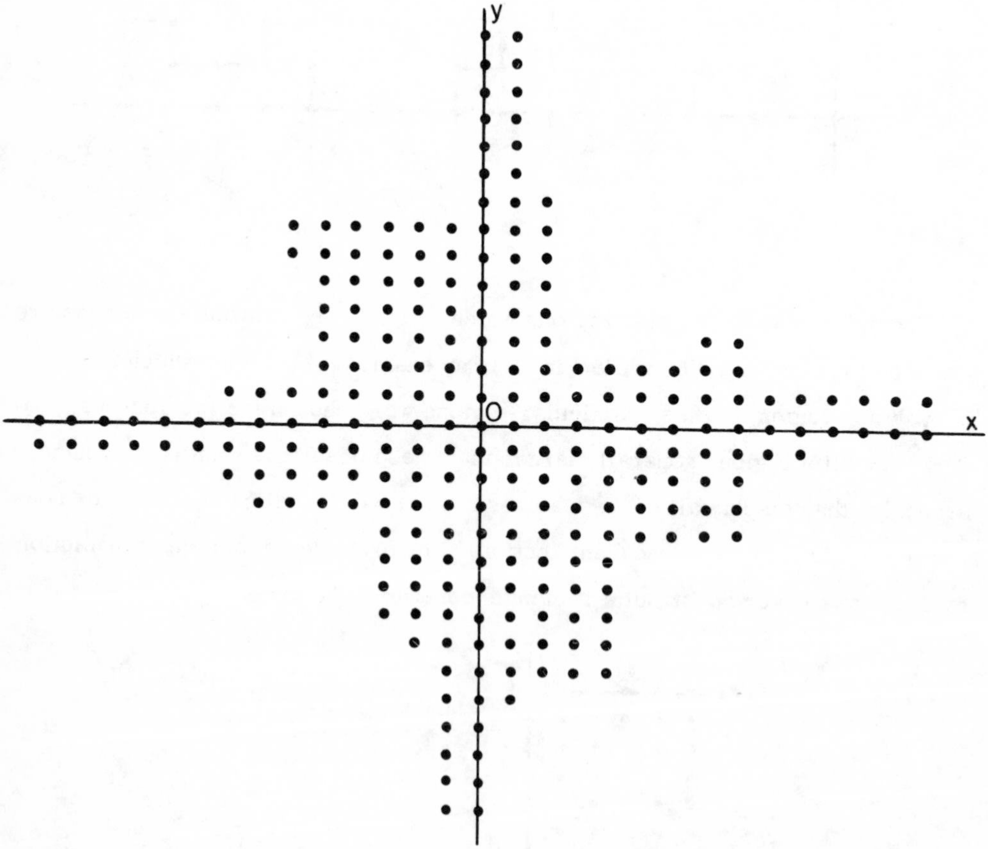

Fig. 4

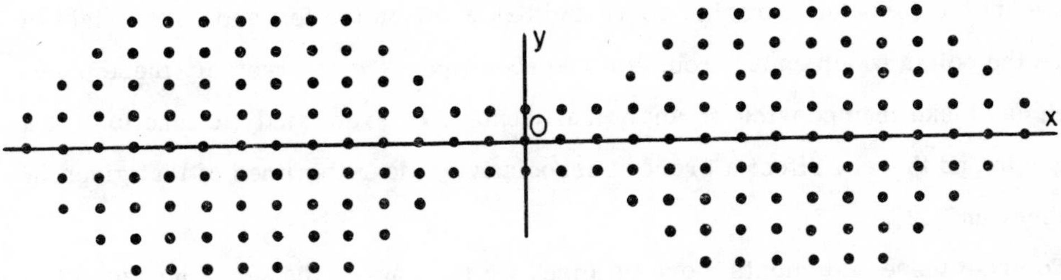

Fig. 5

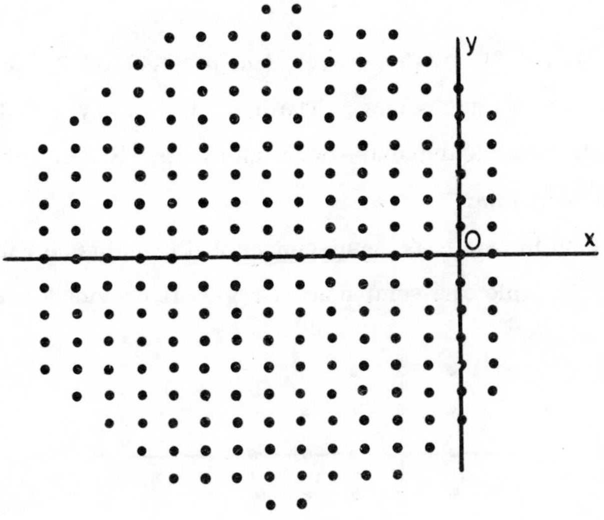

Fig. 6

3. SEMI-CONVEX CONVOLUTION REGIONS

Using Fig. 4 as a base, it is not difficult to realize that the points $z = x + iy$ not in a convolution region but which lie on a vertical or horizontal line which respectively intersects the convolution region above and below (or on the left and on the right of z) are the points which cause trouble for associativity. It is precisely these points which can make it impossible to analytically extend discrete analytic functions to a rectangle and thereby effect a proof of associativity along the lines of that given in [2, theorem 3.2].

To make these statements more precise, we first agree that for a horizontal or vertical line L in the discrete complex plane the positive and negative directions are determined by the axis to which the line is parallel. For a given z_0 on L we let $[z_0, \infty)_L = \{z \in L : z \geq z_0\}$ and $(-\infty, z_0]_L = \{z \in L : z \geq z_0\}$.

<u>Definition 3.1</u>: <u>A gap point relative to a region</u> R <u>is a point</u> $z_0 \notin R$ <u>such that for some horizontal or vertical line</u> L <u>with</u> $z_0 \in L$, $[z_0, \infty)_L \cap R \neq \emptyset$ <u>and</u> $(-\infty, z_0]_L \cap R \neq \emptyset$.

The points (-3, -3), (-6,2), (-6,3), (-6,4), (-6,5), (3,3), (4,3), (5,3), (6,3) in Fig. 4, for example, are all gap points. The line $L = \{x + iy : y = 3\}$ is a horizontal line through the latter four points which intersects R on both the right and left side of any of these points.

A convolution region is said to be <u>semi-convex</u> if it contains no related gap point. For clarity a simple example of a semi-convex region is provided in Fig. 7.

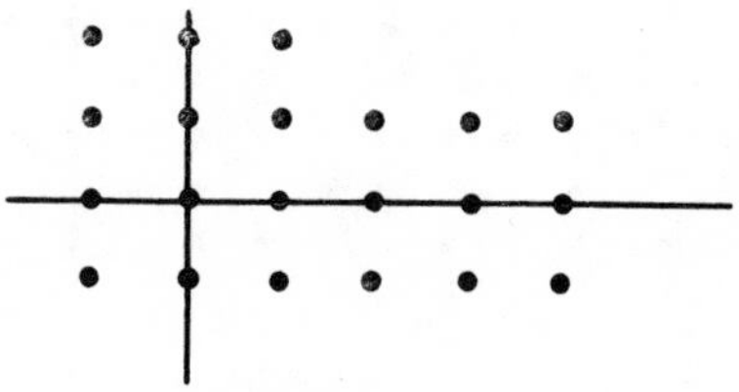

Fig. 7

<u>Theorem 3.2</u>: <u>Any semi-convex convolution region is an associative region.</u>

<u>Proof</u>: Let f, g, and h be discrete analytic functions on R. Using the same extension process employed in [2, theorem 3.2] each of f, g, and h may be

extended to functions f', g', and h' defined and discrete analytic on the entire discrete complex plane. Indeed, by arbitrarily assigning values (complex) to $f'(x)$, $g'(x)$, $h'(x)$, $f'(y)$, $g'(y)$, and $h'(y)$ when $x, y \notin R$ and putting $f'(x) = f(x)$, $g'(x) = g(x)$, $h'(x) = h(x)$, $f'(y) = f(y)$, $g'(y) = g(y)$, and $h'(y) = h(y)$ when $x, y, \in R$ and then extending f', g', and h' to Z_d as in [2, theorem 3.2] the extending algorithm reveals that $f \mid R = f, g' \mid R = g$, and $h' \mid R = h$.

Since $f' * (g' * h') = (f' * g') * h'$ holds by [2, theorem 3.2] and this function as defined is independent of the choice of chain, we must have

$$
\begin{aligned}
f*(g*h) &= f'*(g'*h') \mid R \\
&= (f'*g')*h' \mid R \\
&= (f*g)*h.
\end{aligned}
$$

This completes the proof.

The preceding theorem may be generalized to hold on a much larger collection of convolution regions as follows:

Theorem 3.3: A union of associative regions is an associative region.

Proof: If $R = R_1 \cup R_2 \cup \ldots \cup R_k$ then the point z lies in one of the regions, say R_e. The product is independent of the chain used. Hence, we may use a chain contained in R_e. But the product is associative in R_e. This completes the proof.

All of the regions presented in Figs. 1, 2, 3, 4 and 5 can be shown to be unions of semi-convex convolution subregions. In the absence of a program to check whether possible semi-convex subregions are indeed convolution regions, doing this is admittedly difficult. Fig. 8 shows how this can be done for the region of Fig. 2.

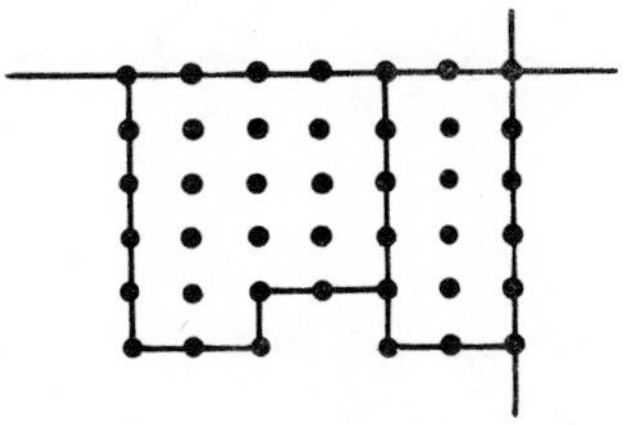

Fig. 8

It is seen that in this case $R = R_1 \cup R_2$.

4. NON-SEMI-CONVEX CONVOLUTION REGIONS

Fig. 9 is an example of a convolution region which is not the union of semi-convex convolution subregions.

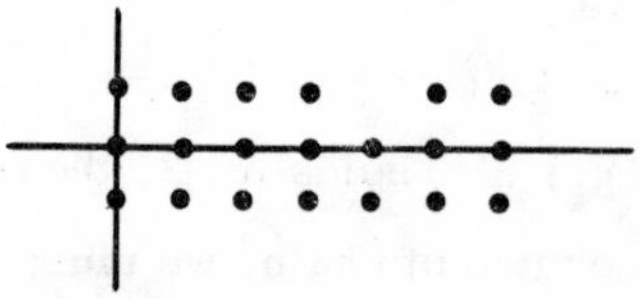

Fig. 9

When we first discovered this region, we thought that it would be possible to easily construct functions which were discrete analytic on R, but for which associativity did not hold. We were quite surprised to discover that the computer said otherwise. All the functions we constructed and then used the program to compute their convolution products with two other discrete analytic functions proved to associate. A closer analysis revealed why.

Lemma 4.1: Let $Z = \{(5,1), (6,1)\}$ and r be the discrete analytic function which is zero on $R \setminus Z$ and for which $r(5,1) = 1$, $r(6,1) = i$. If f is discrete analytic on R then $r*f = \alpha_f r$ where α_f is the complex constant.

$$\alpha_f = \left[\frac{f(t_1) + f(0)}{4}\right] t_1$$

and t_1 is the first point in a chain from 0 to (5,1) which has a counter chain in R from 0 to (5,1).

Proof: That α_f is constant results from the fact that $\int_0^z r(z-t):f(t)\,\delta t$ is independent of chain together with the fact that $r=0$ on $R \setminus Z$. That $r*f=\alpha_f r$ follows because $r*f=0$ on $R \setminus Z$, because $r*f(5,1) = \alpha_f r(5,1)$ and because r is discrete analytic.

In precise terms, if $\{0, t_1, t_2, ---, t_n = (5,1)\}$ is any chain from 0 to (5,1) whose counterchain $\{0, (5,1) - t_{n-1}, ---, (5,1)\}$ belongs to R then by definition when $z = (5,1)$

$$\int_0^z r(r-t):f(t)\,\delta t = \sum_{k=1}^{k=n} \left[\frac{r(z-t_k) + r(z-t_{k-1})}{2}\right] \left[\frac{f(t_k) + (t_{k-1})}{2}\right] (t_k - t_{k-1})$$

$$= \frac{r(t_1) + f(0)}{2} t_1 r(5,1)$$

So chain independence of the integral yields the constancy of α_f.

Now it is clear that $r*f=0$ on $R\setminus Z$ and that $r*f(5,1) = \alpha_f r(5,1)$. The discrete analyticity of $r*f$ then easily yields that $r*f=\alpha_f r$ and completes the proof.

Lemma 4.2: *Let f and g be discrete analytic on R. Then* $r*(f*g) = (\alpha_f\alpha_g)r$.

Proof: This is just a direct computation of α_{f*g}.

Theorem 4.3: *For f, g and h discrete analytic on R,* $f*(g*h) = (f*g)*h$.

Proof: Let $\hat{f}$ be the discrete analytic function obtained by restricting f to $R\setminus Z$ and then extending f discrete analytically to the rectangle R' whose vertices are $i, 6+i, 6-i$, and $-i$. Put $f_e = \hat{f} \mid R\setminus Z$. It is easy to see that $f=f_e+\lambda r$ where $\lambda=f(5,1) - \hat{f}(5,1)$. In a similar fashion, $g=g_e+\alpha r$ and $h=h_e+\rho r$.

Now from the fact that f_e, g_e and h_e are restrictions of functions which are discrete analytic on R', it follows from [2] that $f_e*(g_e*h_e) = (f_e*g_e)*h_e$. Using lemmas 4.1 and 4.2 in conjunction with the preceding fact, a direct computation demonstrates that

$$\begin{aligned} f*(g*h) &= (f_e+\lambda r)*[(g_e+\alpha r)*(h_e+\rho r)] \\ &= [(f_e+\lambda r)*(g_e+\alpha r)]*(h_e+\rho r) \\ &= (f*g)*h. \end{aligned} \tag{4.4}$$

This completes the proof.

It is easy to envision and conjecture an extension of the above type of proof to almost any type of convolution region which is not the union of a finite number of semi-convex convolution regions. The authors believe that any regions of the aforementioned type arise only when there exists a gap point $x+iy$ such that either $|y| = 1$ or $|x| = 1$. A general proof of this fact is presently beyond their grasp.

References

1 R.J. Duffin, Basic properties of discrete analytic functions. Duke Math. Jour. 23 (1956) 335-363.

2 R. J. Duffin and C.S. Duris, A convolution product for discrete function theory. Duke Math. Jour. 31 (1964) 199-220.

3 R.J. Duffin and Elmor L. Peterson, The discrete analogue of a class of entire functions. J.Math. Anal. and Appl. 21 (1968) 619-642.

4 J. Ferrand, Fonctions preharmoniques et fonctions preholomorphes. Bulletin des Sciences Mathematiques, second series, 68 (1944) 152-180.

5 Siro Hayabara, Operational calculus on the discrete analytic functions. Math. Japon. 11(1) (1966) 35-65.

6 R.P. Isaacs, A finite difference function theory. Universidad Nacconal Tucman Revista 2 (1941) 177-201.

J. Bee Bednar
Cynthia English
The University of Tulsa
Tulsa
Oklahoma
USA

H BEGEHR

2 Das Schwarzsche Lemma und verwandte Sätze für pseudoanalytische Funktionen

Die von Bers und Vekua entwickelte Theorie der pseudoanalytischen Funktionen liefert für diese Funktionenklasse eine Reihe von Aussagen, wie sie für analytische Funktionen bekannt sind. Ein grosser Teil dieser Eigenschaften ist eine Folge des Ähnlichkeitsprinzips von Bers. Es gibt aber auch elementare Regeln, die für analytische, nicht aber im allgemeinen für pseudoanalytische Funktionen gelten. So ist zum Beispiel das Produkt pseudoanalytischer Funktionen im Gegensatz zur Summe nicht notwendig wieder pseudoanalytisch. Diesem Mangel kann man aber oft durch die Bemerkung entgehen, dass die pseudoanalytischen Funktionen zur Klasse der approximativ analytischen Funktionen gehören. Diese Klasse ist wiederum gegenüber Produkt und Quotientenbildung abgeschlossen, nicht aber bezüglich Summenbildung. Ihre Funktionen genügen ebenfalls dem Ähnlichkeitsprinzip, zeigen also selbst lokal ein den analytischen Funktionen ähnliches Verhalten. Hier wird diese Bemerkung benutzt, um einige mit dem Schwarzschen Lemma zusammenhängende Sätze für pseudoanalytische Funktionen auszusprechen.

1. GRUNDLEGENDE BEGRIFFE

Eine in einem Gebiet D der komplexen Ebene C definierte komplexe Funktion w heisst approximativ analytisch, wenn sie bis auf isolierte Punkte in D stetige partielle Ableitungen erster Ordnung besitzt und folgenden Bedingungen genügt:

1) Es existieren zwei reelle Konstanten $K \geq 0$ und $\sigma > 1$ derart, dass bis auf die Ausnahmepunkte in D gilt

$$|w_{\bar{z}}(z)| \leq \frac{K}{1+|z|^{\sigma}} |w(z)| \quad (z = x+iy) . \tag{1.1}$$

2) Es existiert eine reelle Konstante $\lambda\,(0 \leq \lambda < 1)$, so dass in der Umgebung eines jeden Ausnahmepunktes z_0 von D gilt

$$|w_{\bar{z}}(z)| \leq \frac{K}{|z-z_0|^{\lambda}} |w(z)| . \tag{1.2}$$

Zu den approximativ analytischen Funktionen gehört die Klasse der pseudoanalytischen Funktionen. Von einem (normalen) erzeugenden Paar (F,G) zweier in D gegebener Funktionen mit den Eigenschaften

i. $M := 1 + \sup_{z\in D} [\, |F(z) - 1|^2 + |G(z) - i|^2]^{\frac{1}{2}} < + \infty,$

ii. $0 < \Delta \leq \mathfrak{I}\mathfrak{m}[\overline{F(z)}\, G(z)] \qquad (z \in D)$,

iii. $F_z(z)$, $F_{\bar{z}}(z)$, $G_z(z)$, $G_{\bar{z}}(z)$ existieren, sind Hölder-stetig in D und es gibt zwei Konstanten $K > 0$ und $\sigma > 1$, so dass in D gilt

$$|F_{\bar{z}}(z)| + |G_{\bar{z}}(z)| \leq \frac{K}{\lambda + |z|^{\sigma}} , \tag{1.3}$$

ausgehend, nennt man eine in D definierte komplexe Funktion w (F,G) - pseudoanalytisch, falls in jedem Punkt z von D der vermittels der wegen ii. eindeutig durch die Gleichung

$$w(z) = \lambda(z) F(z) + \mu(z) G(z) \tag{1.4}$$

gegebenen reellen Funktionen λ und μ gebildete Grenzwert

$$\dot{w}(z) = \frac{d_{(F,G)} w(z)}{dz} := \lim_{\xi \to z} \frac{w(\xi) - (z) F(\xi) - \mu(z) G(\xi)}{\xi - z} \tag{1.5}$$

existiert. Dies ist genau dann der Fall, wenn w in D der Gleichung

$$w_{\bar{z}}(z) = a(z) w(z) + b(z) \overline{w(z)} \tag{1.6}$$

mit den charaktristischen Koeffizienten

$$a := - \frac{\overline{F}G_{\bar{z}} - F_{\bar{z}}\overline{G}}{F\overline{G} - \overline{F}G} , \qquad b := \frac{FG_{\bar{z}} - F_{\bar{z}}G}{F\overline{G} - \overline{F}G}$$

von (F,G) genügt. Die vermittels (1.4) w eindeutig zugeordnete Funktion

$$\omega(z) := \lambda(z) + i\,\mu(z) \qquad (z \in D) \tag{1.7}$$

ist eine quasiholomorphe Funktion, das ist eine Lösung des Beltramischen Differentialgleichungssystems

$$\omega_{\bar z}(z) = \nu(z)\,\overline{\omega_z(z)} \qquad (|\nu(z)| \le k < 1 \tag{1.8}$$

und eine innere Transformation im Sinne von Stöiloff. Man spricht von einer pseudoanalytischen Konstanten, wenn die zugehörige quasiholomorphe Funktion eine (komplexe) Konstante ist. Auf Grund des Ähnlichkeitsprinzips von Bers verhalten sich approximativ analytische Funktionen lokal asymptotish wie analytische Funktionen. Darüber hinaus genügen sie wegen des Ähnlichkeitsprinzips dem Maximumprinzip: Eine in einem beschränkten Gebiet D approximativ analytische Funktion, die auf dem Rand ∂D von D noch stetig ist und dort betragsmässig nicht oberhalb einer Konstanten R liegt, genügt in D der Abschätzung

$|w(z)| \le [\exp 4K]R$

mit der Konstanten K aus (1.1).

Eine quasiholomorphe Funktion ω hat im Inneren des Gebietes kein Maximum, es sei denn, sie ist eine Konstante.

Im folgenden soll D die Kreisscheibe $|z| < 1$ bezeichnen. In diesem Fall kann, da es sich um ein beschränktes Gebiet handelt, die in (1.1) bzw. (1.3) auftretende Konstante σ formal durch null ersetzt werden. Ausserdem wird zusätzlich au i., ii. und iii. noch

iv. $|F_z(z)|^2 + |G_z(z)|^2 + |F_{\bar z}(z)|^2 + |G_{\bar z}(z)|^2 \le M_1^2 < +\infty \; (|z| < 1)$

vorausgesetzt.

2. FOLGERUNGEN AUS DEM MAXIMUMPRINZIP

Eine pseudoanalytische Funktion w soll unimodular beschränkt heissen, wenn stets gilt

$$|\omega(z)|^2 = \lambda^2(z) + \mu^2(z) \le 1. \tag{2.1}$$

Wegen

$$|w(z)| \le |\omega(z)| [1 + (F(z) - 1|^2 + |G(z) - i|^2)^{\frac{1}{2}}] \tag{2.2}$$

gilt dann mit Rücksicht auf i.

$$|w(z)| \le M|\omega(z)| \le M. \tag{2.3}$$

<u>Hilfssatz</u>: Die zu in $|z| < 1$ unimodular beschränkter, nichtkonstanter pseudoanalytischer Funktion w mit festem $z_0 (|z_0| < 1)$, $\lambda_0 = \lambda(z_0)$, $\mu_0 = \mu(z_0)$ gebildete Funktion

$$W(z) := \frac{M[w(z) - \lambda_0 F(z) - \mu_0 G(z)]}{M^2 - (\lambda_0 \overline{F(z)} + \mu_0 \overline{G(z)} w(z)} \tag{2.4}$$

ist in $|z| < 1$ approximativ analytisch und betragsmässig nach oben durch 1 beschränkt.

<u>Beweiss</u>: Die Abschätzung

$$\left|\frac{\partial}{\partial \bar{z}} [\lambda_0 \overline{F(z)} + \mu_0 \overline{G(z)} w(z)]\right| \le \frac{M_1}{M - |\lambda_0 F(z) + \mu_0 G(z)|} \left[\frac{1}{M} + \frac{M}{\Delta} (M+1)^2 - 2\right]^{\frac{1}{2}}].$$

$$|M^2 - \lambda_0 \overline{F(z)} + \mu_0 \overline{G(z)} w(z)|$$

bzw. unter Beachtung von (2.3) und $|\omega_0| < 1$ die Ungleichung

$$\left|\frac{\partial}{\partial \bar{z}} [(\lambda_0 \overline{F(z)} + \mu_0 \overline{G(z)}) w(z)]\right| \le \frac{M_1}{1 - |\omega_0|} \left[\frac{1}{M^2} + \frac{1}{\Delta} [(M+1)^2 - 2]^{\frac{1}{2}}\right].$$

$$|M^2 - (\lambda_0 \overline{F(z)} + \mu_0 \overline{G(z)}) w(z)|$$

bedeutet, dass der Nenner von W approximativ analytisch ist. Da der Zähler pseudoanalytisch ist, gehört W zur Klasse der approximativ analytischen Funktionen. Man bestätigt leicht

$$|W_{\bar{z}}(z)| \leq \frac{kM_1}{1 - |\omega_0|} |W(z)| \qquad (k := \frac{1}{M^2} + \frac{1}{\Delta}[(M+1)^2 - 2]^{\frac{1}{2}}) . \tag{2.5}$$

Naturgemässe gilt in $|z| < 1$

$$|W(z)| \leq 1.$$

Neben $W(z)$ ist auch

$$\hat{W}(z) := W(z) \frac{1 - \bar{z}_0 z}{z - z_0} \qquad (|z| < 1) \tag{2.6}$$

approximativ analytisch, denn $\hat{W}(z)$ hat nur in $z = z_0$ eine (isolierte) Singularität, in deren Umgebung sie wegen der Existenz von $\dot{w}(z_0)$ beschränkt ist. Wendet man also unter Beachtung von

$$\overline{\lim_{|z| \to 1}} \; |W(z) \frac{1 - \bar{z}_0 z}{z - z_0}| \leq 1 \tag{2.7}$$

das Maximumprinzip an, so erhält man

$$|W(z)| \leq \kappa \left|\frac{z - z_0}{1 - \bar{z}_0 z}\right| \tag{2.8}$$

$$\kappa := \exp \frac{4k M_1}{1 - |\omega_0|} \qquad (k := \frac{1}{M^2} + \frac{1}{\Delta}[(M+1)^2 - 2]^{\frac{1}{2}}). \tag{2.9}$$

Die Ungleichung (2.8) gibt nur dann neue Information über $w(z)$, wenn

$$\left|\kappa \frac{z - z_0}{1 - \bar{z}_0 z}\right| < 1, \tag{2.10}$$

d.h., wenn z der in $|z| < 1$ gelegenen Kreisscheibe

$$\left| z - \frac{\kappa^2 - 1}{\kappa^2 - |z_0|^2} z_0 \right| < \frac{\kappa (1 - |z_0|^2)}{\kappa^2 - |z_0|^2} \tag{2.11}$$

angehört. Setzt man noch für $|z_0| < 1$, $|z| < 1$

$$\psi(z, z_0) := \left| \frac{z - z_0}{z - \bar{z}_0 z} \right| \tag{2.12}$$

so folgt aus (2.8) die wiederum in (2.10) interessante Abschätzung

$$|w(z)| \leq M \frac{|\lambda_0 F(z) + \mu_0 G(z)| + \kappa M \psi(z, z_0)}{M + \kappa |\lambda_0 F(z) + \mu_0 G(z)| \psi(z, z_0)} \tag{2.13}$$

oder auch

$$|w(z)| \leq M \frac{|\omega(z_0)| + \kappa \psi(z, z_0)}{1 + \kappa |\omega(z_0)| \psi(z, z_0)} . \tag{2.14}$$

Durch Grenzübergang (z gegen z_0!) folgt aus (2.8) ausserdem, wenn man noch für z_0 nun z schreibt, unter der Voraussetzung $|\omega(z)| < 1$

$$|\dot{w}(z)| \leq \exp\left[\frac{4kM_1}{1 - |\omega(z)|}\right] \frac{M^2 - |w(z)|^2}{M(1 - |z|^2)} , \tag{2.15}$$

oder

$$|\dot{w}(z)| \leq \exp\left[\frac{4kMM_1}{M - |w(z)|}\right] \frac{M^2 - |w(z)|^2}{M(1 - |z|^2)} . \tag{2.16}$$

In (2.8), (2.14) und (2.15) findet man bekannte Abschätzungen für in $|z| < 1$ unimodular beschränkte analytische Funktionen wieder, denn dann gelten $M = 1$ und $M_1 = 0$. Zusammenfassend gilt der

<u>Satz 1</u>: In einer in $|z| < 1$ unimodular beschränkten pseudoanalytischen Funktion w existiert eine vom erzeugenden Paar (F, G) und vom Funktionswert der zu w gehörigen quasiholomorphen Funktion ω im Punkte z_0, $\omega(z_0 = \lambda_0 + \iota\mu_0$, abhängige durch (2.9) gegebene Konstante $\kappa\ (1 \leq \kappa)$, so dass in $|z| < 1$

$$\left| \frac{M(w(z) - \lambda_0 F(z) - \mu_0 G(z))}{M^2 - (\lambda_0 \overline{F_{(z)}} + \mu_0 \overline{G_{(z)}}) w(z)} \right| \leq \kappa \left| \frac{z - z_0}{1 - z_0 z} \right| \tag{2.17}$$

gilt. Die (F, G) - Abteilung von w in z_0 erfüllt

$$|\dot{w}(z_0)| \leq \kappa \frac{M^2 - |w(z_0)|^2}{M(1 - |z_0|^2)} . \tag{2.18}$$

Dabei ist (2.17) nur für die durch (2.11) gegebenen z interessant.

Satz 2: Eine in $|z| < 1$ approximativ analytische Funktion w, für die

$$\overline{\lim_{|z| \to 1}} |w(z)| = M < +\infty ,$$

genügt bei festem z_1, $|z_1| < 1$, in $|z| < 1$ der Ungleichung

$$|w(z)| \leq \kappa M \frac{|w(z_1)| + M\psi(z, z_1)}{M + |w(z_1)| \psi(z, z_1)} \quad (\kappa := \exp 4K \geq 1). \tag{2.19}$$

Der Beweis von (2.19) ergibt sich durch Anwendung von (2.13) auf die auf Grund des Ähnlichkeitsprinzips existierende zu w ähnliche analytische Funktion. Wendet man diese Abschätzung auf die in (2.6) gegebene Funktion $\hat{W}$ im Punkte z_0 an, so folgt mit κ aus (2.9)

$$|\hat{W}(z_0)| \leq \kappa \frac{|\hat{W}(z_1)| + \psi(z_0, z_1)}{1 + |\hat{W}(z_1)| \psi(z_0, z_1)} . \tag{2.20}$$

Hat das pseudoanalytische w nun in $z = 0$ eine Nullstelle, so dass

$$\hat{W}(0) = \frac{\lambda_0 F(0) + \mu_0 G(0)}{Mz_0} ,$$

so ergibt sich mit Rucksicht auf

$$\dot{w}(z_0) = \frac{M^2 - |w(z_0)|^2}{M(1 - |z_0|^2)} \hat{W}(z_0) ,$$

wenn man in (2.20) $z_1 = 0$ setzt und anstelle von z_0 wieder z schreibt, der folgende Satz.

Satz 3: Hat die in $|z| < 1$ unimodular beschränkte, nichtkonstante pseudoanalytische Funktion w in $z = 0$ eine Nullstelle, so gilt mit der zu z durch

$$a M |z| = |\lambda F(0) + \mu G(0)| \quad (w = \lambda F + \mu G)$$

gegebenen Grösse a die Ungleichung

$$|\dot{w}(z)| \leq \exp\left[\frac{4kM_1}{1-|\omega(z)|}\right] \frac{M^2 - |w(z)|^2}{M(1-|z|^2)} \frac{a+|z|}{1+a|z|} . \tag{2.21}$$

Hier kann a auch durch die vermittels

$$\alpha |z| = |\lambda + i\mu|$$

gegebene Grösse α ersetzt werden. (2.21) ist eine Verschärfung von (2.15), wenn a (bzw. α) kleiner als 1 ist.

Wie im analytischen Fall lässt sich Satz 1 durch Betrachtung endlich vieler $(\lambda_0 + i\mu_0)$ - Stellen erweitern.

Satz 4: Hat die in $|z| < 1$ unimodular beschränkte pseudoanalytische Funktion w in z_k $(1 \leq k \leq n)$ ω_0-Stellen, d.h. gilt

$$w(z_k) = \lambda_0 F(z_k) + \mu_0 G(z_k) \quad (1 \leq k \leq n) \text{ mit } \omega_0 = \lambda_0 + i\mu_0 ,$$

so gilt in $|z| < 1$

$$\left| \frac{M(w(z) - \lambda_0 F(z) - \mu_0 G(z)}{M^2 - (\lambda_0 \overline{F(z)} + \mu_0 \overline{G(z)})\, w(z)} \right| \leq \kappa \prod_{k=1}^{n} \left| \frac{z - z_k}{1 - \overline{z}_k z} \right| . \tag{2.22}$$

Ist speziell $\omega_0 = 0$, so erhält man die Jensensche Ungleichung für pseudoanalytische Funktionen in der Form

$$|w(z)| \leq \kappa M \prod_{k=1}^{n} \left| \frac{z - z_k}{1 - z_k z} \right| \tag{2.23}$$

und

$$|w(0)| \leq \kappa M \prod_{k=1}^{n} |z_k| ,$$

wie sie sich übrigens ebenso wie der entsprechende Satz von Blaschke oder der Hadamardsche Dreikreisesatz, auch aus dem Ähnlichkeitsprinzip herleiten lässt.

Beweis: Seien allgemein W_1 und W_2 zwei in $|z| < 1$ beschränkte approximativ analytische Funktionen, so dass W_2 höchstens mit W_1 gemeinsame Nullstellen von nicht höherer Ordnung hat und für die

$$\overline{\lim_{|z|\to 1}}\, |W_1(z)| = R, \qquad \overline{\lim_{|z|\to 1}}\, |W_2(z)| = 1$$

gelten. Dann ist die durch $W_1 + WW_2$ in $|z| < 1$ gegebene Funktion W approximativ analytisch, so dass nach dem auf W angewandten Maximumprinzip mit einer geeigneten von W_1 und W_2 abhängigen Konstanten κ $(1 \leq \kappa)$

$$|W_1(z)| \leq \kappa\, R\, |W_2(z)|$$

gilt. Durch geeignete Festlegung von W_1 und W_2 folgt (2.22). Der in den gegebenen Abschätzungen auftretende Faktor κ $(1 \leq \kappa)$ verhindert Aussagen in der Nähe der Peripherie von $|z| < 1$. Dadurch sind etwa Ergebnisse, wie sie im analytischen Fall $(\kappa = 1\,!)$ der Satz von Julia-Carathéodory liefern, im allgemeinen nicht zu erhalten. Die Ursache für das Auftreten von κ ist darin zu sehen, dass eine approximativ analytische wie auch eine pseudoanalytische Funktion im allgemeinen Fall ihr Maximum in einem inneren Punkt des Gebietes annehmen kann. Ein Beispiel für diese Tatsache ist etwa die Funktion

$$w(z) = 4\, \frac{1 - |z|^{10}}{12 + 5|z|^8 z} + z \qquad (|z| < 1).$$

Sie ist mit

$$\nu(z) = \frac{5|z|^8 z}{12 + 5|z|^8 z}$$

$((1 - \nu),\, i(1 + \nu))$-pseudoanalytisch. Dass es auch (F,G)-pseudoanalytische Funktionen gibt, die nicht analytisch sind und dem Maximumprinzip genügen, mag die einfache Funktion

$$w(z) = z^{\alpha}\,\overline{z}^{\beta} \qquad (0 < \beta < \alpha;\; 0 < \rho < |z| < 1,\; |\arg z| < \eta < \pi)$$

mit

$$F = 1 - \nu, \quad G = i(1 + \nu), \quad \nu(z) = \frac{\beta}{\alpha}\,(z\,\overline{z}^{-1})^{\alpha - \beta}$$

belegen.

References

1 H. Begehr, Zur Wertverteilung approximativ analytischer Funktionen. Arch. Math. (Basel) 23 (1972) 41-49.

2 H. Begehr, Die logarithmische Methode in der Wertverteilungstheorie pseudoanalytischer Funktionen. Ann. Acad. Sci. Fenn. A.I. 549 (1973).

3 H. Begehr, Über beschränkte verallgemeinerte analytische Funktionen. Analele Stiintificevale Universitatii 'Al. I Cuza' din Iasi 20 (1974) 295-303.

4 H. Begehr, Eine Bemerkung zum Maximumprinzip für morphe Funktionen mehrerer komplexer Veränderlichen. Math. Nachricht. (erscheint demnächst).

5 L. Bers, Theory of pseudo-analytic functions Hektographierte Vorlesungsausarbeitung. Institute for Mathematics and Mechanics, New York University, New York (1953).

6 L. Bers and L. Nierenberg, On linear and non-linear elliptic boundary value problems in the plane. Conv. intern. equazioni lineari alle derivate parziali, Trieste (1954). Edizioni Cremonese, Rom (1955) S. 141-167.

7 L. Bieberbach, Lehrbuch der Funktionentheorie, Band II, 2. Auflage, Chaelsea Publ. Comp., New York (1945).

8 C. Carathéodory, Funktionentheorie, Band I, 2. Auflage, Birkhäuser Verlag, Basel-Stuttgart (1960).

9 A. Dinghas, Vorlesungen über Funktionentheorie. Die Grundlehren der Mathematischen Wissenschaften in Einzeldarstellungen, Band 110. Springer-Verlag, Berlin-Göttingen-Heidelberg (1961).

10 K. Habetha, Über die Wertwerteilung pseudoanalytischer Funktionen. Ann. Acad. Sci. Fenn. A.I. (1967) 406.

11 K. Habetha, Zum Phragmén-Lindelöfschen Prinzip bei quasiholomorphen und pseudoanalytischen Funktionen. Appl. Analysis 2 (1972) 169-185.

12 P. Künzi, Quasikonforme Abbildungen. Ergebnisse der Mathematik und ihrer Grenzgebiete. Springer-Verlag, Berlin-Göttingen-Heidelberg (1960).

13 O. Letho and K.I. Virtanen, Quasikonforme Abbildungen. Die Grundlehren der Math. Wiss. in Einzeldarstellungen, Band 126. Springer-Verlag, Berlin-Göttingen-Heidelberg (1965).

14 I.N. Vekua, Verallgemeinerte analytische Funktionen. Mathematische Lehrbücher und Monographien, II Abt. Band 15. Akademie Verlag Berlin (1963).

Heinrich Begehr
Institut für Mathematik I
Freie Universität Berlin
1 Berlin 33
Hüttenweg 9
GFR

F BRACKX

3 On (k)-monogenic functions of a quaternion variable

Abstract

The theory of holomorphic functions of a complex variable has been subjected to various generalizations, which are mainly of two kinds: on the one hand enlargement of domain and co-domain, where we can cite a.o.: J.A. Ward [9], R.Fueter [3] and [4], H.Snyder [7], R.Delanghe [1], and on the other hand extension of the notion of holomorphy as in the works of a.o. A. Kriszten [6], R.P. Gilbert [5]. In this paper we have joined both ways by considering (k)-monogenic functions with values in the algebra $\mathbb{H}$ of real quaternions.

In the third section we prove Stokes' Theorem which leads us to Cauchy's Formula (section 4) and a mean-value theorem (section 5).

In paragraph 8 we study some special homogeneous (k)-monogenic polynomials which will be the basic elements for the expansion in Taylor series established in section 10.

Acknowledgement

We wish to thank Prof. Dr. R.Delanghe for his kind suggestions and various comments in connection with this work.

1. PRELIMINARY DEFINITIONS AND RESULTS

Let M be a four-dimensional, differentiable, oriented manifold, contained in some open non-void subset Ω of $\mathbb{R}^4$, and C a p-chain on M (p fixed and $1 \leq p \leq 4$). If ω_α ($\alpha = 0,1,2,3$) are real p-forms on M, i.e.

$$\omega_\alpha = \sum_h \eta_{\alpha,h}\, dx^h$$

- where $h = (h_1,\ldots,h_p) \in \{0,1,2,3\}^p$, $0 \leq h_1 < h_2 < \ldots < h_p \leq 3$;

$\eta_{\alpha,h} \in C_r(\Omega)$ $(r \geq 1)$, $\eta_{\alpha,h} : \mathbb{R}^4 \to \mathbb{R}$ for all α and h;

$dx^h \in \Lambda^p w$, w being the four-dimensional real vector space with basis $\{dx_0, dx_1, dx_2, dx_3\}$ and $\{e_0, e_1, e_2, e_3\}$ denotes the basis of the algebra $\mathbb{H}, +, \times$ of real quaternions, then we can compose the quaternion -p-form

$$\omega = \sum_{\alpha=0}^{3} e_\alpha \, \omega_\alpha \quad ,$$

and define

$$\int_C \omega = \sum_{\alpha=0}^{3} e_\alpha \int_C \omega_\alpha \; .$$

Next we consider functions with argument in $\mathbb{R}^4$ and values in $\mathbb{H}$. Such a function can be represented by

$$f = \sum_{\alpha=0}^{3} e_\alpha f_\alpha , \; x = (x_0, x_1, x_2, x_3) \in \mathbb{R}^4 \rightarrow f(x) = \sum_{\alpha=0}^{3} e_\alpha f_\alpha(x)$$

where the f_α $(\alpha = 0,1,2,3)$ are real valued.

The concept of (k)-monogenity $(k \in \mathbb{N})$ is now defined by means of the differential operator

$$D = \sum_{\beta=0}^{3} e_\beta \frac{\partial}{\partial x_\beta}$$

which, in view of the non-commutativity of the multiplication in $\mathbb{H}$, can act from the left or the right upon a function f:

$$Df = \sum_{\alpha,\beta=0}^{3} e_\beta \, e_\alpha \frac{\partial f_\alpha}{\partial x_\beta} \quad ; \quad fD = \sum_{\alpha,\beta=0}^{3} e_\alpha e_\beta \frac{\partial f_\alpha}{\partial x_\beta} \; .$$

<u>Definition 1.1</u>: A function $f: \mathbb{R}^4 \rightarrow \mathbb{H}$ is called left (res. right) - (k)-monogenic in Ω iff

i. $f \in C_k(\Omega)$, this means $f_\alpha \in C_k(\Omega)$ for all $\alpha = 0,1,2,3$;

ii. $D^k f = D(D^{k-1} f) = 0$ in Ω (resp. $f D^k = 0$ in Ω).

Remark: If $k = 1$, resp. $k = 2$, f is called regular, resp. areolar-monogenic in Ω.

Theorem 1.1: If $f \in C_{2k}(\Omega)$ is a left (or right) - (k)-monogenic function in Ω, then f is polyharmonic in Ω.

Proof:

Let $\bar{D}$ be the operator $\bar{D} = \sum_{\beta=0}^{3} \epsilon_\beta e_\beta \frac{\partial}{\partial x_\beta}$, $\epsilon_0 = +1$, $\epsilon_\beta = -1$ if $\beta = 1,2,3$.

Since $D\bar{D} = \bar{D}D = \Delta e_0$ $\left(\Delta = \sum_{\alpha=0}^{3} \frac{\partial^2}{\partial x_\alpha^2}\right)$ it follows from $D^k f = 0$ in Ω that

$\bar{D}^k D^k f = \Delta^k f = 0$ in Ω. Q.E.D.

2. STOKES' THEOREM AND GREEN'S IDENTITY

In this section we make use of the following notations:

$$d\hat{x}_0 = dx_1 \wedge dx_2 \wedge dx_3 , \qquad d\hat{x}_1 = dx_0 \wedge dx_2 \wedge dx_3 ,$$

$$d\hat{x}_2 = dx_0 \wedge dx_1 \wedge dx_3 , \qquad d\hat{x}_3 = dx_0 \wedge dx_1 \wedge dx_2 ,$$

$$d\sigma_x = \sum_{\alpha=0}^{3} (-1)^\alpha e_\alpha d\hat{x}_\alpha \quad \text{and} \quad dx^4 = dx_0 \wedge dx_1 \wedge dx_2 \wedge dx_3 .$$

Theorem 2.1: Let Ω be an open, non empty subset of $\mathbb{R}^4$, M a 4-dimensional, differentiable, oriented manifold contained in Ω, C a 4-chain on M and f, $g \in C_1(\Omega)$. Then

$$\int_{\partial C} f\, d\sigma_x\, g = \int_C (fD.g + f.Dg)\, dx^4 . \tag{2.1}$$

Proof: As

$$\int_{\partial C} f\, d\sigma_x\, g = \int_{\partial C} \sum_{\alpha,\beta,\gamma=0}^{3} (-1)^\alpha e_\beta e_\alpha e_\gamma f_\beta g_\gamma \, d\hat{x}_\alpha$$

and, by Stokes' Theorem for real-valued functions [2],

$$\int_{\partial C} f_\beta\, g_\gamma\, d\hat{x}_\alpha = (-1)^\alpha \int_C \frac{\partial}{\partial x_\alpha} (f_\beta \cdot g_\gamma)\, dx^4 ,$$

we get

$$\int_{\partial C} f\, d\sigma_x\, g = \sum_{\alpha,\beta,\gamma=0}^{3} e_\beta\, e_\alpha\, e_\gamma \int_C \left(\frac{\partial f_\beta}{\partial x_\alpha} . g_\gamma + f_\beta . \frac{\partial g_\gamma}{\partial x_\alpha}\right) dx^4$$

$$= \int_C (fD.g + f.Dg)\, dx^4 \qquad \text{Q.E.D.}$$

Remark: Putting alternatively $f = 1$ and $g = 1$ in the above formula, we obtain the so-called Stokes' formula for functions of a quaternion variable;

$$\int_{\partial C} d\sigma_x\, g = \int_C Dg\, dx^4 \quad ; \quad \int_{\partial C} f\, d\sigma_x = \int_C fD\, dx^4 .$$

Theorem 2.2 (Green's identity): If f and $g \in C_k(\Omega)$, then for any 4-chain C on $M \subset \Omega$,

$$\int_{\partial C} \sum_{j=0}^{k-1} (-1)^j\, fD^{k-1-j}\, d\sigma_x\, D^j g = \int_C (fD^k.g + (-1)^{k-1} f.D^k g)\, dx^4 .$$

Proof: In (2.1), substitute f and g respectively by fD^{k-1-j} and $D^j g$, and sum up over j from 0 to $k-1$; then

$$\int_{\partial C} \sum_{j=0}^{k-1} (-1)^j\, fD^{k-1-j}\, d\sigma_x\, D^j g =$$

$$= \sum_{j=0}^{k-1} (-1)^j \int_C fD^{k-j}.D^j g\, dx^4 + \sum_{j=1}^{k} (-1)^{j-1} \int_C fD^{k-j}.\, D^j g\, dx^4$$

$$= \int_C fD^k.\, g\, dx^4 + \sum_{j=1}^{k-1} ((-1)^j + (-1)^{j-1}) \int_C fD^{k-j}.\, D^j g\, dx^4 + (-1)^{k-1} .$$

$$. \int_C f.\, D^k g\, dx^K$$

$$= \int_C (fD^k.g + (-1)^{k-1} f.\, D^k g)\, dx^4. \qquad \text{Q.E.D.}$$

Corollary 2.2: If f is right-(k)-monogenic and g is left-(k)-monogenic in Ω, then for any 4-chain C on $M \subset \Omega$,

$$\int_{\partial C} \sum_{j=0}^{k-1} (-1)^j \; f D^{k-1-j} \; d\sigma \; D^j g = 0 \; .$$

Remark: As it is well known (see [8]) all these theorems remain valid if the 4-chain C (resp. ∂C) is replaced by a 4-dimensional, compact, differentiable, oriented manifold-with-boundary S (resp. ∂S).

3. FUNDAMENTAL SOLUTIONS OF THE OPERATORS D^k ($k \in \mathbb{N}$)

Of special interest in developing the theory of (k)-monogenic functions of a quaternion variable are the functions g_k, defined by

$$g_k(x) = \frac{1}{\omega_4} \cdot \frac{\bar{x}}{\rho_x^4} \cdot \frac{x_0^{k-1}}{(k-1)!} \; , \quad x \in \mathbb{R}^4 \setminus \{0\} \quad , \quad k = 1,2,\ldots$$

where ρ_x is the euclidean distance between x and the origin:

$$\rho_x^2 = |x|^2 = \sum_{\alpha=0}^{3} x_\alpha^2 \; ,$$

$$x = \sum_{\alpha=0}^{3} e_\alpha \, x_\alpha \; , \; \bar{x} = \sum_{\alpha=0}^{3} \epsilon_\alpha \, e_\alpha \, x_\alpha \quad \text{and} \quad \omega_4 = 2\pi^2 \; ,$$

the surface-area of the four-dimensional unit sphere.

They have the following properties:

i. each g_k is analytic in co $\{0\}$; this means that the components of g_k are real-analytic in co $\{0\}$ (see also definition 8.1 in section 8);

ii. for $k > 1$, $Dg_k = g_k D = g_{k-1}$ in co $\{0\}$ which implies that $D^k g_k = g_k D^k = 0$ in co $\{0\}$;

iii. for each $k \in \mathbb{N}$, $g_k \in L_1^{loc}(\mathbb{R}^4, \mathbb{H})$; the functions g_k can thus be interpreted as quaternion distributions:

$$g_k[\phi] = \int g_k(x) \, . \, \phi(x) \; dx^4$$

where ϕ belongs to $\mathscr{D}$, the class of the indefinitely continuously differentiable real-valued functions with compact support in $\mathbb{R}^4$.

Using the distribution-derivative, we get

$$D^k g_k [\phi] = (-1)^k [D^k \phi] g_k = (-1)^k \int D^k \phi(x) \,.\, g_k(x) \; dx^4$$

$$= \lim_{\epsilon \to +0} (-1)^k \int_{Co\mathring{B}(o,\epsilon)} \phi(x) D^k \,.\, g_k(x) \; dx^4$$

where $B(0,\epsilon)$ denotes the sphere with center at the origin and radius ϵ.

Application of Theorem 2.2 yields

$$D^k g_k [\phi] = \lim_{\epsilon \to +0} \int_{\partial B(0,\epsilon)} \sum_{j=0}^{k-1} (-1)^{k+1-j} \,.\, \phi(x) D^{k-1-j} \,.\, d\sigma_x \cdot D^j g_k(x) \,,$$

or, putting $x = \epsilon u$,

$$D^k g_k [\phi] = \lim_{\epsilon \to +0} \int_{\partial B(0,1)} \sum_{j=0}^{k-2} (-1)^{k+1-j} \,.\, \phi(\epsilon u) D^{k-1-j} \,.\, \epsilon^3 \,.\, d\sigma_u \,.\, g_{k-j}(\epsilon u)$$

$$+ \lim_{\epsilon \to +0} \int_{\partial B(0,1)} \phi(\epsilon u) \,.\, \epsilon^3 \,.\, d\sigma_u \,.\, g_1(\epsilon u) \,.$$

As $d\hat{u}_\alpha = (-1)^\alpha n_\alpha \, dS$ $(\alpha = 0,1,2,3)$ where n_α is the α-th component of the external surface-normal and dS the elementary surface-element, the 3-form $d\sigma_u$ for the unit sphere can be written as

$$d\sigma_u = \sum_{\alpha=0}^{3} e_\alpha \; u_\alpha \; dS = u \; dS \,.$$

Since $\phi \in \mathscr{D}$ there exist $M_j \in \mathbb{R}^+$ $(j = 0,1,\ldots k-2)$ so that on $B(0,1)$,

$$| \phi D^{k-1-j} | \leq M_j \,.$$

Consequently,

$$0 \leq \Big| \int_{\partial B(0,1)} \phi(\epsilon u) \; D^{k-1-j} \,.\, \epsilon^3 \,.\, d\sigma_u \,.\, g_{k-j}(\epsilon u) \Big| \leq \frac{1}{\omega^4} \cdot \frac{\epsilon^{k-1-j}}{(k-j-1)!} \cdot M_j \,.$$

$$\int_{\partial B(0,1)} | u o |^{k-j-1} \; dS \leq \epsilon^{k-1-j} \,.\, \frac{M_j}{(k-j-1)!}$$

and

$$\lim_{\epsilon \to +0} \int_{\partial B(0,1)} \sum_{j=0}^{k-2} (-1)^{k+1-j} . \phi(x) D^{k-1-j} . d\sigma_x . D^j g_k(x) = 0 .$$

For ϵ sufficiently small, we know that

$$\phi(\epsilon u) = \phi(0) + Q(\epsilon) \quad \text{with} \quad \lim_{\epsilon \to +0} Q(\epsilon) = 0 .$$

Hence,

$$\int_{\partial B(0,1)} \phi(\epsilon u) . \epsilon^3 . d\sigma_u . g_1(\epsilon u) = \frac{1}{\omega_4} \int_{\partial B(0,1)} [\phi(0) + Q(\epsilon)] \, dS$$

and

$$\lim_{\epsilon \to +0} \int_{\partial B(0,1)} \phi(x) . d\sigma_x . g_1(x) = \phi(0).$$

Combining these results we obtain

$$D^k g_k [\phi] = (-1)^k \int D^k \phi(x) . g_k(x) . dx^4 = \phi(0) .$$

Analogous computations show that

$$g_k D^k [\phi] = \phi(0) .$$

So we have proved that $D^k g_k = g_k D^k = \delta$, δ being the Dirac-measure.

Conclusion: In view of the properties established in (i), (ii) and (iii) we can call the function g_k a two-sided fundamental solution of the operator D^k.

4. CAUCHY'S FORMULA

The next theorem gives us an expression for the value of a (k)-monogenic function f at an interior point x of a manifold S, by means of an integral taken over the boundary ∂S. This formula is therefore an extension of the classical Cauchy-formula for holomorphic functions of a complex variable.

Theorem 4.1: Let Ω be an open, non-empty subset of $\mathbb{R}^4$, f a left-(k)-monogenic function in Ω. $S \subset \Omega$ a 4-dimensional, compact, differentiable, oriented manifold-with-boundary. Then for every $x \in \mathring{S}$,

$$f(x) = \int_{\partial S} \sum_{j=0}^{k-1} (-1)^j . g_{j+1}(u-x). d\sigma_u . D^j f(u)$$

$$= \frac{1}{\omega_4} \int_{\partial S} \sum_{j=0}^{k-1} (-1)^j . \frac{\overline{u-x}}{\rho^4} . \frac{(u_0 - x_0)^j}{j!} . d\sigma_u \, D^j f(u)$$

where ρ is the euclidean distance between u and x.

Proof: With x fixed in $\mathring{S}$, choose a $t \in \mathbb{R}^+$ so that $\overline{B}(x,t) \subset \mathring{S}$. Application of corollary 2.2 to the domain $S \setminus \overline{B}(x,t)$ with the functions g_k and f, results into

$$\int_{\partial S} \sum_{j=0}^{k-1} (-1)^j \, g_{j+1}(u-x). d\sigma_u . D^j f(u) =$$

$$= \frac{1}{\omega_4} \int_{\partial B(x,t)} \sum_{j=0}^{k-1} (-1)^j \frac{\overline{u-x}}{\rho^4} . \frac{(u_0 - x_0)^j}{j!} . d\sigma_u . D^j f(u)$$

$$= \frac{1}{\omega_4} \sum_{j=0}^{k-1} \frac{(-1)^j}{j! \, t^4} \int_{\partial B(x,t)} \sum \epsilon_\gamma \, e_\gamma \, e_i \, e_{\beta 1} \ldots e_{\beta j} \, e_\alpha (-1)^i \tag{4.1}$$

$$\frac{\partial^j f_\alpha}{\partial u_{\beta 1} \ldots \partial u_{\beta j}} (u_\gamma - x_\gamma)(u_0 - x_0)^j \, d\hat{u}_i \, .$$

By Stokes' theorem,

$$(-1)^i \int_{\partial B} \frac{\partial^j f_\alpha}{\partial u_{\beta 1} \ldots \partial u_{\beta j}} (u_\gamma - x_\gamma)(u_0 - x_0)^j \, d\hat{u}_i = \int_B \frac{\partial^{j+1} f_\alpha}{\partial u_i \, \partial u_{\beta 1} \ldots \partial u_{\beta j}} (u_\gamma - x_\gamma) \, .$$

$$. (u_0 - x_0)^j \, du^4 + \int_B \frac{\partial^j f_\alpha}{\partial u_{\beta 1} \ldots \partial u_{\beta j}} (u_\gamma - x_\gamma) \, j \, (u_0 - x_0)^{j-1} \delta_{0i} \, du^4 +$$

$$\int_B \frac{\partial^j f \alpha}{\partial u_{\beta 1} \ldots \partial u_{\beta j}} \delta_{i\gamma} (u_0 - x_0)_j \, du^4.$$

Since $f \in C_k(\Omega)$ we have on S:

$$\left| \frac{\partial^{j+1} f_\alpha(x)}{\partial u_i \, \partial u_{\beta 1} \dots \partial u_{\beta j}} \right| \leq M_{\alpha, i, \beta 1, \dots, \beta j} \quad \text{for all} \quad \alpha, i, \beta 1, \dots \beta j \qquad \text{and}$$

$$\left| \frac{\partial^{j} f_\alpha(x)}{\partial u_{\beta 1} \dots \partial u_{\beta j}} \right| \leq M_{\alpha, \beta 1, \dots, \beta j} \quad \text{for all} \quad \alpha, \beta 1, \dots \beta j \, .$$

Hence, with $M_1 = \max M_{\alpha, i, \beta 1, \dots, \beta j}$ and $M_2 = \max M_{\alpha, \beta 1, \dots, \beta j}$ we have the following inequalities:

$$0 \leq \left| \int_B \frac{\partial^{j+1} f_\alpha}{\partial u_i \, \partial u_{\beta 1} \dots \partial u_{\beta j}} (u_\gamma - x_\gamma)(u_0 - x_0)^j \, du^4 \right|$$

$$\leq M_1 \int_B |u_\gamma - x_\gamma| . |u_0 - x_0|^j \, du^4 \leq M_1 \, . \, t^{j+5} . \, V_4, \tag{4.2}$$

$$0 \leq \left| \int_B \frac{\partial^{j} f_\alpha}{\partial u_{\beta 1} \dots \partial u_{\beta j}} (u_\gamma - x_\gamma) \, j \, (u_0 - x_0)^{j-1} \, du^4 \right|$$

$$\leq j \, M_2 \int_B |u_\gamma - x_\gamma| . |u_0 - x_0|^{j-1} \, du^4 \leq j . M_2 \, . \, t^{j+4} . \, V_4 \tag{4.3}$$

and

$$0 \leq \left| \int_B \frac{\partial^{j} f_\alpha}{\partial u_{\beta 1} \dots \partial u_{\beta j}} (u_0 - x_0)^j \, du^4 \right|$$

$$\leq M_2 \int_B |u_0 - x_0|^j \, du^4 \leq M_2 . \, t^{j+4} . \, V_4 \qquad (j > 0) \tag{4.4}$$

where $V_4 = \frac{1}{4} \omega_4$ represents the volume of the four-dimensional unit sphere.

For $j = 0$ the last integral reduces to $\int_B f_\alpha \, du^4$, and as for sufficiently small t,

$$f_\alpha(u) = f_\alpha(x) + Q_\alpha(t) \quad \text{with} \quad \lim_{t \to +0} Q_\alpha(t) = 0 \qquad (\alpha = 0,1,2,3) \, ,$$

$$\int_B f_\alpha \, du^4 = t^4 . \, V_4 . \, f_\alpha(x) + t^4 . \, V_4 . Q_\alpha(t) .$$

Finally, passing to the limit for $t \to +0$ in (4.1) and taking into account the inequalities (4.2), (4.3) and (4.4), we obtain:

$$\int_{\partial S} \sum_{j=0}^{k-1} (-1)^j g_{j+1}(u-x).d\sigma_u.D^j f(u) = \frac{1}{\omega_4} \sum_{\alpha,i=0}^{3} \epsilon_i e_i e_i e_\alpha V_4 . f_\alpha(x)$$

$$= \frac{4V_4}{\omega_4} \sum_{\alpha=0}^{3} e_\alpha f_\alpha(x) = f(x) \qquad \text{Q.E.D.}$$

Corollary 4.1: If f is left-(k)-monogenic in Ω, then $f \in C_\infty(\Omega)$.

5. THE MEAN-VALUE THEOREM

In the special case that the compact manifold mentioned in Theorem 4.1 is a closed sphere, we can derive from Cauchy's formula a mean-value theorem, by transforming the integral over the boundary of the manifold to an integral over the sphere itself.

Theorem 5.1: If f is left-(k)-monogenic in Ω and $a \in \Omega$, then

$$f(a) = \frac{4}{t^4 . V_4} \int_{B(a,t)} \sum_{j=0}^{k-1} (-1)^j \frac{(u_0 - a_0)^j}{j!} D^j f(u) \, du^4$$

for any sphere $B(a,t)$ so that $\overline{B}(a,t) \subset \Omega$.

Proof: Choose $t \in \mathbb{R}^+$ so that $\overline{B}(a,t) \subset \Omega$ and apply Cauchy's formula (Theorem 4.1). Then

$$f(a) = \frac{1}{\omega_4} \int_{\partial B(a,t)} \sum_{j=0}^{k-1} (-1)^j \frac{\overline{u} - \overline{a}}{\rho^4} \frac{(u_0 - a_0)^j}{j!} . d\sigma_u D^j f(u) .$$

By means of Theorem 2.1 we obtain

$$t^4 \omega_4 f(a) = \sum_{j=0}^{k-1} (-1)^j \int_B [(\overline{u} - \overline{a}) . \frac{(u_0 - a_0)^j}{j!} D^{j+1} f(u) + 4 \frac{(u_0 - a_0)^j}{j!} D^j f(u)] \, du^4$$

$$+ \sum_{j=1}^{k-1} (-1)^j \int_B \frac{(\overline{u} - \overline{a})(u_0 - a_0)^{j-1}}{(j-1)!} . D^j f(u) . du^4 ,$$

so that indeed

$$f(a) = \frac{4}{t^4 \omega_4} \int_{B(a,t)} \sum_{j=0}^{k-1} (-1)^j \cdot \frac{(u_0 - a_0)^j}{j!} \cdot D^j f(u) \; du^4 .$$

Notice that this formula can be rewritten as

$$f(a) = \left[\int_{B(a,t)} \sum_{j=0}^{k-1} (-1)^j . \frac{(u_0 - a_0)^j}{j!} . D^j f(u) . du^4 \right] . \left[\int_{B(a,t)} du^4 \right]^{-1} \qquad \text{Q.E.D.}$$

6. HOMOGENEOUS (k)-MONOGENIC POLYNOMIALS

Consider the function

$$Q_n^{(0)} : \mathbb{R}^4 \to \mathbb{H} , \; x \to \sum_{\alpha=0}^{3} e_\alpha \, Q^{(0)}_{n,\alpha} (x) , \quad \text{where for all} \quad \alpha = 0,1,2,3$$

$Q^{(0)}_{n,\alpha} : \mathbb{R}^4 \to \mathbb{R}$ are homogeneous polynomials of a fixed degree n in $x_0, x_1, x_2,$ x_3.

If $D^k Q_n^{(0)} = 0$ in a certain open neighbourhood Ω of the origin, then we call $Q_n^{(0)}$ a homogenous left-(k)-monogenic polynomial of degree n in Ω.

If $a = a_0, a_1, a_2, a_3) \neq 0$ then we use the notation $Q_n^{(a)}$ for the quaternion polynomial $Q_n^{(a)}(x) = \sum_{\alpha=0}^{3} e_\alpha \, Q^{(a)}_{n,\alpha} (x)$, where for all $a = 0,1,2,3$ $Q^{(a)}_{n,\alpha} : \mathbb{R}^4 \to \mathbb{R}$ are homogeneous polynomials of degree n in $x_0 - a_0, \; x_1 - a_1, \; x_2 - a_2, \; x_3 - a_3$.

From now on we study only the polynomials $Q_n^{(0)}$, because the results can be proved analogously for the polynomials $Q_n^{(a)}$. Furthermore we drop the superscript (0) to avoid too complicated notations, and also admit tacitly that $n \geq k$, because $D^k Q_n = 0$ is automatically fulfilled if $n < k$.

Of special interest are the homogeneous regular (this means (1)-monogenic) polynomials of degree $(n-m)$, $p_{1_1 \ldots 1_{n-m}}$, which were already introduced by R. Fueter [4] and are defined by

$$p_{1_1 \ldots 1_{n-m}} (x) = \frac{1}{(n-m)!} \sum_{\pi(1_1, \ldots, 1_{n-m})} z_{1_1} \cdot z_{1_2} \cdots z_{1_{n-m}}$$

where the sum runs over all possible permutations with repetition of the sequence $(1_1, 1_2, \ldots 1_{n-m})$, and the z_1 $(1=1,2,3)$ are so-called hypercomplex variables defined by

$$z_1 = x_1 \, e_0 - x_0 \, e_1 \qquad (1 = 1,2,3)$$

Theorem 6.1: If Q_n is a homogeneous left-(k)-monogenic polynomial of degree n in the open neighbourhood Ω of the origin, then in Ω,

$$Q_n(x) = \sum_{m=0}^{k-1} \sum_{(1_1,\ldots,1_{n-m})} \frac{x_0^m}{m!} \cdot p_{1_1 \ldots 1_{n-m}} \cdot \frac{\partial^{n-m} \, D^m Q_n}{\partial x_{1_1} \ldots \partial x_{1_{n-m}}}$$

where the second sum runs over all possible combinations with repetition of $(1,2,3)$ in sets of $(n-m)$ elements.

Proof (by induction):

i.   The theorem is true for $k=1$. Indeed, for a homogeneous regular polynomial Q_n, the formula reduces to

$$Q_n = \sum_{(1_1,\ldots,1_n)} p_{1_1 \ldots 1_n} \frac{\partial^n Q_n}{\partial x_{1_1} \ldots \partial x_{1_n}} \, .$$

This formula was already shown by R. Fueter [4].

ii.   Suppose now that the formula holds for $(k-1)$. As Q_n is homogeneous of degree n we have on one hand, by Euler's formula, that

$$nQ_n = x_0 \frac{\partial Q_n}{\partial x_0} + \sum_{i=1}^{3} x_i \frac{\partial Q_n}{\partial x_i} \, . \tag{6.1}$$

On the other hand the left-(k)-monogenity of Q_n implies that DQ_n is a homogeneous left-(k-1)-monogenic polynomial of degree $(n-1)$, and thus by our hypothesis,

$$DQ_n = \sum_{m=0}^{k-2} \sum_{(1_1,\ldots,1_{n-1-m})} \frac{x_0^m}{m!} \cdot p_{1_1 \ldots 1_{n-1-m}} \cdot \frac{\partial^{n-1-m} \, D^{m+1} Q_n}{\partial x_{1_1} \ldots \partial x_{1_{n-1-m}}}$$

33

But

$$DQ_n = e_0 \frac{\partial Q_n}{\partial x_0} + \sum_{\beta=1}^{3} e_\beta \frac{\partial Q_n}{\partial x_\beta}$$

so that (6.1) turns into

$$nQ_n = \sum_{i=1}^{3} z_i \frac{\partial Q_n}{\partial x_i}$$

$$+ \sum_{m=0}^{k-2} \sum_{l_1=1}^{3} \cdots \sum_{l_{n-1-m}=1}^{3} \frac{x_0^{m+1}}{m!(n-1-m)!} z_{l_1} \ldots z_{l_{n-1-m}} \cdot \frac{\partial^{n-1-m} D^{m+1} Q_n}{\partial x_{l_1} \ldots \partial x_{l_{n-1-m}}} . \tag{6.2}$$

But the functions $Q_{n,\alpha}$ are all of the class C_∞ in $\mathbb{R}^4$, so that in Ω:

$$\frac{\partial}{\partial x_i} (D^k Q_n) = D^k \left(\frac{\partial Q_n}{\partial x_i}\right) \qquad (i = 1,2,3) .$$

Hence, since Q_n is left-(k)-monogenic in Ω, so are also $\frac{\partial Q_n}{\partial x_i}$ $(i = 1,2,3)$ and by means of analogous reasonings as made above, we obtain for $i = 1,2,3$:

$$(n-1) \frac{\partial Q_n}{\partial x_i} = \sum_{j=1}^{3} z_j \cdot \frac{\partial^2 Q_n}{\partial x_j \partial x_i}$$

$$+ \sum_{m=0}^{k-2} \sum_{s_1,\ldots,s_{n-2-m}=1}^{3} \frac{x_0^{m+1}}{m!(n-2-m)!} \cdot z_{s_1} \ldots z_{s_{n-2-m}} \frac{\partial^{n-2-m} D^{m+1} \left(\frac{\partial Q_n}{\partial x_i}\right)}{\partial x_{s_1} \ldots \partial x_{s_{n-2-m}}}$$

Substitution of this expression for $\frac{\partial Q_n}{\partial x_i}$ in (6.2) leads to

$$n(n-1) Q_n = \sum_{i,j=1}^{3} z_i z_j \frac{\partial^2 Q_n}{\partial x_j \partial x_i}$$

$$+ \sum_{m=0}^{k-2} \frac{1}{m!} \cdot \frac{2n-2-m}{(n-1-m)!} \sum_{l_1,\ldots,l_{n-1-m}=1}^{3} x_0^{m+1} z_{l_1} \ldots z_{l_{n-1-m}} \cdot \frac{\partial^{n-1-m} D^{m+1} Q_n}{\partial x_{l_1} \ldots \partial x_{l_{n-1-m}}} .$$

Proceeding in the same manner we arrive after all at

$$n!\, Q_n = \sum_{j_1,\dots,j_n=1}^{3} z_{j_1}\dots z_{j_n} \frac{\partial^n Q_n}{\partial x_{j_1}\dots \partial x_{j_n}}$$

$$+ \sum_{m=1}^{k-1} \frac{1}{(n-m)!} \sum_{l_1,\dots,l_{n-m}=1}^{3} \frac{x_0^m}{m!} z_{l_1}\dots z_{l_{n-m}} \frac{\partial^{n-m} D^m Q_n}{\partial x_{l_1}\dots \partial x_{l_{n-m}}}$$

or

$$Q_n = \sum_{m=0}^{k-1} \sum_{(l_1,\dots,l_{n-m})} \frac{x_0^m}{m!} p_{l_1\dots l_{n-m}} \frac{\partial^{n-m} D^m Q_n}{\partial x_{l_1}\dots \partial x_{l_{n-m}}} .$$

iii. The theorem is true for $k = 1$, thus for $k = 2,3,\dots$ Q.E.D.

7. THE BASIC POLYNOMIALS $\frac{1}{m!} x_0^m p_{l_1\dots l_{n-m}}$

Lemma 7.1: The polynomials $\frac{1}{m!} x_0^m p_{l_1\dots l_{n-m}}$, $(l_1,\dots,l_{n-m}) \in \{1,2,3\}^{n-m}$, of degree n, are left- and right-$(m+1)$-monogenic in the whole $\mathbb{R}^4$.

Proof: Taking into account the left- as well as right-regularity of the polynomials $p_{l_1\dots l_{n-m}}$ in the whole $\mathbb{R}^4$, we have successively

$$D^{m+1}\left(\frac{x_0^m}{m!} p_{l_1\dots l_{n-m}}(x)\right) = D^m\left(\frac{x_0^{m-1}}{(m-1)!} p_{l_1\dots l_{n-m}}(x)\right) = \dots =$$

$$= D p_{l_1\dots l_{n-m}}(x) = 0 , \qquad \forall x \in \mathbb{R}^4 .$$

Analogously, for all $x \in \mathbb{R}^4$, $\left(\frac{x_0^m}{m!} p_{l_1\dots l_{n-m}}(x)\right) D^{m+1} = 0.$ Q.E.D.

Corollary 7.1: Any $\left\{\begin{matrix}\text{right}\\ \text{left}\end{matrix}\right\}$ linear combination (with constants in $\mathbb{H}$) of polynomials $\frac{1}{m!} x_0^m p_{l_1\dots l_{n-m}}$, $(l_1,\dots,l_{n-m}) \in \{1,2,3\}^{n-m}$ is $\left\{\begin{matrix}\text{left}\\ \text{right}\end{matrix}\right\}$-$(m+1)$-monogenic in $\mathbb{R}^4$.

Theorem 7.1: If a homogeneous polynomial Q_n of degree n in x_0, x_1, x_2, x_3 is left-(k)-monogenic in an open neighbourhood Ω of the origin, then it is left-(k)-monogenic in the whole $\mathbb{R}^4$.

Proof: As Q_n is homogeneous of degree n in x_0, x_1, x_2, x_3, $D^k Q_n$ is homogeneous and of degree $(n-k)$ in x_0, x_1, x_2, x_3, so that $D^k Q_n$ can be written as

$$D^k Q_n = \sum_{(l_1,\ldots,l_{n-k})} x_{l_1} \ldots x_{l_{n-k}} \cdot c_{l_1 \ldots l_{n-k}}$$

where the sum runs over all possible combinations with repetition of $(0,1,2,3)$ in sets of $(n-k)$ elements. But Q_n is left-(k)-monogenic in Ω, so in Ω:

$$\sum_{(l_1,\ldots,l_{n-k})} x_{l_1} \ldots x_{l_{n-k}} \cdot c_{l_1 \ldots l_{n-k}} = 0$$

from which it follows that all $c_{l_1 \ldots l_{n-k}} = 0$ and $D^k Q_n = 0$ in $\mathbb{R}^4$. Q.E.D.

We now consider the set of all homogeneous polynomials of degree n in x_0, x_1, x_2, x_3 which are left-(k)-monogenic in a certain open neighbourhood of the origin and thus in the whole $\mathbb{R}^4$; we denote this set by ${}^k[Q_n]$. Clearly $({}^k[Q_n], + ; \mathbb{H}, +, \times)$ is a right vector space over $\mathbb{H}$. If we put

$${}^kB_n = \{\frac{1}{m!} x_0^m \, p_{l_1 \ldots l_{n-m}} : m = 0,1,2,\ldots,k-1 \, ; \, (l_1,\ldots l_{n-m}) \in \{1,2,3\}^{n-m}\}$$

we can prove

Lemma 7.2: ${}^kB_n$ is a set of generators for the vector space ${}^k[Q_n]$.

Proof: By Lemma 7.1 each element of ${}^kB_n$ is left-(k)-monogenic in $\mathbb{R}^4$. Moreover, we have proved in Theorem 6.1 that for any element $Q_n \in {}^k[Q_n]$,

$$Q_n(x) = \sum_{m=0}^{k-1} \sum_{(l_1,\ldots l_{n-m})} \frac{x_0^m}{m!} \, p_{l_1 \ldots l_{n-m}} \, \frac{\partial^{n-m} D^m Q_n}{\partial x_{l_1} \ldots \partial x_{l_{n-m}}},$$

so that indeed ${}^kB_n$ is a generating set for ${}^k[Q_n]$. Q.E.D.

Lemma 7.3: ${}^kB_n$ is right-$\mathbb{H}$-free.

Proof: Consider the linear combination

$$\sum_{m=0}^{k-1} \sum_{(l_1,\dots,l_{n-m})} \frac{1}{m!} \cdot x_0^m \cdot p_{l_1 \dots l_{n-m}}(x) \cdot a^{(m)}_{l_1 \dots l_{n-m}} = 0$$

or more explicitly

$$\sum_{(l_1,\dots l_n)} p_{l_1 \dots l_n} \, a^{(0)}_{l_1 \dots l_n} + \sum_{(l_1,\dots,l_{n-1})} x_0 p_{l_1 \dots l_{n-1}} \, a^{(1)}_{l_1 \dots l_{n-1}} + \dots$$

$$\dots + \sum_{(l_1,\dots,l_{n-k+1})} \frac{1}{(k-1)!} \, x_0^{k-1} \, p_{l_1 \dots l_{n-k+1}} \, a^{(k-1)}_{l_1 \dots l_{n-k+1}} = 0 ,$$

all $a^{(m)}_{l_1 \dots l_{n-m}} \in \mathbb{H}$.

An arbitrary term of the first sum can be written as

$$\frac{1}{n!} \sum_{\pi(l_1',\dots,l_n')} z_{l_1'} \cdot z_{l_2'} \dots z_{l_n'} \cdot a^{(0)}_{l_1' \dots l_n'}$$

or as

$$\frac{1}{n!} \sum_{\pi(l_1',\dots,l_n')} x_{l_1'} \dots x_{l_n'} \cdot a^{(0)}_{l_1' \dots l_n'} + \text{terms in which } x_0 \text{ appears},$$

or still as

$$\frac{1}{n!} \cdot \frac{n!}{s_1!s_2!s_3!} \, x_1^{s_1} x_2^{s_2} x_3^{s_3} \cdot a^{(0)}_{l_1' \dots l_n'} + \text{terms in which } x_0 \text{ appears}.$$

Hereby s_i represents the number of times that i appears in the sequence $(l_1',\dots,l_n')$. So (7.1) reads also as

$$\sum_{(l_1,\dots,l_n)} \frac{1}{s_1!s_2!s_3!} \, x_1^{s_1} x_2^{s_2} x_3^{s_3} \, a^{(0)}_{l_1 \dots l_n} = \text{sum of terms in which } x_0$$

appears,

from which it easily follows that all $a^{(0)}_{l_1 \dots l_n} = 0$ and (7.1) reduces to

$$\sum_{m=1}^{k-1} \sum_{(l_1,\dots,l_{n-m})} \frac{1}{m!} x_0^m \, p_{l_1 \dots l_{n-m}} \, a^{(m)}_{l_1 \dots l_{n-m}} = 0.$$

By a similar argument - separating the terms containing x_0^2 - one can show that all $a^{(1)}_{l_1 \dots l_n} = 0$. Working further in the same way we obtain $a^{(m)}_{l_1 \dots l_{n-m}} = 0$ for all $m = 0,1,\dots,k-1$ and all $(l_1,\dots,l_{n-m}) \in \{1,2,3\}^{n-m}$. Q.E.D.

Theorem 7.2: ${}^k B_n$ is a basis for the right vector space ${}^k[Q_n]$, and this for all $n \in \mathbb{N}$.

Proof: This statement follows immediately from Lemmas 7.2 and 7.3. Q.E.D.

Remarks:

1) If $Q_n^{(a)}$ is left-(k)-monogenic in a certain open neighbourhood Ω_a of a, then in $\mathbb{R}^4$,

$$Q_n^{(a)} = \sum_{m=0}^{k-1} \sum_{(l_1,\dots,l_{n-m})} \frac{(x_0 - a_0)^m}{m!} \cdot p^{(a)}_{l_1 \dots l_{n-m}} \frac{\partial^{n-m} D^m Q_n^{(a)}}{\partial x_{l_1} \dots \partial x_{l_{n-m}}}$$

with $p^{(a)}_{l_1 \dots l_{n-m}}(x-a) = \dfrac{1}{(n-m)!} \displaystyle\sum_{\pi(l_1,\dots,l_{n-m})} (z_{l_1} - a'_{l_1}) \dots (z_{l_{n-m}} - a'_{l_{n-m}})$

and $z_l - a'_l = (x_l - a_l)\, e_0 - (x_0 - a_0)\, e_l$, $l = 1,2,3$.

2) ${}^k B_n(a) = \{\frac{1}{m!}(x_0 - a_0)^m \, p^{(a)}_{l_1 \dots l_{n-m}}(x-a) : m = 0,1,\dots,k-1 ;$

$(l_1,\dots,l_{n-m}) \in \{1,2,3\}^{n-m}\}$

is a basis for the right vector space ${}^k[Q_n^{(a)}]$ of homogeneous left-(k)-monogenic polynomials of degree n in the $x_\alpha - a_\alpha$, $\alpha = 0,1,2,3$.

3) If $n = 0$, ${}^k[Q_0]$ consists of the constant functions, with basis ${}^k B_0 = \{e_0\}$

8. THE FIRST TAYLOR EXPANSION FOR AN ANALYTIC FUNCTION

We now introduce a new class of functions of a quaternion variable, namely the analytic functions.

Definition 8.1: A function $f : \mathbb{R}^4 \to \mathbb{H}$ is called analytic in some open, non-void subset $\Omega \subset \mathbb{R}^4$ iff the four components f_α, $\alpha = 0,1,2,3$, are real-analytic in Ω, in other words iff

i. $f \in C_\infty(\Omega)$;

ii. for each compact subset $k \subset \Omega$ there exist constants $C_1(k)$ and $C_2(k)$ so that for all $n \in \mathbb{N}$:

$$\sup_{x \in k} \left| \frac{\partial^n f_\alpha(x)}{\partial x_0^{n_0}\, \partial x_1^{n_1}\, \partial x_2^{n_2}\, \partial x_3^{n_3}} \right| \leq n!\; C_1(k).C_2(k)^n , \quad \sum_{i=0}^{3} n_i = n , \quad \forall_\alpha = 0,1,2,3.$$

Let us now assume that f is analytic in a certain region Ω containing the origin. Then it is well known that there exists an open neighbourhood of the origin $\Omega_0 \subset \Omega$ in which each component f_α of $f\,(\alpha = 0,1,2,3)$ can be expanded in a Taylor series. This series, considered as a multiple power series, converges absolutely and uniformly on each compact subset of Ω_0. Let us agree once and for all that, any time we mention the uniform convergence of such a series in an open region, we mean: "the multiple power series converges uniformly on the compact subsets of that region". So, for each $\alpha = 0,1,2,3$,

$$f_\alpha(x) = \sum_{n=0}^{\infty} \frac{1}{n!} \sum_{l_1,\dots,l_n=0}^{3} x_{l_1} \dots x_{l_n} \left. \frac{\partial^n f_\alpha}{\partial x_{l_1} \dots \partial x_{l_n}} \right|_{x=0}$$

where the right-hand side converges absolutely and uniformly in Ω_0.

Introducing the notation

$$\left. \frac{\partial^n f}{\partial x_{l_1} \dots \partial x_{l_n}} \right|_{x=0} = \sum_{\alpha=0}^{3} e_\alpha \left. \frac{\partial^n f_\alpha}{\partial x_{l_1} \dots \partial x_{l_n}} \right|_{x=0}$$

we obtain the first Taylor expansion for the analytic function f in the open neighbourhood Ω_0 of the origin:

$$f(x) = \sum_{n=0}^{\infty} \frac{1}{n!} \sum_{l_1=0}^{3} \dots \sum_{l_n=0}^{3} x_{l_1} \dots x_{l_n} \left. \frac{\partial^n f}{\partial x_{l_1} \dots \partial x_{l_n}} \right|_{x=0}$$

This series naturally remains absolutely and uniformly convergent in Ω_0, and its homogeneous part of degree n is denoted by $T_n^{(0)}$:

$$T_n^{(0)} = \frac{1}{n!} \sum_{l_1,\dots,l_n=0}^{3} x_{l_1}\dots x_{l_n} \left.\frac{\partial^n f}{\partial x_{l_1}\dots\partial x_{l_n}}\right|_{x=0} .$$

9. THE SECOND TAYLOR EXPANSION FOR A (K)-MONOGENIC FUNCTION

We now build a bridge between analytic and (k)-monogenic functions.

Theorem 9.1: If f is left-(k)-monogenic in an open, non-void subset $\Omega \subset \mathbb{R}^4$, then

i. f is analytic in Ω;

ii. for each $a \in \Omega$ there exists an open neighbourhood $\Omega_a \subset \Omega$ of a in which

$$f(x) = \sum_{n=0}^{\infty} \sum_{m=0}^{k-1} \frac{1}{m!(n-m)!} \sum_{l_1=1}^{3} \dots \sum_{l_{n-m}=1}^{3} (x_0-a_0)^m .$$

$$(z_{l_1}-a'_{l_1})\dots(z_{l_{n-m}}-a'_{l_{n-m}}) . \left.\frac{\partial^{n-m} D^m f}{\partial x_{l_1}\dots\partial x_{l_{n-m}}}\right|_{x=a} .$$

Proof:

i. Since f is left-(k)-monogenic in Ω, f belongs to $C_\infty(\Omega)$, so that in Ω, in view of Theorem 1.1, the components f_α are solutions of the elliptic differential equation $\Delta^k = 0$. Hence the f_α $(\alpha = 0,1,2,3)$ are real-analytic.

ii. Without loose of generality we may assume that Ω contains the origin. It follows from (i) and section 8, that there exists an open neighbourhood Ω_0 of the origin in which f can be expanded in its first Taylor series. First of all we proof that each homogeneous polynomial T_n - again we drop the superscript (0) - appearing in this expansion, is left-(k)-monogenic in $\mathbb{R}^4$. If $n < k$, this statement is trivial; from now on we assume $n \geq k$.

$D^k T_n$ is a homogeneous polynomial of degree $(n-k)$, thus it can be written as

$$D^k T_n = \sum x_0^{l_0} x_1^{l_1} x_2^{l_2} x_3^{l_3} a_{l_0 l_1 l_2 l_3}$$

with all $a_{l_0 l_1 l_2 l_3} \in \mathbb{H}$ and $\sum_{i=0}^{3} l_i = n-k$.

Choose arbitrarily a sequence $(t_1, t_2, \ldots, t_{n-k}) \in \{0,1,2,3\}^{n-k}$ and consider the real differential operator

$$D' = \frac{\partial^{n-k}}{\partial x_{t_1} \cdots \partial x_{t_{n-k}}} = \frac{\partial^{n-k}}{\partial x_0^{s_0}\, \partial x_1^{s_1}\, \partial x_2^{s_2}\, \partial x_3^{s_3}}$$

where s_i equals the number of times that i appears in the sequence $(t_1, \ldots t_{n-k})$. In $\Omega, D^k f = 0$, so that $D''f = D'D^k f = 0$. As D'' is a differential operator of the n-th degree,

$$0 = D''f = D'(D^k T_n) + \text{homogeneous terms of degree } q \geq 1,$$

or $0 = s_0!\, s_1!\, s_2!\, s_3!.\ a_{s_0 s_1 s_2 s_3}$ + homogeneous terms of degree $q \geq 1$.

In particular, if $x = 0$ this expression results into

$$0 = s_0!\, s_1!\, s_2!\, s_3!.\ a_{s_0 s_1 s_2 s_3} \quad ;$$

hence $a_{s_0 s_1 s_2 s_3} = 0$.

As the sequence $(t_1 \ldots t_{n-k})$ and thus (s_0, s_1, s_2, s_3) too, were arbitrary, we may conclude that all $a_{s_0 s_1 s_2 s_3} = 0$, which implies that $D^k T_n = 0$ in $\mathbb{R}^4$.

By Theorem 6.1 we obtain for the homogeneous left-(k)-monogenic polynomial T_n:

$$T_n = \sum_{m=0}^{k-1} \sum_{(l_1, \ldots, l_{n-m})} \frac{1}{m!} \cdot x_0^m \cdot p_{l_1 \ldots l_{n-m}} \frac{\partial^{n-m} D^m T_n}{\partial x_{l_1} \cdots \partial x_{l_{n-m}}} .$$

Now it is easy to check, starting with the original definition of T_n, that

$$\frac{\partial^{n-m} D^m T_n}{\partial x_{l_1} \cdots \partial x_{l_{n-m}}} = \left. \frac{\partial^{n-m} D^m f}{\partial x_{l_1} \cdots \partial x_{l_{n-m}}} \right|_{x=0}$$

So in Ω,

$$T_n = \sum_{m=0}^{k-1} \sum_{(l_1,\ldots,l_{n-m})} \frac{1}{m!} \cdot x_0^m \cdot p_{l_1 \ldots l_{n-m}} \cdot \left. \frac{\partial^{n-m} D^m f}{\partial x_{l_1} \ldots \partial x_{l_{n-m}}} \right|_{x=0},$$

and the first Taylor expansion gives rise to

$$f(x) = \sum_{n=0}^{\infty} \sum_{m=0}^{k-1} \sum_{(l_1,\ldots,l_{n-m})} \frac{1}{m!} \cdot x_0^m . p_{l_1 \ldots l_{n-m}} \left. \frac{\partial^{n-m} D^m f}{\partial x_{l_1} \ldots \partial x_{l_{n-m}}} \right|_{x=0} \tag{9.1}$$

where the third sum, as before, is taken over all possible combinations with repetition of $(1,2,3)$ in sets of $(n-m)$ elements. The expression (9.1) is called the second Taylor expansion of the left-(k)-monogenic function f in Ω_0. Considered as a multiple power series it converges absolutely and uniformly on the compact subsets of Ω_0. In view of the definition of the polynomials $p_{l_1 \ldots l_{n-m}}$, it can be rewritten as

$$f(x) = \sum_{n=0}^{\infty} \sum_{m=0}^{k-1} \frac{1}{m!(n-m)!} \sum_{l_1=1} \sum_{l_{n-m}=1} x_0^m . z_{l_1} \ldots z_{l_{n-m}} \cdot \left. \frac{\partial^{n-m} D^m f}{\partial x_{l_1} \ldots \partial x_{l_{n-m}}} \right|_{x=0}$$

Q.E.D.

Theorem 9.2: If for each point a in an open, non-void subset Ω of $\mathbb{R}^4$, there exists an open neighbourhood $\Omega_a \subset \Omega$ of a, in which f can be represented by a uniformly convergent series of the form

$$f(x) = \sum_{n=0}^{\infty} \sum_{m=0}^{k-1} \sum_{l_1=1}^{3} \cdots \sum_{l_{n-m}=1}^{3} (x_0 - a_0)^m (z_{l_1} - a'_{l_1}) \ldots (z_{l_{n-m}} - a'_{l_{n-m}}) .$$

$$. c^{(m)}_{l_1 \ldots l_{n-m}}$$

with all $c^{(m)}_{l_1 \ldots l_{n-m}} \in \mathbb{H}$, then f is left-(k)-monogenic in Ω.

Proof: Obviously the components of f are real-analytic in Ω_a and the cited series may be derived termswise. So in Ω_a,

$$D^k f = \sum_{n=0}^{\infty} \sum_{m=0}^{k-1} \sum_{(1_1,\ldots,1_{n-m})} [\, D^k \sum_{\pi(1_1,\ldots,1_{n-m})} (x_0 - a_0)^m \,.$$

$$(z_{1_1} - a'_{1_1}) \ldots (z_{1_{n-m}} - a'_{1_{n-m}}) \,] \,.\, c_{1_1 \ldots 1_{n-m}}^{(m)} \,.$$

But

$$\sum_{\pi(1_1,\ldots,1_{n-m})} (x_0 - a_0)^m (z_{1_1} - a'_{1_1}) \ldots (z_{1_{n-m}} - a'_{1_{n-m}}) =$$

$$m!(n-m)! \quad \frac{(x_0 - a_0)^m}{m!} \quad p_{1_1 \ldots 1_{n-m}}^{(a)} \quad (x - a) \in {}^k B_n (a) .$$

Consequently

$$D^k \,[\sum_{\pi(1_1,\ldots,1_{n-m})} (x_0 - a_0)^m . (z_{1_1} - a'_{1_1}) \ldots (z_{1_{n-m}} - a'_{1_{n-m}}) \,] = 0$$

for all $m = 0, 1, \ldots, k-1$ and all $(1_1, \ldots, 1_{n-m}) \in \{1,2,3\}^{n-m}$, and $D^k f = 0$ in Ω_a. As a was arbitrarily chosen in Ω, f is left-(k)-monogenic in Ω.

Remark: From Theorems 9.1 and 9.2 and the linear independance of the basic poly-nomials $(m!)^{-1} x_0^m p_{1_1 \ldots 1_{n-m}} (x)$, it follows that the second Taylor expansion for a left-(k)-monogenic function is unique.

References

1   R. Delanghe,   On regular-analytic functions with values in a Clifford algebra, Math. Ann. 185 Heft 2 (1970) 91-111.

2   H. Flanders,   Differential forms with applications to the physical sciences, Academic Press, New York (1963).

3   R. Fueter,   Die Funktionentheorie der Differentialgleichungen $\Delta u = 0$ und $\Delta \Delta u = 0$ mit vier reellen Variablen, Comm. Math. Helv. 7 (1934) 307-330.

4 R. Fueter, Uber die analytische Darstellung der regulären Funktionen einer Quaternionenvariablen, Comm. Math. Helv. 8 (1935) 371-378.

5. R.P. Gilbert, Pseudohyperanalytic function theory, Gesellschaft für Mathematik und Datenverarbeitung, Bonn, 77 (1973).

6 A. Kriszten, Areolar monogene und polyanalytische Funktionen, Comm. Math. Helv. 21 (1948) 73-78.

7 H.H. Snijder, A hypercomplex function theory associated with Laplace's equation, V.E.B. Deutscher Verlag der Wissenschaften, Berlin, (1968).

8 M. Spivak, Calculus on manifolds, Benjamin Inc., New York (1965).

9 J.A. Ward, A theory of analytic functions in linear associative algebras, Duke Math. Journal 7 (1940) 233-248.

Freddy Brackx
Seminarie voor Wiskundige Analyse
Rijksuniversiteit te Gent
J. Plateaustraat 22
B-9000 Gent
Belgium

K HABETHA

4 On zeros of elliptic systems of first order in the plane

1. INTRODUCTION

This paper deals with zeros of linear elliptic systems of partial differential equations of first order in the plane. Such a system can be written in the form

$$A_0 \cdot u_x + B_0 \cdot u_y + C_0 \cdot u = 0 ,$$

where A_0, B_0, C_0 are real nxn-matrices; u is a real (column-) vector with n components and the dot means matrix multiplication. The system is defined in a domain G of the plane; the continuity assumptions will be specified later on. The system is called elliptic if

$$\det (A_0 \xi + B_0 \eta) \neq 0$$

for all $(\xi, \eta) \in \mathbb{R}^2$, $\xi^2 + \eta^2 > 0$. Therefore n has to be even, $n = 2r$, and because of $\det A_0 \neq 0$ one can assume A_0 to be the unit matrix:

$$u_x + B_0 \cdot u_y + C_0 \cdot u = 0. \qquad (1)$$

In the case $r = 1$ Carleman [4] proved, that all zeros of the solutions are isolated and of finite order; furthermore, the solutions behave near zeros like holomorphic functions, and there is also a global similarity principle between solutions of (1) and holomorphic functions (Bers [1], Vekua [19]). These three problems are investigated in the following three sections for $r > 1$. When possible normal forms of (1) are avoided. They were introduced by Douglis [6]; however, the existence of normal forms requires strong assumptions on B_0 in (1). (Such assumptions may be found in Petrowski [17].)

2. ISOLATED ZEROS

Theorem 1: If the system (1) has locally Hölder continuous coefficients in G, all zeros of finite order are isolated.

Proof: Bers [2] proved for linear elliptic equations $Lu = 0$ with Hölder continuous coefficients that

$$u(x,y) = p_N(x,y) + 0([(x-x_0)^2 + (y-y_0)^2]^{(N+\epsilon)/2})$$

for $(x,y) \to (x_0, y_0)$, (x_0,y_0) a zero of u of finite order. p_N is a homogeneous polynomial of degree N and a solution of the osculating equation $L_0 u = 0$, where L_0 is the leading part (only the terms of highest order) of L with coefficients taken at (x_0,y_0). As Bers pointed out, [2] p.475, his proof carries over to the case of linear elliptic systems; the necessary fundamental solutions for homogeneous systems with constant coefficients may be found in John [11].

Therefore the question whether a zero of finite order of a solution of (1) is isolated can be decided for systems of the form $u_x + B_0 \cdot u_y = 0$ with a constant matrix B_0. If $B_1 = p^{-1} \cdot B_0 \cdot P$ is the Jordan normal form of B_0, the transformation $u = P \cdot v$ gives us the equation $v_x + B_1 \cdot v_y = 0$. Douglis proved in [6] that this can be written with complex valued functions $w_1, \ldots, w_r$ in the form

$$\frac{\partial w_1}{\partial \bar{z}} = 0, \quad \frac{\partial w_j}{\partial \bar{z}} + a \frac{\partial w_{j-1}}{\partial \bar{z}} + b \frac{\partial w_{j-1}}{\partial z} = 0$$

for $j = 2, \ldots, r$. The first function w_j which does not vanish identically is therefore holomorphic and has only isolated zeros. So v and also u have only isolated zeros if these are of finite order.

There is an immediate

Corollary 2: If equation (1) has (real)analytic coefficients in G, all zeros of the solutions of (1) are of finite order and isolated.

Proof: It is known (e.g. John [11]) that all solutions of an elliptic system with analytic coefficients are themselves analytic. Therefore they have only zeros of finite order and Theorem 1 gives the result.

A proof of this result for the special system dealt with by Douglis was given recently by Gilbert and Wendland [10]. The next theorem determines another special class of systems (1) for which all zeros of solutions are isolated and of finite order.

Unfortunately this is not true in general. Pliś [18] showed that there are elliptic equations of fourth order possessing solutions which are zero for $y < 0$ and not identically zero in every neighbourhood of the origin. Such an equation is equivalent to a system of four complex equations of first order. As this relationship between system and equation is by no means trivial, it is desirable to have an explicit example of a system with nonisolated zeros. Furthermore, it would be of interest to know whether zeros of solutions of (1) are in general isolated only for the case $r = 1$ or also for $r = 2,3$.

Theorem 3: Let the matrices B_0 and C_0 of (1) be locally Hölder continuous in G and let P be a matrix with locally Hölder continuous first derivatives such that $P^{-1} . B_0 . P$ is diagonal in G. Then all solutions of (1) have only isolated zeros of finite order.

It is clear, that P has complex valued elements. The existence of P cannot be proved from continuity assumptions for B_0 alone. But for two extremal cases the condition is obvious, namely if B_0 is Hölder continuous differentiable and (i) has $2r$ distinct eigenvalues at each point of G (see Kato [12] Ch. II, section 5.4) or (ii) is diagonal in the following sense:

$$B_0 = \begin{pmatrix} D_1 & & 0 \\ & \ddots & \\ 0 & & D_r \end{pmatrix}$$

with real 2×2-matrices D_j.

Proof: If $P^{-1} . B_0 . P = B_1$ is diagonal let $u = Pv$ and it follows that

$$v_x + B_1 . v_y + C_1 . v = 0 \tag{2}$$

with $C_1 = P^{-1} . (P_x + B_0 . P_y + C_0 . P)$. As all eigenvalues of B_0 are complex, we can assume

$$B_1 = \begin{pmatrix} \lambda_1 & & & & & 0 \\ & \ddots & & & & \\ & & \lambda_r & & & \\ & & & \bar{\lambda}_1 & & \\ & & & & \ddots & \\ 0 & & & & & \bar{\lambda}_r \end{pmatrix},$$

with $\operatorname{Im}\lambda_j > 0$, $j = 1,\ldots, r$. As $P = (p_1 \ldots p_{2r})$ with the (column) eigenvectors p_j of B_0, and as B_0 is real and B_1 has the above form it follows that

$$P = \begin{pmatrix} P_{11} & \overline{P_{11}} \\ P_{21} & \overline{P_{21}} \end{pmatrix} .$$

Here the P_{jk} are $r \times r$-matrices, and $\overline{P_{jk}}$ is the complex conjugate matrix of P_{jk}. Obviously P_x and P_y are also of this form, and it is easy to see that

$$P^{-1} = \begin{pmatrix} Q_{11} & Q_{12} \\ \overline{Q_{11}} & \overline{Q_{12}} \end{pmatrix} .$$

B_0 and C_0 being real, it follows that

$$C_1 = \begin{pmatrix} C_{11} & C_{12} \\ \overline{C_{12}} & \overline{C_{11}} \end{pmatrix} .$$

Putting

$$B_{11} = \begin{pmatrix} \lambda_1 & & 0 \\ & \ddots & \\ 0 & & \lambda_r \end{pmatrix}, \quad w = \begin{pmatrix} v_1 \\ \vdots \\ v_r \end{pmatrix}, \quad \text{and} \quad \tilde{w} = \begin{pmatrix} v_{r+1} \\ \vdots \\ v_{2r} \end{pmatrix}$$

we get (2) in the form

$$\begin{aligned} & w_x + B_{11} \cdot w_y + C_{11} \cdot w + C_{12} \cdot \tilde{w} = 0 , \\ & \tilde{w}_x + \bar{B}_{11} \cdot \tilde{w}_y + \overline{C_{12}} \cdot w + \overline{C_{11}} \cdot \tilde{w} = 0 . \end{aligned} \tag{3}$$

So it is only necessary to look for (complex) solutions of

$$w_x + B_{11} \cdot w_y + C_{11} \cdot w + C_{12} \cdot \bar{w} = 0 \tag{4}$$

since setting $\tilde{w} = \bar{w}$ the solutions of (3) and (4) are equivalent.

Using the complex differentiations

$$\frac{\partial}{\partial \bar z} = \tfrac{1}{2}\left(\frac{\partial}{\partial x} + i\,\frac{\partial}{\partial y}\right), \quad \frac{\partial}{\partial z} = \tfrac{1}{2}\left(\frac{\partial}{\partial x} - i\,\frac{\partial}{\partial y}\right),$$

we have

$$(E - i\,B_{11}) \, . \, w_{\bar z} + (E + i\,B_{11}) \, . \, w_z + C_{11} \, . \, w + C_{12} \, . \, \bar w = 0 ,$$

where E is the identity matrix. As the elements of B_{11} have positive imaginary part, $(E - i\,B_{11})$ is invertible and with

$$Q = (E - i\,B_{11})^{-1} \, . \, (E + i\,B_{11}) = \begin{pmatrix} \mu_1 & & 0 \\ & \ddots & \\ 0 & & \mu_r \end{pmatrix}, \quad \mu_j = \frac{1 + i\,\lambda_j}{1 - i\,\lambda_j} ,$$

$$A = (E - i\,B_{11})^{-1} \, . \, C_{11} , \qquad B = (E - i\,B_{11})^{-1} \, . \, C_{12} ,$$

we get

$$w_{\bar z} + Q \, . \, w_z + A \, . \, w + B \, . \, \bar w = 0 \tag{5}$$

where the elements of Q, A and B are Hölder continuous, $|\mu_j| < 1$. To prove the theorem by contradiction we assume the origin to be a zero of w of infinite order (isolated or not). We fix a small circle K with center O and radius R. For each j, $j = 1, \ldots, r$, we define a quasiconformal mapping ζ_j of $\bar K$ onto $\bar G_j$, where G_j is a domain, and ζ_j is a solution of

$$\frac{\partial \zeta_j}{\partial \bar z} + \mu_j \frac{\partial \zeta_j}{\partial z} = 0 .$$

Here we have assumed $\zeta_j(0) = 0$. Because of the Hölder continuity of μ_j we have with a constant c_0

$$\frac{1}{c_0} \le \left|\frac{\zeta_j(z)}{z}\right| \le c_0 \tag{6}$$

uniformly for $z \in K$ and $j = 1, \ldots, r$; also the Jacobian J_j of the mapping is bounded away from zero uniformly in K and $j = 1,\dots,r$ (see Lehto-Virtanen [14] Ch. V, section 7.2).

Introducing this new variable in the equation with number j

$$\frac{\partial w_j}{\partial \bar{z}} + \mu_j \frac{\partial w_j}{\partial z} + \sum_{k=1}^{r} (a_{jk} w_k + b_{jk} \bar{w}_k) = 0 \tag{7}$$

we get

$$\frac{\partial w_j}{\partial \bar{\zeta}_j} + \sum_{k=1}^{r} (\tilde{a}_{jk} w_k + \tilde{b}_{jk} \bar{w}_k) = 0 \tag{8}$$

with

$$\tilde{a}_{jk} = a_{jk} [(1 - |\mu_j|^2) \frac{\partial \bar{\zeta}_j}{\partial \bar{z}}]^{-1} , \quad \tilde{b}_{jk} = b_{jk} [(1 - |\mu_j|^2 \frac{\partial \bar{\zeta}_j}{\partial \bar{z}}]^{-1} .$$

$\tilde{a}_{jk}$ and $\tilde{b}_{jk}$ have the same continuity properties as a_{jk} and b_{jk}, in particular they are bounded in G_j uniformly with respect to ζ_j and j. Putting

$$W_j = w_j \zeta_j^{-m}$$

for $m \in \mathbb{N}$ it follows from (8)

$$\frac{\partial W_j}{\partial \bar{\zeta}_j} + \sum_{k=1}^{r} (\alpha_{jk} W_k + \beta_{jk} \bar{W}_k) = 0 \tag{9}$$

with

$$\alpha_{jk} = \tilde{a}_{jk} \zeta_k^{m} \zeta_j^{-m} , \quad \beta_{jk} = \tilde{b}_{jk} \zeta_k^{-m} \zeta_j^{-m} .$$

α_{jk} and β_{jk} are Hölder continuous outside 0 and bounded in G_j uniformly in j. Green's formula for G_j gives us ($d\sigma$ is the area element)

$$W_j(\zeta_j) = \frac{1}{2\pi i} \int_{\partial G_j} \frac{W_j(\tau_j)}{\tau_j - \zeta_j} d\tau_j + \frac{1}{\pi} \int_{G_j} \frac{\sum_{k=1} \alpha_{jk}(\tau_j) W_k(\tau_j) + \beta_{jk}(\tau_j) \bar{W}_k(\tau_j)}{\tau_j - \zeta_j} d\sigma(\tau_j)$$

from which we have

$$|W_j(\zeta_j)| \le \frac{1}{2\pi} \int_{\partial G_j} \frac{|W_j(\tau_j)|}{|\tau_j - \zeta_j|} |d\tau_j| + c_1 \int_{G_j} \frac{\sum_{k=1}^{r} |W_k(\tau_j)|}{|\tau_j - \zeta_j|} d\sigma(\tau_j)$$

with a suitable constant c_1. We integrate with respect to ζ_j over G_j and sum over j;

$$\sum_{j=1}^{r} \int_{G_j} |W_j(\zeta_j)| \, d\sigma(\zeta_j) \le \frac{1}{2\pi} \sum_{j=1}^{r} \int_{G_j} \int_{\partial G_j} \frac{|W_j(\tau_j)| \, |d\tau_j|}{|\tau_j - \zeta_j|} d\sigma(\zeta_j)$$

$$+ c_1 \sum_{j,k=1}^{r} \int_{G_j} |W_k(\tau_j)| \int_{G_j} \frac{d\sigma(\zeta_j)}{|\tau_j - \zeta_j|} d\sigma(\tau_j) .$$

From E. Schmidt's inequality

$$\int_G \frac{d\sigma(\zeta)}{|\zeta - z|} \le 2\pi \{ \frac{1}{\pi} \int_G d\sigma(\zeta) \}^{1/2}$$

we deduce

$$\int_{G_j} \frac{d\sigma(\zeta_j)}{|\tau_j - \zeta_j|} \le 2\pi \{ \frac{1}{\pi} \int_K J_j(z) \, d\sigma(z) \}^{1/2} \le c_2 R$$

with a suitable constant c_2. Hence for all sufficiently small R one has

$$\sum_{j=1}^{r} \int_{G_j} |W_j(\zeta_j)| \, d\sigma(\zeta_j) \le c_3 R \sum_{j=1}^{r} \int_{\partial G_j} |W_j(\zeta_j)| \, |d\zeta_j| .$$

As the Jacobian is bounded away from zero we can transform this back to z getting

$$\sum_{j=1}^{r} \int_K |W_j(z)| \, d\sigma(z) \le c_4 R \sum_{j=1}^{r} \int_{\partial K} |W_j(z)| \, |dz| .$$

We use (6), with

$$\|w\| = \sum_{j=1}^{r} |w_j|$$

we have

$$\int_K \frac{\|w(z)\|}{|z|^m}\, d\sigma(z) \leq c_5 R^{1-m} \int_{\partial K} \|w(z)\|\, |dz|\, .$$

Now it is easy to reproduce the argument by Carleman [5]. w is continuous and not identically zero, therefore we have a $z_0 \in K$ and $\delta > 0$ with $0 < \delta < |z_0| < |z_0| + \delta < R$ and $\mu = \min_{|z-z_0| \leq \delta} \|w(z)\| > 0$, so

$$\frac{\mu\, \pi\, \delta^2}{(|z_0| + \delta)^m} \leq c_6 R^{2-m}\, .$$

Letting $m \to \infty$ we get $\mu = 0$, and hence a contradiction.

3. BEHAVIOUR NEAR A ZERO OF FINITE ORDER

For this section we assume (1) to be written in the following normal form:

$$w_{\bar{z}} + Q \cdot w_z + A \cdot w + B \cdot \bar{w} = 0 \tag{10}$$

with a complex r-vector w and $r \times r$-matrices Q, A, B,

$$Q = \begin{pmatrix} \mu_1 & & \\ q_{21} & \mu_2 & 0 \\ \vdots & & \\ q_{r1} & \cdots\; q_{r(r-1)} & \mu_r \end{pmatrix}, \quad |\mu_j| < 1 \text{ for } j=1,\ldots,r. \tag{11}$$

All coefficients should be Hölder continuous. It is difficult to describe conditions for B_0 in (1) under which the transformation to the normal form (10) along the lines sketched in the proof of Theorem 3 is possible. Conditions of such a type can be found in Petrowski [17]. The form (10), (11) is more general than the one used by Douglis [6] and also the one used by Bojarski [3] and Kühn [13]. In all these cases it is at least assumed, that all μ_j are equal and $q_{jk} = q_{j+m,k+m}$ $(m = 1,\ldots,r-j)$. Then the system (10) can be written in the algebra introduced by Douglis.

For $r = 1$ each solution has the following behaviour near a zero

$$w(z) = a(z-z_0)^n (1+0(1)) \quad \text{for} \quad z \to z_0\, , \tag{12}$$

which means w behaves like a homomorphic function. The theorem by Bers used in the proof of Theorem 1 gives some information in this direction:

$$w(z) = p_N(x,y) + 0(|z - z_0|^{N+\epsilon}) \tag{13}$$

with a homogeneous polynomial p_N of degree N which is a solution of the osculating equation

$$w_{\bar{z}} + B(z_0) \cdot w_z = 0 .$$

This can be sharpened as follows.

Theorem 4: Let w be a solution of (10) with Q given by (11) and Q, A, B locally Hölder continuous, z_0 a zero of w of order N. If w_{j_0} is the first component of w having a zero exactly of order N we have for $z \to z_0$

$$w_j(z) = 0(|z - z_0|^{N+1}) , \quad j = 1, \dots, j_0 - 1 ,$$

$$w_{j_0}(z) = a_{j_0}(\zeta_{j_0} - z_0)^N + 0(|\zeta_{j_0} - z_0|^{N+1}) \tag{14}$$

$$w_j(z) = p_{j\,N}(z - z_0, \overline{z - z_0}) + 0(|z - z_0|^{N+\epsilon}) .$$

Here $a_{j_0} \neq 0$, p_{j_N} is a homogeneous polynomial of degree N and ζ_j is a quasiconformal transformation of a neighbourhood of z_0, $\zeta_j(z_0) = z_0$, which is a solution of

$$\frac{\partial \zeta_j}{\partial \bar{z}} + \mu_j \frac{\partial \zeta_j}{\partial z} = 0 .$$

The equation with index j_0 in (14) is the remaining part from (12) in case $r > 1$. Theorem 4 gives us only some information about the components of w which have a zero of lowest possible order. For all other components nothing can be said, their zeros may be of arbitrary order and not isolated. This is shown by a simple example (oral communication by E. Kühn)

$$\frac{\partial w_1}{\partial \bar{z}} = 0, \qquad \frac{\partial w_2}{\partial \bar{z}} = a w_1 \;;$$

if $w_1 = z$, we have $a = \frac{1}{z} \frac{\partial w_2}{\partial \bar{z}}$ and all orders greater than 1, of the zero of w_2 at 0 are possible. Indeed, w_2 may have a zero of infinite order.

Proof: We assume $z_0 = 0$ and introduce as in the proof of Theorem 3 new variables ζ_j getting

$$\frac{\partial w_j}{\partial \bar{\zeta}_j} + \sum_{k=1}^{j-1} \gamma_{jk} \frac{\partial w_k}{\partial \bar{\zeta}_k} + \sum_{k=1}^{r} (\alpha_{jk} w_k + \beta_{jk} \bar{w}_k) = 0, \quad j = 1, \ldots, r,$$

with suitable coefficients $\gamma_{jk}, \alpha_{jk}, \beta_{jk}$. It follows from Green's formula for a small circle K_j with centre $\zeta_j = 0$, that

$$w_j(\zeta_j) = \frac{1}{2\pi i} \int_{\partial K_j} \frac{w_j(\tau_j)}{\tau_j - \zeta_j} d\tau_j + \frac{1}{\pi} \sum_{k=1}^{j-1} \int_{K_j} \frac{\gamma_{jk}(\tau_j) \frac{\partial w_k}{\partial \bar{\zeta}_k}(\tau_j)}{\tau_j - \zeta_k} d\sigma(\tau_j)$$

$$+ \frac{1}{\pi} \sum_{k=1}^{r} \int_{K_j} \frac{\alpha_{jk}(\tau_j) w_k(\tau_j) + \beta_{jk}(\tau_j) \bar{w}_k(\tau_j)}{\tau_j - \zeta_j} d\sigma(\tau_j) . \tag{15}$$

The proof is based on three simple facts: The boundary integral in (15) is holomorphic in ζ_j, secondly

$$\int_{K_j} \frac{P_N(\tau,\bar{\tau})}{\tau - \zeta} d\sigma(\tau) = q_N(\zeta) + r_{N+1}(\zeta,\bar{\zeta}) \tag{16}$$

where q_N is a polynomial of degree N and r_{N+1} a homogeneous polynomial of degree $N+1$. (16) follows from

$$\frac{1}{\tau - \zeta} = \frac{1}{\tau} + \frac{\zeta}{\tau^2} + \ldots + \frac{\zeta^N}{\tau^{N+1}} + \frac{\zeta^{N+1}}{\tau^{N+1}} \frac{1}{\tau - \zeta}, \tag{17}$$

and (for a circle K with centre 0)

$$\int_K \frac{1}{\tau(\tau - \zeta)} \left(\frac{\bar{\tau}}{\tau}\right)^k d\sigma(\tau) = -\frac{\pi}{k+1} \left(\frac{\bar{\zeta}}{\zeta}\right)^{k+1}$$

(see Kühn [13]. p.36). The third fact is

$$\int_K \frac{0(|\tau|^{N+\epsilon})}{\tau-\zeta}\, d\sigma(\tau) = s_{N+1}(\zeta) + 0(|\zeta|^{N+1+\epsilon}) \tag{18}$$

$$\frac{\partial}{\partial\zeta}\int_K \frac{0(|\tau|^{N+\epsilon})}{\tau-\zeta}\, d\sigma(\tau) = s'_{N+1}(\zeta) + 0(|\zeta|^{N+\epsilon}) \tag{19}$$

with a polynomial s_{N+1} of degree $N+1$. This follows also from (17):

$$\int_K \frac{0(|\tau|^{N+\epsilon})}{\tau-\zeta}\, d\sigma(\tau) = s_{N+1}(\zeta) + \zeta^{N+1}\int_K \frac{0(|\tau|^{n+\epsilon})}{\tau^{N+1}}\left(\frac{1}{\tau-\zeta} - \frac{1}{\tau}\right) d\sigma(\tau) .$$

The last integral is $0(|\zeta|^{\epsilon})$ as can easily be seen by dividing K in $D_1 = \{\tau | \, |\tau-\zeta| < \frac{1}{2}|\zeta|\}$, $D_2 = \{\tau | \, |\tau| < 2|\zeta|\} - D_1$, $D_3 = K - (D_1 \cup D_2)$.
For the derivative we get

$$\frac{\partial}{\partial\zeta}\int_K \frac{0(|\tau|^{N+\epsilon})}{\tau-\zeta}\, d\sigma(\tau) = s'_{N+1}(\zeta) + N\zeta^{N}\int_K \frac{0(|\tau|^{N+\epsilon})}{\tau^{N+1}}\left(\frac{1}{\tau-\zeta} - \frac{1}{\tau}\right) d\sigma(\tau)$$

$$+ \zeta^{N}\frac{\partial}{\partial\zeta}\int_K \frac{0(|\tau|^{N+\epsilon})}{\tau^{N}}\left(\frac{1}{\tau-\zeta} - \frac{1}{\tau} - \frac{\zeta}{\tau^2}\right) d\sigma(\tau)$$

$$= s'_{N+1}(\zeta) + 0(|\zeta|^{N+\epsilon}) + \zeta^{N}\Big\{\int_K \Big[\frac{0(|\tau|^{N+\epsilon})}{\tau^{N}} - \frac{0(|\zeta|^{N+\epsilon})}{\zeta^{N}}\Big]\frac{d\sigma(\tau)}{(\tau-\zeta)^2}$$

$$- \int_K \frac{0(|\tau|^{N+\epsilon})}{\tau^{N+2}}\, d\sigma(\tau)\Big\} .$$

Here the differentiation formula for the singular integral may be found in Bers [1], p.7, together with the Hölder continuity of the expression in the last brackets. This gives the desired result

$$= s'_{N+1}(\zeta) + 0(|\zeta|^{N+\epsilon}) .$$

Now we are able to prove the theorem. We start with w_1, and note that in this case in (15) there is no second term on the right side. The Hölder continuity of the coefficients, the result (13) by Bers, and (16), (18), (19) give us

$$w_1(\zeta_1) = h_1(\zeta_1) + r_{N+1}(\zeta_1,\overline{\zeta}_1) + 0(|\zeta_1|^{N+1+\epsilon})$$

$$\frac{\partial w_1}{\partial \zeta_1}(\zeta_1) = h_1'(\zeta) + \frac{\partial}{\partial \zeta_1} r_{N+1}(\zeta_1,\overline{\zeta}_1) + 0(|\zeta_1|^{N+\epsilon}) , \tag{20}$$

with a homomorphic function h_1 and a homogeneous polynomial r_{N+1} of degree $N+1$.

If w_1 has no zero of order N, we have

$$w_1(\zeta_1) = \tilde{r}_{N+1}(\zeta_1,\overline{\zeta}_1) + 0(|\zeta_1|^{N+1+\epsilon})$$

$$\frac{\partial w_1}{\partial \zeta_1}(\zeta_1) = \tilde{r}_N(\zeta_1,\overline{\zeta}_1) + 0(|\zeta_1|^{N+\epsilon}) ; \tag{21}$$

this gives us the same starting point for w_2 as before and we have the same results for w_2. Repetition proves the theorem for $j = 1,\ldots,j_0$. In each step we have to change the variables from ζ_1 to ζ_2 and so on, but because of the differentiability of the quasiconformal transformations we have the same expression in (21) also for the variables with higher index (also for z).

If w_1 has a zero of order N, we have from (20)

$$w_1(\zeta_1) = a_1\zeta_1^{N} + r_{N+1}(\zeta_1,\overline{\zeta}_1) + 0(|\zeta_1|^{N+1+\epsilon})$$

$$\frac{\partial w_1}{\partial \zeta_1}(\zeta_1) = Na_1\zeta_1^{N-1} + r_N(\zeta_1,\overline{\zeta}_1) + 0(|\zeta_1|^{N+\epsilon}) . \tag{22}$$

Changing the variable to ζ_2 gives us

$$w_1(\zeta_2) = p_N(\zeta_2,\overline{\zeta}_2) + 0(|\zeta_2|^{N+\epsilon})$$

$$\frac{\partial w_1}{\partial \zeta_1}(\zeta_2) = p_{N-1}(\zeta_2,\overline{\zeta}_2) + 0(|\zeta_2|^{N-1+\epsilon}). \tag{23}$$

This is not the same starting point as before and we get only from (15), (16), (18),

(19), (23) and (13)

$$w_2(\zeta_2) = h_2(\zeta_2) + r_N(\zeta_2, \overline{\zeta}_2) + 0(|\zeta_2|^{N+\epsilon}) ,$$

$$\frac{\partial w_2}{\partial \zeta_2}(\zeta_2) = h_2'(\zeta_2) + \frac{\partial}{\partial \zeta_2} r_N(\zeta_2, \overline{\zeta}_2) + 0(|\zeta_2|^{N-1+\epsilon}) .$$

The holomorphic function h_2 has a zero at least of order N at $\zeta = 0$, and so this provides the same information as (23) about w_1. Repetition proves the theorem completely. Better results are available, if all μ_j are equal, so that they can be assumed to be zero, and if Q has Hölder continuously differentiable elements.

4. SIMILARITY PRINCIPLE

For $r = 1$ Bers proved the so-called similarity principle: For every solution of $w_{\overline{z}} + Aw + B\overline{w} = 0$ in G there is a bounded, continuous function s and a holomorphic function f such that in G

$$w = fe^s .$$

For $r > 1$ similarity principles have been proved in special cases by Gilbert [9], Gilbert and Hile [7], Kühn [13], Pascali [15, 16]. In these cases there exists for every solution of the given elliptic system a regular matrix T and a homomorphic (or hyperanalytic) f, such that $w = T \cdot f$ in G.

Here it will be shown that this is not true in general. For this purpose we restrict ourselves to the relatively simple system

$$w_{\overline{z}} + A \cdot w + B \cdot \overline{w} = 0. \tag{24}$$

Pascali [15,16] proved a similarity principle for this system. Unfortunately his proof is incorrect and ensures the existence of T only for sufficiently small domains. In fact not much more can be proved. For clarity we write down this in a theorem, a) is partly the result by Pascali.

Theorem 5:

a) Let the coefficients of (24) be bounded and measurable in the bounded domain G and let w be a continuously differentiable solution of (24), let

$$C = A + B \cdot \begin{pmatrix} \bar{w}_1/w_1 & & 0 \\ & \ddots & \\ 0 & & \bar{w}_r/w_r \end{pmatrix}$$

with $\bar{w}_j(z)/w_j(z) = 0$ for $w_j(z) = 0$. Let M be the Banachspace of $r \times r$-matrices within $\hat{C}$ continuous elements and norm

$$\|T\| = \max_{i,j} \max_{z \in \hat{C}} |t_{ij}(z)|$$

for $T = (t_{ij})$. $J : M \to M$ may be defined by

$$J(T)(z) = \frac{-1}{\pi} \int_G \frac{T(\zeta) \cdot C(\zeta)}{\zeta - c} \, d\sigma(\zeta) .$$

J is a compact operator and if 1 is not an eigenvalue of J there is a unique solution of $T = J(T) + E$. For this solution $\det T \neq 0$ in $\hat{C}$ and $f = T \cdot w$ is a holomorphic vector in G, so that w is similar to $f = T \cdot w$. 1 is not an eigenvalue if J is contracting.

b) In the closure of unit circle D there exists a continuously differentiable matrix A such that there is no regular matrix T whose elements are continuous in $\bar{D}$, for which $T \cdot w$ is a holomorphic vector, and w is a solution of $w_{\bar{z}} + A \cdot w = 0$.

We see that there is, in general, no similarity principle in the class of continuous matrices in $\bar{G}$. It is more or less trivial that one has such a matrix T with $(T \cdot w)$ holomorphic if T is only assumed to be continuous in G, so that it may be unbounded (for the example of b) we give such a matrix). But such an unbounded matrix is not of very much interest as the importance of the similarity principle is that nearly all function-theoretic properties of w may be found from the holomorphic f. For an unbounded T at least the boundary properties of w are transferred to T and not to f.

Proof:

a) If $T.w$ has to be holomorphic, we have to solve $T_{\bar{z}}.w + T.w_{\bar{z}} = (T_{\bar{z}} - T.C).w = 0$. So it is sufficient to solve $T_{\bar{z}} - T.C = 0$. It is known that a solution of $T = J(T) + H$ with a holomorphic matrix H is a solution of $T_{\bar{z}} + T.C = 0$. J is compact in M (see Vekua [19], Ch. I, sections 5,6), so that it has only enumerably many eigenvalues λ_n with zero as convergence point. For each other $\lambda \neq \lambda_n$ $T = \lambda J(T) + T_0$ is solvable for each $T_0 \in M$. If $\lambda = 1$ is not an eigenvalue, $T = J(T) + E$ has a unique solution. This solution is continuous in $\hat{C}$, holomorphic outside G and a solution of $T_{\bar{z}} - T.C = 0$ in G. Moreover $(\det T)_{\bar{z}} = (\operatorname{tr} C)\det T$ as is easily seen, so $\det T = g\,e^s$ with a holomorphic g and a continuous s in $\hat{C}$ because of the similarity principle for $r = 1$. As $\det T(\infty) = \det E = 1$ we have $\det T = e^s \neq 0$ in $\hat{C}$. It is clear that 1 is not an eigenvalue if J is contracting, especially if diameter G or $\|C\|$ are sufficiently small. This is the case proved by Pascali.

b) Let D be the unit circle,

$$A(z) = \begin{pmatrix} 0 & \dfrac{6z^2}{3 - z^2\bar{z}^2} \\ -1 & 0 \end{pmatrix}$$

is continuously differentiable in D and

$$w(z) = \begin{pmatrix} 1 - z^2\bar{z}^2 \\ \bar{z} - \frac{1}{3} z^2\bar{z}^3 \end{pmatrix} \tag{25}$$

is a solution of $w_{\bar{z}} + A.w = 0$. Let S be a matrix, such that $(S.w)$ is holomorphic, (i.e.)

$$(S_{\bar{z}} - S.A).w = 0 .$$

If u is a (column-) vector and $Lu = u_{\bar{z}}^T - u^T.A$ (T means transposition), we have with $L^*v = -v_{\bar{z}}^T - v^T . A^T$ by Green's formula

$$\int_G [(Lu)\,.\,v - (L^*v)\,.\,u]\,d\sigma = \int_G (u^T.\,v)_{\bar z} = \frac{1}{2i}\int_{\partial G} u^T.\,v\,dz\,.$$

In particular, for $G = D$, u a row of the matrix S, and the above solution w, we obtain

$$0 = \int_{|z|=1} u^T.\,w\,dz = \frac{2}{3}\int_{|z|=1} u_2\bar z\,dz\,.$$

Moreover for all $n \in \mathbb{N}$ $z^n w$ is also a solution of $w_{\bar z} + A\,.\,w = 0$, and $(Lu)\,.\,wz^n = 0$, so that

$$\int_{|z|=1} u_2\, z^{n-1}\,dz = 0 \qquad n = 0,1,2,\ldots\,. \tag{26}$$

Now we assumed u_2 to be continuous on ∂D; the Fourier series of u_2

$$\sum_{n\in Z} a_n e^{in\varphi}$$

has no terms with $n \le 0$ because of (26). Therefore u is holomorphic in D and has a zero at 0. From $u_{2\bar z} = \dfrac{6z^2}{3 - z^2\bar z^2}\,u_1$ it follows $u_1 = 0$ in D and therefore from $u_{1\bar z} = -u_2$ also $u_2 = 0$.

Therefore S has to be zero and the theorem is proved.

At last an example will be given for an unbounded T, such that $T\,.\,w = \binom{1}{1}$ which is holomorphic: With $w = \binom{w_1}{w_2}$ from (25)

$$T(z) = \begin{pmatrix} \dfrac{1}{w_1(z)} & 0 \\ 1 - \dfrac{\bar z}{1 - z^2\bar z^2} & \dfrac{1 + z^2\bar z}{1 - \frac{1}{3} z^2\bar z^2} \end{pmatrix}$$

$$\det T(z) = \frac{1}{w_1(z)}\;\frac{1 + z^2\bar z^2}{1 - \frac{1}{3} z^2\bar z^2} \ne 0$$

for $z \in D$.

References

1 L. Bers, Theory of pseudo-analytic functions. Lecture Notes, New York University (1953).

2 L. Bers, Local behaviour of solutions of general linear elliptic equations. Comm. Pure Appl. Math. 8 (1955) 473-496.

3 B.B. Bojarski, Theory of the generalized analytic vector. Annales Polon. Math. 17 (1966) 281-320.

4 T. Carleman, Sur les systèmes linéaires aux dérivées partielles du premier ordre à deux variables. C.R. Acad. Sci. Paris 197 (1933) 471-474.

5 T. Carleman, Sur une problème d'unicité pour les systèmes d'équations aux derivées partielles à deux variables indépendantes. Ark. Mat. Astr. Fys. 26B, Nr.17 (1939) 1-9.

6 A. Douglis A function-theoretic approach to elliptic systems of equations in two variables. Comm. Pure Appl. Math. 6 (1953) 259-289.

7 R.P. Gilbert and G.N. Hile, Generalized hyperanalytic function theory. Bull. Amer. Math. Soc. 78 (1972) 998-1001; also Trans. Amer. Math. Soc. 195 (1974) 1-29.

8 R.P. Gilbert, Constructive methods for elliptic partial differential equations. Lecture Notes in Mathematics 365, Berlin-Heidelberg-New York (1974).

9 R.P. Gilbert, Pseudohyperanalytic function theory. Ber. Ges. Math. Datenverarb. Bonn 77 (1973) 53-63.

10 R.P. Gilbert and W.L. Wendland, Analytic, generalized, hyperanalytic function theory and an application to elasticity. Proc. Royal Soc. Edinburgh, 73A, 22 (1974/75)

11 F. John, Plane waves and spherical means applied to partial equations, Interscience tracts in pure and applied mathematics, 2.London-New York (1955).

12 T. Kato, Perturbation theory for linear operators. Grundlehren der math. Wissenschaften 132. Berlin-Heidelberg-New York (1966).

13 E. Kühn, Über die Funktionentheorie und das Ähnlichkeitsprinzip einer Klasse elliptischer Differentialgleichungssysteme in der Ebene. Dissertation, Universität Dortmund (1974).

14 O. Lehto and K.I. Virtanen, Quasikonforme Abbildungen. Grundlehren der math. Wissenschaften 126. Berlin-Heidelberg-New York (1965).

15 D. Pascali, Vecteurs analytiques généralisés. Revue Roumaine Math. Pure Appl. 10 (1965) 779-808.

16 D. Pascali, Sur la représentation de première espèce des vecteurs analytiques généralisés. Revue Roumaine Math. Pure Appl. 12 (1967) 685-689.

17 I.G. Petrowski, Vorlesungen über partielle Differentialgleichungen. Leipzig (1955).

18 A. Pliś, A smooth linear elliptic differential equation without any solution in a sphere. Comm. Pure. Appl. Math. 14 (1961) 599-617.

19 I.N. Vekua, Verallgemeinerte analytische Funktionen. Berlin (1963).

Klaus Habetha
Abteilung Mathematik
Universität Dortmund
46 Dortmund-Hombruch
Postfach 500
GFR

G JANK and K-J WIRTHS

5 Generalized maximum principles in certain classes of pseudoanalytic functions

In his basic work on pseudo-analytic functions [3] Bers proved that for these functions the following maximum principle is valid: Let D be a bounded domain, (F,G) a generating pair in D and $w(z)$ (F,G)-pseudo-analytic in D, continuous in $\bar{D}$. If $|w(z)| \leq M$ on ∂D, then $|w(z)| \leq K^2 M$ in D, K being a constant depending only on D and (F,G).

In general, the constant K^2 involved in this theorem is greater than unity since there are pseudo-analytic functions that attain their maximum value at an interior point of D.

Further discussions concerning maximum principles and the growth of pseudo-analytic functions can be found for instance in [1], [2] and [4].

In this paper we shall show that it is possible to prove for certain classes of pseudo-analytic functions a generalized 'sharp' maximum principle resp. a generalized 'sharp' Schwarz' Lemma. Our estimates in the following theorem turn out to be sharp since we compare the pseudo-analytic functions with appropriate generators and not with constants or powers of z.

Theorem: Let g be holomorphic in a domain $D \subset C$, $g(z) \neq 0$ in D and $\gamma \in C^2(D)$ real-valued in D, γ^{-2} subharmonic in D. If $w \in C^2(D)$ is a solution of

$$w_{\bar{z}} = \frac{g\,\gamma_{\bar{z}}}{\bar{g}\,\gamma}\,\bar{w}, \tag{1}$$

such that

$$\overline{\lim_{z \to \eta}} \left|\frac{w(z)}{g(z)\,\gamma(z)}\right| \leq 1 \tag{2}$$

for any $\eta \in \partial D$, then

$$|w| \leq |g\,\gamma| \quad \text{in } D. \tag{3}$$

If for some $z_0 \in D$ $|w(z_0)| = |g(z_0)\,\gamma(z_0)|$ then $|w| = |g\,\gamma|$ in D.

Proof: Let $w \not\equiv 0$ and

$$N := \{z \mid z \in D,\ w(z) = 0\}, \quad \dot{D} = D/N .$$

We consider the function

$$U := \log(w\overline{w}) - \log(g\overline{g}\gamma^2),$$

which belongs to the class $C^2(\dot{D})$.
The functions

$$b := \frac{g\,\gamma_{\overline{z}}}{\overline{g}\,\gamma}$$

and

$$b\,\frac{\overline{w}}{w} \quad \text{resp.} \quad \overline{b}\,\frac{w}{\overline{w}}$$

belong to the class $C^1(\dot{D})$. Therefore they are bounded in each compactum in $\dot{D}$.
A computation using (1) yields

$$L[U] := U_{z\overline{z}} + b\,\frac{\overline{w}}{w}\,U_z + \overline{b}\,\frac{w}{\overline{w}}\,U_{\overline{z}}$$

$$= 4b\overline{b} - (\log(g\overline{g}\gamma^2))_{z\overline{z}} + 2\,\mathrm{Re}\,(\frac{\overline{w}}{w}(b_z - b(\log(g\overline{g}\gamma^2))_z))$$

$$= (4b\overline{b} - (\log(\gamma^2))_{z\overline{z}})\ \mathrm{Re}\ (1 - \frac{\overline{g}}{g}\,\frac{\overline{w}}{w})$$

$$= \gamma^2(\gamma^{-2})_{z\overline{z}}\ \mathrm{Re}\ (1 - \frac{\overline{g}}{g}\,\frac{\overline{w}}{w}).$$

As γ^{-2} is subharmonic in D this implies

$$L[U] \geq 0 \text{ in } \dot{D}.$$

Under these circumstances the maximum principle of E. Hopf (cf., e.g. [5] p.61) asserts that $U \equiv M$ in $\dot{D}$ if the function U attains its maximum value M in a point of $\dot{D}$. In this case the assertion of the theorem is an immediate consequence of (2), for the points of N (3) is trivial.

If the function U does not attain its maximum value in D, then

$$U(z) \leq \sup_{\eta \in \partial D} \{ \overline{\lim_{z \to \eta}} U(z) \}$$

and (3) follows from (2) and

$$\lim_{z \to \eta} U(z) = -\infty \quad \text{for} \quad \eta \in N .$$

In this case equality cannot occur for (3) for any point of D. Further we want to study the question whether $|w| = |g\gamma|$ in D implies $w = \pm g\gamma$ in D. This is wrong if there is a function $\varphi \in C^2(D)$, $\varphi\overline{\varphi} = 1$, for which $\varphi g \gamma$ is a solution of (1), i.e.

$$((\varphi^2 - 1)\gamma^2)_{\overline{z}} = 0 \quad \text{in} \quad D .$$

Thus we conclude the existence of a holomorphic function h, $\operatorname{Re} h > 0$ in D, such that

$$\gamma = (h + \overline{h})^{-\frac{1}{2}}. \tag{4}$$

Then $\varphi^2 = -\dfrac{\overline{h}}{h}$.

In all other cases $w = \pm g\gamma$ if equality holds in (3) for some point in D. Only if γ is chosen as in (4) we obtain $L[U] \equiv 0$ for each solution $w \in C^2(D)$ of (1).

The pair $(g\gamma, \frac{ig}{\gamma})$ forms a generating pair for the differential equation (1). That there may exist more than one such pair which can be used to state a generalized maximum principle as in the above theorem will be demonstrated by the example

$$w_{\overline{z}} = 2z\overline{w} . \tag{5}$$

This differential equation has the solutions

$$w_1 = i\, C_1\, e^{-2z\bar{z}}, \quad w_2 = C_2\, z\, e^{z^2+\bar{z}^2}, \quad C_1, C_2 \in \mathbb{R}.$$

The solutions of (5) are analytic. Therefore we can replace the condition (2) by the condition $w(0) = 0$, if we use w_2 as a test function in $D_2 = \{z \mid 0 < |z| < r\}$. In doing so we obtain two possibilities for the function U defined above: Either $\overline{\lim}_{z\to 0} U(z) = -\infty$ or there is an analytic completion U' of U in $D_2 \cup \{0\}$ so that $L[U'] \geq 0$ in $D_2 \cup \{0\}$. In both cases U cannot attain its maximum value at the origin unless it vanishes identically, if $\overline{\lim}_{|z|\to r} U(z) \leq 0$.

So we have two corollaries to the theorem.

Corollary 1: Let w be a solution of (5) in $D_1 = \{z \mid |z| < r\}$ and $\overline{\lim}_{z\to\eta} |w(z)| \leq M$, $\eta \in \partial D_1$. Then

$$|w(z)| \leq M\, e^{-2z\bar{z}+2r^2} \quad \text{for} \quad z \in D_1. \tag{6}$$

If for some $z_0 \in D_1$ equality holds in (6) then

$$w(z) = \pm M\, e^{-2z\bar{z}+2r^2}. \quad \text{('Maximum principle')}$$

Corollary 2: Let w be a solution of (5) in D_2, $w(0) = 0$ and

$$\overline{\lim_{z\to\eta}}\, |w(z)| \leq M\, r\, e^{\eta^2+\bar{\eta}^2} \quad \text{for} \quad \eta \in \partial D_2.$$

Then

$$|w(z)| \leq M\, |z|\, e^{z^2+\bar{z}^2} \quad \text{for} \quad z \in D_2. \tag{7}$$

If for some $z_0 \in D_2$ equality holds in (7) then

$$w(z) = \pm M\, z\, e^{z^2+\bar{z}^2} \quad \text{('Schwarz' Lemma')}$$

References

1 H. Begehr, Das Schwarzsche Lemma und verwandte Sätze für pseudoanalytische Funktionen. Applicable Analysis (to appear).

2 H. Begehr, Über beschränkte verallgemeinerte analytische Funktionen. Analele Stiintifice ale Universitatii 'Al. I. Cuza' din Iasi (to appear).

3 L. Bers, Theory of pseudo-analytic functions. New York University, New York (1953).

4 K. Habetha, Zum Phragmen-Lindelöfschen Prinzip bei quasiholomorphen und pseudoanalytischen Funktionen. Applicable Analysis 2 (1972) 169-185.

5 M.H. Protter and H.F. Weinberger, Maximum Principles in differential equations. Englewood Cliffs, N.J. (1967).

Gerhard Jank
Karl-Joachim Wirths
Abt. Mathematik der Universität Dortmund
D-46 Dortmund 50
Postfach 500500
GFR

S RUSCHEWEYH

6 Hardy spaces of λ-harmonic functions

1. INTRODUCTION AND BASIC RESULTS

In [10] B. Muckenhoupt and E.M. Stein studied Hardy classes of ultraspherical expansions and their related 'Poisson integrals' $(\zeta = \rho e^{i\theta})$:

$$u(\zeta) = \sum_{j=0}^{\infty} a_j \rho^j P_j^{\lambda} (\cos\theta), \ \rho < 1, \ \lambda > 0, \tag{1}$$

where P_μ^ν denote the Gegenbauer polynomials. If (1) is compact convergent in $|\zeta| < 1$, then $u(\zeta)$ is a solution of

$$(\zeta - \bar{\zeta}) u_{\zeta\bar{\zeta}} - \lambda (u_\zeta - u_{\bar{\zeta}}) = 0 \tag{2}$$

in that disc. Note that the singular real axis intersects the domain of regularity of u.

For $z \in \Delta = \{z \mid |z| < 1\}$ let

$$k_\lambda(z) = [\mathrm{Re}\, \frac{1+z}{1-z}]^\lambda \,/\, {}_2F_1(1-\lambda, \lambda; 1; \frac{-|z|^2}{1-|z|^2}), \tag{3}$$

where ${}_2F_1$ stands for the hypergeometric function (the denominator is nonvanishing in Δ, compare Lemma 3), and put $\zeta = i(1+z)/(1-z)$. A straightforward calculation shows that $w(z)$ is a solution of

$$D_\lambda w := w_{z\bar{z}} + \frac{1}{2|z|} (\frac{1}{\partial |z|} \, {}_2F_1(1-\lambda, \lambda; 1; \frac{-|z|^2}{1-|z|^2})) (z w_z + \bar{z} w_{\bar{z}}) = 0 \tag{4}$$

in Δ if and only if $w(z) = k_\lambda(z) u(\zeta)$, where $u(\zeta)$ is a solution of (2) in the upper halfplane.

In the present paper we shall study Hardy spaces of the solutions of (4) in Δ. Although the boundary of Δ is singular for (4) and $\lambda \neq 1$, many basic theorems of the classical theory of h^p - spaces ($\lambda = 1$) can be extended to the general case. A striking difference, however, appears in the discussion of the Hardy classes of univalent solutions (see section 5).

First of all we need a suitable representation theorem for all solutions of (4). Clearly we can restrict ourselves to the case $\lambda \geq \frac{1}{2}$. Let

$$d_{\lambda j}(r) = \frac{\Gamma(\lambda+j)}{\Gamma(\lambda)\Gamma(j+1)}\, r^j\, \frac{{}_2F_1(1-\lambda,\lambda;j+1;-r^2/(1-r^2))}{{}_2F_1(1-\lambda,\lambda;1;-r^2/(1-r^2))}, \quad j = 0,1,\ldots,$$

$d_{\lambda,-j}(r) = d_{\lambda j}(r)$, $j = 1,2,\ldots$. It is easily seen that the functions $e^{ij\varphi}d_{\lambda j}(r)$ are solutions of (4) ($z = re^{i\varphi}$) and the following theorem which is a special case of [6, Th. 1] shows the completeness of this system.

Theorem 1: Let w be a complexvalued function in $|z| < R \leq 1$. Then w is a solution of (4) in $|z| < R$ if and only if it has an expansion

$$w = \sum_{j=-\infty}^{\infty} a_j d_{\lambda j}(r)\, e^{ij\varphi}, \quad z = re^{i\varphi}, \tag{5}$$

which is absolutely (and thus compact) convergent in this disc. The series (5) is uniquely determined for a given solution w and the coefficients can be expressed by the 'Fourier integrals'

$$a_j = \frac{1}{2\pi d_{\lambda j}(r)} \int_{-\pi}^{\pi} w(re^{i\varphi})\, e^{-ij\varphi}\, d\varphi, \quad 0 < r < R. \tag{6}$$

A detailed study of the behaviour of the functions $d_{\lambda j}(r)$ will be fundamental for the investigation of (4). These results are collected in the Appendix to this paper and will be referred to as Lemmata A1-A3.

By h_λ we denote the set of all solutions of (4) in Δ. For convenience and in view of their close connection to the harmonic functions we shall call these solutions λ-harmonic' (there will be no confusion with a similar notation in [10]). Theorem 2 is an immediate consequence of Lemma A3.

Theorem 2: There exists a natural isomorphism between the spaces h_λ for different $\lambda \geq \frac{1}{2}$, generated by $w_{\lambda_1} \to w_{\lambda_2}$ with

$$w_\lambda = \sum_{j=-\infty}^{\infty} a_j d_{\lambda j}(r) e^{ij\varphi} . \tag{7}$$

Actually the main portion of this paper is devoted to the study of this isomorphism. Theorem 3 is a first hint for the strong relations between the functions w_λ for different λ. We shall write clco A for the closed convex hull of a set A.

Theorem 3: Let $w_{\lambda_0} \in h_{\lambda_0}$ for a certain $\lambda_0 \geq \frac{1}{2}$. Then $\text{clco}\,(w_\lambda(\Delta)) = \text{clco}\,(w_{\lambda_0}(\Delta))$ for all $\lambda \geq \frac{1}{2}$.

Proof: By assumption and Theorem 2 we have $w_\lambda \in h_\lambda$, $\lambda \geq \frac{1}{2}$. Thus it will be sufficient to show that every separating straight line for $\text{clco}\,(w_\mu(\Delta))$ has the same property for $\text{clco}\,(w_\nu(\Delta))$ where $\mu, \nu \geq \frac{1}{2}$ are arbitrary. From the invariance properties of h_λ it becomes clear that it will be enough to prove: $\text{Re}\, w_\mu(z) > 0$, $z \in \Delta$, implies $\text{Re}\, w_\nu(z) > 0$, $z \in \Delta$. Let

$$w_\mu = \sum_{j=-\infty}^{\infty} a_j d_{\mu j}(r) e^{ij\varphi} ,$$

and

$$F(r,\rho,e^{i\varphi}) = \sum_{j=-\infty}^{\infty} a_j d_{\mu j}(r) d_{\nu j}(\rho) e^{ij\varphi} .$$

Since $d_{\lambda j}(t)$ is nondecreasing (Hopf's maximum principle) with $\lim_{t \to 1-0} d_{\lambda j}(t) = 1$ (Lemma A2) Theorem 1 implies $F(r,\rho,e^{i\varphi}) \in h_\nu$ $(z = \rho e^{i\varphi})$ for any fixed $r < 1$. The boundary values $(\rho = 1)$ of this function are $w_\mu(re^{i\varphi})$ and thus contained in the right halfplane. The range of a λ-harmonic function is contained in the convex hull of its boundary values (see [13, Th.1] combined with [6, Th.1]). Thus $\text{Re}\, F(r,\rho,e^{i\varphi}) > 0$ for $r < 1$, $\rho < 1$, $\varphi \in \mathbb{R}$, and by continuity our assertion follows from $F(1,\rho,e^{i\varphi}) = w_\nu(\rho e^{i\varphi})$.

Section 2 is devoted to the study of some general properties of L^p-norms of λ-harmonic functions while a generalization of the Poisson-Stieltjes integral is discussed in section 3. In section 4 we investigate the 'λ-analytic' functions which are defined as follows: A λ-harmonic function is called λ-analytic if and only if it has an expansion

$$w = \sum_{j=0}^{\infty} a_j d_{\lambda j}(r) e^{ij\varphi}, \quad z = re^{i\varphi} \in \Delta .$$

This, of course, is only one of the possibilities to introduce λ-analytic functions. Another one would be to consider pseudo-analytic functions whose real parts are λ-harmonic. Our definition, however, leads to some immediate generalizations of classical H^p-theorems.

One final remark should be made concerning the apparently vast structure of equation (4). It is easily seen, that $w \in h_\lambda$ if and only if

$$W(z) = w(z)\,{}_2F_1(1-\lambda,\lambda;1;\frac{-|z|^2}{1-|z|^2}) \tag{8}$$

is a solution of the 'nicer' equation

$$(1 - z\bar{z})^2 W_{z\bar{z}} - \lambda(\lambda - 1) W = 0, \; z \in \Delta . \tag{9}$$

This equation has been studied in many papers by several authors (cf. K.W. Bauer [1], I. Haeseler and St. Ruscheweyh [4], M. Kracht and E. Kreyszig [7], H. Maass [8], St. Ruscheweyh [14], I.N. Vekua [15]) and all results of this paper are easily transformed into theorems concerning (9).

2. PROPERTIES OF THE L^p-MEANS

Let $0 < p \leq \infty$. For a given λ-harmonic function w we define $M_p(r,w) := \|w(re^{i\varphi})\|_p$, where $\| \; \|_p$ denote the common L^p-norms over $[-\pi,\pi]$.

Lemma 1: Let w be λ-harmonic, $\lambda \geq \frac{1}{2}$, and let $1 \leq p \leq \infty$. Then $M_p(r,w)$ is nondecreasing for $0 < r < 1$.

Proof: Hopf's maximum principle takes care of the case $p = \infty$ (compare [13, Th.1]). Now let $1 \leq p < \infty$. A straightforward calculation shows that $|w|^p$ is a subsolution of (4), i.e. $D_\lambda |w|^p > 0$, $z \in \Delta$. Let $0 < s < t < 1$. Since Dirichlet's problem is solvable for equation (4) with respect to the circle $|z| = t$, we find a function v such that

i. $D_\lambda v = 0$, $|z| < t$,

ii. v is continuous in $|z| \leq t$,

iii. $v = |w|^p$ on $|z| = t$.

Thus $D_\lambda(|w|^p - v) \geq 0$, $|z| < t$, and by Hopf's maximum principle we can conclude $|w|^p \leq v$, $|z| \leq t$. By Theorem 1

$$v = \sum_{j=-\infty}^{\infty} b_j d_{\lambda j}(r) e^{ij\varphi},$$

where the series is compact convergent for $|z| < t$. Hence

$$\int_{-\pi}^{\pi} v(\rho e^{i\theta})\, d\theta = 2\pi b_0, \quad 0 < \rho < t, \tag{10}$$

and (10) remains valid for $\rho = t$ by continuity. Thus

$$(M_p(s,w))^p = \frac{1}{2\pi} \int_{-\pi}^{\pi} |w(se^{i\theta})|^p\, d\theta$$

$$\leq \frac{1}{2\pi} \int_{-\pi}^{\pi} v(se^{i\theta})\, d\theta$$

$$= b_0$$

$$= \frac{1}{2\pi} \int_{-\pi}^{\pi} v(te^{i\theta})\, d\theta$$

$$= \frac{1}{2\pi} \int_{-\pi}^{\pi} |w(te^{i\theta})|^p\, d\theta$$

$$= (M_p(t,w))^p,$$

which settles the assertion.

Lemma 2: Let $w = w_{\lambda_0}$ be λ_0-harmonic, and let w_λ be defined as in (7). Then for any $\lambda \geq \frac{1}{2}$, $1 \leq p \leq \infty$, we have

$$\lim_{r\to 1-0} M_p(r,w_\lambda) = \lim_{r\to 1-0} M_p(r,w_{\lambda_0}).$$

Proof: The argument is similar to that one we have used for the proof of Theorem 3. Let

$$w_{\lambda_0} = \sum_{j=-\infty}^{\infty} a_j d_{\lambda_0 j}(r) e^{ij\varphi},$$

and let $\mu,\nu \geq \frac{1}{2}$. Then for fixed $\rho \in (0,1]$

$$\sum_{j=-\infty}^{\infty} a_j d_{\mu j}(\rho) d_{\nu j}(r) e^{ij\varphi} \in h_\nu .$$

Thus, by Lemma 1,

$$\| \sum_{j=-\infty}^{\infty} a_j d_{\mu j}(\rho) d_{\nu j}(r) e^{ij\varphi} \|_p \leq \| \sum_{j=-\infty}^{\infty} a_j d_{\mu j}(\rho) e^{ij\varphi} \|_p = M_p(\rho,w_\mu).$$

$\rho \to 1-0$ gives $M_p(r,w_\nu) \leq \lim_{\rho\to 1-0} M_p(\rho,w_\mu)$ and hence $\lim_{r\to 1-0} M_p(r,w_\nu) \leq \lim_{r\to 1-0} M_p(r,w_\mu)$. Interchanging the roles of μ,ν gives our assertion.

Let $0 < p \leq \infty$. The Hardy class $h_\lambda{}^p \subset h_\lambda$ is the collection of those λ-harmonic functions w for which $\sup_{0<r<1} M_p(r,w) < \infty$. In the case $p \geq 1$ it follows from Lemma 1 that $\sup_{0<r<1} M_p(r,w) = \lim_{r\to 1-0} M_p(r,w)$, and this functional introduces a norm $\| \ \|_p'$ in the linear space $h_\lambda{}^p$. Theorem 4 below is an immediate consequence of Lemmata 1,2.

Theorem 4: Let $1 \leq p \leq \infty$. Then, with the notation of (7), $w_{\lambda_0} \in h_{\lambda_0}{}^p$ for a certain $\lambda_0 \geq \frac{1}{2}$ if and only if $w_\lambda \in h_\lambda{}^p$ for all $\lambda \geq \frac{1}{2}$. Furthermore $\|w_\lambda\|_p' = \|w_{\lambda_0}\|_p'$ for $\lambda \geq \frac{1}{2}$.

Corollary 1: Let $p \geq 1$. Then $h_\lambda{}^p$, for different $\lambda \geq \frac{1}{2}$, are isomorphic Banach spaces.

3. GENERALIZED POISSON-STIELTJES INTEGRALS

In the present section we shall generalize the classical representation theorem for h^p functions by means of the Poisson integrals $(p > 1)$ and Poisson-Stieltjes integral $(p = 1)$ respectively to $h_\lambda{}^p$. This theory rests on the following explicit solution of Dirichlet's problem for (4) with respect to the singular line $|z| = 1$.

Theorem 5: Let $f(e^{i\theta})$ be a real valued, continuous function on $[-\pi, \pi]$. Then there exists a unique function $w(z)$, continuous in $|z| \leq 1$, λ-harmonic in Δ and with $w(e^{i\theta}) = f(e^{i\theta})$, $\theta \in \mathbb{R}$. This function can be represented by

$$w(z) = \frac{1}{2\pi} \int_{-\pi}^{\pi} k_\lambda(ze^{i\theta}) f(e^{i\theta})\, d\theta \tag{11}$$

where k_λ is defined by (3).

Lemma 3: $k_\lambda(z)$ is an 'approximate identity' for L^1, i.e.

i. $k_\lambda(z) > 0, \quad z \in \Delta,$

ii. $\int_{-\pi}^{\pi} k_\lambda(re^{i\theta})\, d\theta = 2\pi, \quad 0 < r < 1,$

iii. for all $\delta > 0$: $\lim_{r \to 1-0} \sup_{\delta \leq |\theta| \leq \pi} k_\lambda(re^{i\theta}) = 0.$

Proof: From Laplace's second integral (cf. [16, p.314]) for Legendre functions and the representation of Legendre functions by means of the hypergeometric function we obtain

$${}_2F_1(1-\lambda, \lambda; 1; -r^2/(1-r^2)) = \frac{1}{2\pi} \int_{-\pi}^{\pi} \left(\operatorname{Re} \frac{1+re^{i\theta}}{1-re^{i\theta}}\right)^\lambda d\theta\,, \tag{12}$$

which takes care of both (i) and (ii). (iii) is easily deduced from the asymptotic behaviour of the function (12) for $r \to 1-0$ as described in Lemma A1.

Proof of Theorem 5: We have $k_\lambda(ze^{i\theta}) \in h_\lambda$ for arbitrary $\theta \in [-\pi, \pi]$, $\lambda \leq \frac{1}{2}$. Thus $w(z)$ as defined by (11) is λ-harmonic. A general property of approximate identities (cf. [5, p.18]) shows that w has the desired boundary values. An obvious application of the maximum principle proves the uniqueness of this solution of

Dirichlet's problem.

Corollary 2: For $\lambda \geq \frac{1}{2}$, $z = re^{i\varphi}$, we have

$$k_\lambda(z) = \sum_{j=-\infty}^{\infty} d_{\lambda j}(r) e^{ij\varphi} .$$

Proof: Since $k_\lambda \in h_\lambda$, it has an expansion (5) with coefficients (6). $e^{ij\varphi} d_{\lambda j}(r)$ is the unique solution of the complex Dirichlet problem with the boundary values $e^{ij\varphi}$, thus we can use (11) to evaluate (6).

Corollary 2 combined with Theorem 1 shows that for every (finite) complex Baire measure μ and for every $L^1[-\pi, \pi]$ function f the integrals

$$Q_\lambda[d\mu] := \frac{1}{2\pi} \int_{-\pi}^{\pi} k_\lambda(ze^{-i\theta}) \ d\mu ,$$

$$Q_\lambda[f] = \frac{1}{2\pi} \int_{-\pi}^{\pi} k_\lambda(ze^{-i\theta}) f(e^{i\theta}) \ d\theta$$

represent λ-harmonic functions.

Theorem 6:

1) $w \in h_\lambda{}^p$ for a certain $p > 1$, $\lambda \geq \frac{1}{2}$, if and only if there exists a function $f \in L^p[-\pi, \pi]$ such that $w = Q_\lambda[f]$.

2) $w \in h_\lambda{}^1$ for a certain $\lambda \geq \frac{1}{2}$ if and only if there exists a complex Baire measure on $[-\pi, \pi]$ such that $w = Q_\lambda[d\mu]$.

3) $w \in h_\lambda$ with $\operatorname{Re} w > 0$ in Δ for a certain $\lambda \geq \frac{1}{2}$ if and only if there exists a positive Baire measure μ on $[-\pi, \pi]$ such that $w = Q_\lambda[d\mu]$.

Proof: This theorem is well known for $\lambda = 1$ and can be reduced to this case. To give an example we prove 3). Put $w = w_\lambda$ and, according to (7), consider $w_1 \in h_1$. By Theorem 3 we have $\operatorname{Re} w_1 > 0$ in Δ and thus the existence of a positive Baire measure μ on $[-\pi, \pi]$ such that $w_1 = Q_1[d\mu]$. Corollary 2 can be used to show $w = Q_\lambda[d\mu]$ with the same μ. The other direction and Parts 1) 2) follow likewise. The following results (analoga to Hardy and Littlewood's maximal theorem (cf. [2, p.11]) and Fatou's theorem (cf. [5, p.34])) can be proved along the common lines, using some obvious properties of the kernels $k_\lambda(z)$.

Theorem 7: To every $p \in (1,\infty]$ there exists a constant A_p with the following property: Let $f \in L^p[-\pi,\pi]$, $w_\lambda = Q_\lambda[f]$ and $U_\lambda(\theta) := \sup_{0<r<1} |w_\lambda(re^{i\theta})|$. Then, for $\lambda \geq \frac{1}{2}$, $U_\lambda \in L^p[-\pi,\pi]$ and $\|U_\lambda\|_p \leq A_p \|f\|_p$.

Theorem 8: Let $w_\lambda \in h_\lambda^p$, $1 \leq p \leq \infty$. Then w_λ has nontangential limits a.e. on $|z| = 1$. These limits form a function $f_\lambda(e^{i\theta}) \in L^p[-\pi,\pi]$ and we have $f_{\lambda_1}(e^{i\theta}) = f_{\lambda_2}(e^{i\theta})$ a.e. for $\lambda_1, \lambda_2 \geq \frac{1}{2}$.

A property of approximate identities (cf. [5, p.18]) and Theorem 6 imply the following extension of the second part of Theorem 4.

Theorem 9: Let $w_{\lambda_0} \in h_{\lambda_0}^p$, $1 < p < \infty$. Then for any $\lambda \geq \frac{1}{2}$ we have

$$\lim_{r\to 1-0} \| w_\lambda(re^{i\theta}) - w_{\lambda_0}(re^{i\theta}) \|_p = 0.$$

One remark concerning the equation (9)

$$(1 - z\bar{z})^2 w_{z\bar{z}} - \lambda(\lambda - 1) w = 0$$

should be made. From (8) and Theorem 6 we have that every positive solution of (9) in Δ can be represented by

$$w = \frac{1}{2\pi} \int_{-\pi}^{\pi} \left(\frac{1 - r^2}{1 + r^2 - 2r\cos(\varphi - \theta)} \right)^\lambda d\mu(\theta), \quad z = re^{i\varphi}, \tag{13}$$

with a positive Baire measure μ, which is also formally a direct extension of Herglotz' formula. (13) contains the following estimate for the positive solutions w of (9) which generalizes a well known and highly useful result of C. Carathéodory on positive harmonic functions in Δ:

$$\left(\frac{1 - r}{1 + r}\right)^\lambda \leq \frac{w(re^{i\theta})}{w(0)} \leq \left(\frac{1 + r}{1 - r}\right)^\lambda, \quad z = re^{i\theta} \in \Delta.$$

4. λ-ANALYTIC FUNCTIONS

A λ-harmonic function is called λ-analytic if it has an expansion

$$\sum_{j=0}^{\infty} a_j d_{\lambda j}(r) e^{ij\varphi}.$$

Obviously $\mathrm{Re}\, w \in h_\lambda$ if $w \in H_\lambda$ where H_λ denotes the class of λ-analytic functions in Δ. On the other hand, if u is a real valued λ-harmonic function, there exists a second real valued λ-harmonic function v (uniquely determined up to an arbitrary additive real constant) such that $w = u + iv \in H_\lambda$. v is called conjugate to u. By H_λ^p, $p > 0$, we denote the subset of λ-analytic functions in $h_\lambda{}^p$. Using the F. and M. Riesz theorem we can strengthen Theorems 6 and 8 for H_λ^1.

Theorem 10: Let $w_{\lambda_0} \in H_{\lambda_0}^1$ for a certain $\lambda_0 \geq \frac{1}{2}$, and let $f(e^{i\theta})$ be the function formed by the boundary values of w_{λ_0}. Then we have for all $\lambda \geq \frac{1}{2}$:

i. $w_\lambda = Q_\lambda[f]$,

ii. $\lim_{r \to 1-0} \| w_\lambda(re^{i\theta}) - w_{\lambda_0}(re^{i\theta}) \|_1 = 0$,

iii. If $f(e^{i\theta}) = 0$ on a set of positive measure then $w_\lambda \equiv 0$.

The next theorem extends the well known results of M. Riesz and A. Kolmogoroff (cf. [2, p.53]) on conjugate functions.

Theorem 11: Let $u \in h_\lambda$ be real valued, and let $v \in h_\lambda$ be that conjugate function to u which fulfils $v(0) = 0$.

1) Let $1 < p < \infty$. Then there exists a constant A_p such that $M_p(r,v) \leq A_p M_p(r,u)$, $0 < r < 1$, for $u \in h_\lambda^p$.

2) Let $0 < p < 1$. Then there exists a constant B_p such that $M_p(r,v) \leq B_p M_p(r,u)$, $0 < r < 1$, for $u \in h_\lambda^1$. In particular $v \in h_\lambda^p$ for every $p \in (0,1)$.

Proof:

1) Let

$$w = u + iv = \sum_{j=0}^{\infty} a_j d_{\lambda j}(r) e^{ij\varphi},$$

and for a fixed $r \in (0,1)$ let

$$f_r(z) = u_r(z) + i v_r(z) = \sum_{j=0}^{\infty} a_j d_{\lambda j}(r) z^j .$$

Then $f_r \in H_1{}^p$ for every $p > 0$ and by the M. Riesz theorem we have

$$M_p(r,v) = \lim_{\rho\to 1-0} M_p(\rho,v_r) \leq \lim_{\rho\to 1-0} A_p M_p(\rho,u_r) = A_p M_p(r,u).$$

2) follows likewise.

The following corollaries are obvious.

Corollary 3:

1) $F \in H_\lambda{}^p$, $p \in (1,\infty)$ if and only if $\text{Re } F\ h_\lambda{}^p$.

2) Let $F \in H_\lambda$, $\text{Re } F \in h_\lambda{}^1$. Then $F \in H_\lambda{}^p$ for every $p \in (0,1)$.

Corollary 4: Let $F \in H_\lambda$, $\text{Re } F > 0$ in Δ. Then $F \in H_\lambda{}^p$ for every $p \in (0,1)$.

Remark: Corollary 4 holds especially for

$$K_\lambda(z) = 1 + 2 \sum_{j=1}^{\infty} d_{\lambda j}(r) e^{ij\varphi} \in H_\lambda , \tag{14}$$

since $\text{Re } K_\lambda(z) = k_\lambda(z)$.

5. UNIVALENT λ-ANALYTIC FUNCTIONS

It is a classical result essentially due to H. Prawitz [12] that every conformal mapping of Δ belongs to $H_1{}^p$ for every $p \in (0,\frac{1}{2})$. The situation is quite different if we consider univalent λ-analytic functions, at least for $\lambda \geq 3/2$.

Theorem 12: Let $w \in H_\lambda$, $\lambda \geq 3/2$, be univalent in Δ. Then $w \in H_\lambda{}^p$ for every $p \in (0,1)$.

The main portion of the proof is contained in the following lemma.

Lemma 4: Let $w_\lambda \in H_\lambda$ be univalent in Δ, $\lambda \geq 3/2$. Then w_1 (see (7)) is univalent in Δ and maps Δ onto a convex domain.

Proof: We recall that an analytic function $f(z)$ maps Δ univalently onto a convex domain if and only if

$$\text{Re } \left(\frac{z f''(z)}{f'(z)} + 1\right) > 0, \quad z \in \Delta. \tag{15}$$

Now let

$$w_\lambda = \sum_{j=0}^{\infty} a_j d_{\lambda j}(r) e^{ij\varphi}$$

and put

$$f_t(z) = \sum_{j=0}^{\infty} a_j d_{\lambda j}(t) z^j, \qquad 0 < t < 1.$$

Obviously $f_t(e^{i\varphi}) = w_\lambda(te^{i\varphi})$, and since this, for $-\pi < \varphi < \pi$, represents a closed Jordan curve we may conclude that $f_t(z)$ is a univalent function in Δ. Furthermore we have $f_{t_1}(\Delta) \subset f_{t_2}(\Delta)$ whenever $t_1 < t_2$ by the univalence of w_λ.

In the language of the geometric theory of functions this situation is described as: $f_t(z)$ forms a subordination chain over $(0,1)$. Since this chain is differentiable with respect to t we are entitled to use a result of Ch. Pommerenke [11] which characterizes such subordination chains and leads in this case to the relation

$$\operatorname{Re} \frac{\frac{\partial}{\partial t} f_t(z)}{z f_t'(z)} \frac{d_{\lambda 1}(t)}{d_{\lambda 1}'(t)} > 0, \; z \in \Delta, \quad 0 < t < 1.$$

If we let $t \to 1-0$ and make use of

i. $\displaystyle \lim_{t \to 1-0} \frac{d_{\lambda j}'(t)}{d_{\lambda 1}'(t)} = j^2, \quad j = 1, 2, \ldots, \quad \lambda \geq 3/2,$ (Lemma A2),

ii. $\displaystyle \lim_{t \to 1-0} f_t(z) = w_1(z)$ uniformly on compacta in Δ,

we obtain $\operatorname{Re}\left(\frac{z w_1''(z)}{w_1'(z)} + 1\right) > 0$, $z \in \Delta$, and according to (15) the assertion follows.

Proof of Theorem 12: Let $f(z)$ be analytic in Δ and map Δ univalently onto a convex domain. P.J. Eenigenburg and F.R. Keogh [3] proved that $f \in H^1$ unless $f = a + b(1 - ze^{i\tau})^{-1}$ where $a, b \in \mathbb{C}$, $b \neq 0$, $\tau \in \mathbb{R}$. Now we let $w_\lambda \in H_\lambda$, $\lambda \geq 3/2$, be univalent. Then, by Lemma 4, w_1 is univalent with a convex image. Thus $w_1 \in H_1^1$ (and consequently $w_\lambda \in H_\lambda^1$) unless $w_1 = a + b(1 - ze^{i\tau})^{-1}$. But in the latter case we find

$$\frac{2w_\lambda(e^{-i\tau}z) - 2a - b}{b} = K_\lambda(z) ,$$

and $K_\lambda \in H_\lambda^p$ for all $p \in (0,1)$ by the remark after Corollary 4.

There seems to be a much stronger connection between univalent λ-analytic functions and convex univalent analytic functions. Actually we conjecture that the inverse of Lemma 4 is true. The key problem is the univalence of $K_\lambda(z)$ which by now has been settled only in the cases $\lambda = 2$ and $\lambda = 3$ (compare [14]).

6. APPENDIX

Lemmata A1-A2 follow from lengthy but straightforward calculations using the expansions of the hypergeometric function near infinity. See for instance [9].

Lemma A1: For $x \to \infty$:

$$ {}_2F_1(1-\lambda,\lambda;1;-x) = \begin{cases} \dfrac{\Gamma(2\lambda-1)}{\Gamma^2(\lambda)} x^{\lambda-1}(1+0(1)), & \lambda > \frac{1}{2}, \\[2ex] \dfrac{\log x}{\pi} x^{-\frac{1}{2}}(1+0(1)), & \lambda = \frac{1}{2}. \end{cases} $$

Lemma A2: For $t = (1-r^2)/r^2 \to 0+0$, $j = 1,2,\ldots$:

$$ d_{\lambda j}(r) = \begin{cases} 1 + \dfrac{j^2}{8(3-2\lambda)} t^2(1+0(1)), & \lambda > 3/2, \\[2ex] 1 + \dfrac{1}{8} j^2 t^2 (\log t)(1+0(1)), & \lambda = 3/2, \\[2ex] 1 + \dfrac{\Gamma(1-2\lambda)\Gamma(\lambda)}{\Gamma(2\lambda-1)\Gamma(1-\lambda)}\left[\dfrac{\Gamma(\lambda+j)}{\Gamma(1-\lambda+j)} - \dfrac{\Gamma(\lambda)}{\Gamma(1-\lambda)}\right] t^{2\lambda-1}(1+0(1)), & \\ & 3/2 > \lambda > 1/2,\ \lambda \neq 1, \\[2ex] 1 + (\psi(j+\frac{1}{2}) - \psi(\frac{1}{2}))(\log t)^{-1} + 0(t), & \lambda = 1/2. \end{cases} $$

$$d'_{\lambda j}(r) = \begin{cases} \dfrac{-j^2 t}{2(3-2\lambda)} \quad (1 + 0(1)), & \lambda > 3/2 , \\ -\tfrac{1}{2} j^2 t (\log\, t)(1 + 0(1)), & \lambda = 3/2 . \end{cases}$$

<u>Lemma A3</u>: For $r \in (0,1)$, $j = 1,2,\ldots,$ we have

$$1 \leq \frac{d_{\lambda j}(r)}{r^j} \leq \frac{\Gamma(\lambda+j)}{j!\Gamma(\lambda)} , \qquad \lambda > 1 ,$$

$$\frac{\Gamma(\lambda+j)}{j!\Gamma(\lambda)} \leq \frac{d_{\lambda j}(r)}{r^j} \leq 1 \qquad \tfrac{1}{2} \leq \lambda < 1 .$$

In particular $\lim\limits_{j\to\infty} (d_{\lambda j}(r))^{1/j} = r, \ 0 < r < 1, \ \lambda \geq \tfrac{1}{2}$.

<u>Proof</u>: $w_0 = {}_2F_1(1-\lambda,\lambda;1;-|z|^2/(1-|z|^2))$ is a solution of equation (9) and positive in Δ (see (12)). Then by the common maximum-minimum principles for elliptic equations we deduce that w_0 is increasing for $\lambda > 1$ and decreasing for $\tfrac{1}{2} \leq \lambda < 1$. Thus

$$A_\lambda(r) := \frac{\partial}{\partial r} (\log\, {}_2F_1(1-\lambda,\lambda;1;-r^2/(1-r^2))$$

is positive for $\lambda > 1$ and negative for $\tfrac{1}{2} \leq \lambda < 1$. The function $v(r) := r^{-j} d_{\lambda j}(r)$ is a positive solution of

$$v'' + (\frac{2j+1}{r} + 2A_\lambda(r))v' + \frac{2j}{r} A_\lambda(r) v = 0,$$

and thus cannot take a positive minimum $(\lambda > 1)$ or a positive maximum $(\tfrac{1}{2} \leq \lambda < 1)$. In a neighbourhood of $r = 0$ we have

$$v(r) = \frac{\Gamma(\lambda+j)}{j!\Gamma(\lambda)} (1 - \frac{\lambda(\lambda-1)j}{j+1} r^2 + 0(r^2)) ,$$

and it is easily seen that

$$\frac{\Gamma(\lambda+j)}{j!\Gamma(\lambda)} \begin{cases} > 1, & \lambda > 1, \\ < 1, & \tfrac{1}{2} \leq \lambda < 1. \end{cases}$$

Thus $v(r)$ has the following properties:

1) $\lambda > 1$: $v(0) > 1$, $v'(0) < 0$. $v(r)$ positive and without minimum in $0 < r < 1$. $\lim_{r \to 1-0} v(r) = 1$.

2) $\frac{1}{2} \leq \lambda < 1$: $v(0) < 1$, $v'(0) > 0$. $v(r)$ positive and without maximum in $0 < r < 1$. $\lim_{r \to 1-0} v(r) = 1$.

By continuity this implies $v(0) \geq v(r) \geq 1$ $(\lambda > 1)$ and $v(0) \leq v(r) \leq 1$ $(\frac{1}{2} \leq \lambda < 1)$ for $0 < r < 1$.

After [9, p. 12]

$$\left(\frac{\Gamma(\lambda+j)}{\Gamma(1+j)}\right)^{1/j} = j^{(\lambda-1)/j}\,(1 + 0(1)), \qquad j \to \infty,$$

such that

$$\lim_{j \to \infty} \left(\frac{\Gamma(\lambda+j)}{j!\,\Gamma(\lambda)}\right)^{1/j} = 1, \qquad \lambda > \tfrac{1}{2}.$$

Lemma A3 is established.

References

1 K.W. Bauer, Über die Lösungen der elliptischen Differentialgleichung $(1 \pm z\bar{z})^2 w_{z\bar{z}} + \lambda w = 0$, J. reine angew. Math. 221 (1966) Teil I: 48-84, Teil II: 176-196.

2 P.L. Duren, Theory of H^p spaces. Academic Press (1970).

3 P.J. Eenigenburg and F.R. Keogh, On the Hardy class of some univalent functions and their derivatives. Michigan Math. J. 17 (1970) 335-346.

4 I. Haeseler and St. Ruscheweyh, Singuläre Eichlerintegrale und verallgemeinerte Eisensteinreihen. Math. Ann. 203 (1973) 251-259.

5 K. Hoffman, Banach spaces of analytic functions. Prentice Hall (1965).

6 G. Jank and St. Ruscheweyh, Funktionenfamilien mit einem Maximumprinzip und elliptische Differentialgleichung II. Monatsch.Math.79 (1975) 103-113.

7 M. Kracht and E. Kreyszig, Bergman-Operatoren mit Polynomen als Erzeugenden. Manuscripta Math. 1 (1969) 369-376.

8 H. Maass, Lectures on modular functions of one complex variable. Tata Institut of Fundamental Research (1964).

9 W. Magnus and F. Oberhettinger and R.P. Soni, Formulas and theorems for the special functions of mathematical physics, Third edition. Springer (1966).

10 B. Muckenhoupt and E.M. Stein, Classical expansions and their relation to conjugate harmonic functions. Trans.Amer.Math.Soc.118 (1965) 17-92.

11 Ch. Pommerenke, Über die Subordination analytischer Funktionen. J.reine agnew.Math. 218 (1965) 159-173.

12 H. Prawitz, Über Mittelwerte analytischer Funktionen. Arkiv Mat.Astr. Fys. 20 (1927-28) No.6, 1-12.

13 St. Ruscheweyh, Funktionenfamilien mit einem Maximumprinzip und elliptische Differentialgleichungen I. Monatsh.Math 78. (1974) 246-255.

14 St. Ruscheweyh, Geometrische Eigenschaften der Lösungen der Differentialgleichung $(1 - z\overline{z})^2 w_{z\overline{z}} - n(n+1)w = 0$. J.reine angew. Math. 270 (1974) 143-157.

15 I N. Vekua, New methods for solving elliptic equations. North Holland (1968).

16 E.T. Whittaker and G.N. Watson, Modern analysis, Fourth edition. Cambridge Univ. Press (1927).

St. Ruscheweyh
Abteilung f. Mathematik
Universität Dortmund
46 Dortmund-Hombruch
Postfach 500
GFR

K W BAUER and H FLORIAN

7 Bergman-Operatoren mit Polynomerzeugenden

SUMMARY

The representation of solutions of partial differential equations by simple operators is of great importance in the study of their function theoretical properties. In the application of Bergman integral operators for instance it therefore is essential to select particularly simple generating functions (kernels). So Florian [3] and Kreyszig [7] investigated operators whose generating functions satisfy ordinary differential equations. Moreover in recent years Kracht and Kreyszig [6,8] just as Florian and Jank [4] have considered generating functions that are polynomials of the variable of integration. These operators generally can be represented in a form free of integrals. (Generally such a form of representation is especially appropriate for investigations of the function theoretic properties of solutions.) Kreyszig [9] has determined a class of differential equations in an explicit fashion which possesses generating functions of a very simple form. The present paper considers the differential equation

(A) $\omega^2 RSw + \lambda\omega Sw - n(n+1-\lambda)w = 0$

where λ any constant, n any natural number,

$$\omega = \varphi(z_1) + \psi(z_2), \quad R = \frac{1}{\varphi'}\frac{\partial}{\partial z_1}, \quad S = \frac{1}{\psi'}\frac{\partial}{\partial z_2},$$

and

i. $\varphi(z_1)$, $\psi(z_2)$ holomorphic in the simple connected domains G_1 resp. G_2,

ii. $\omega\varphi'\psi' \neq 0$ in $G_1 \times G_2$.

At first Bergman operators corresponding to (A) are determined whose generating functions are polynomials. The solutions of (A) obtained by means of these operators are converted to a form free of integrals applying a theorem which was proved in [6]. By the aid of the operators R and S the solutions can be represented in a particularly simple form (Theorem 1).

In the special case where the differential equation (A) takes the form

$$\omega^2 RSw + (n - n^*)\omega Sw - n(n^* + 1)w = 0 \qquad (n, n^* \text{ any natural numbers})$$

by the discussed method solutions are determined that contain two arbitrary holomorphic functions $g(z_1)$ and $h(z_2)$. (Theorem 2.) Finally it is pointed to some important differential equations which are included in (A).

ZUSAMMENFASSUNG

Für die Untersuchung der funktionentheoretischen Eigenschaften der Lösungen partieller Differentialgleichungen ist deren Darstellung durch einfache Operatoren von grosser Bedeutung. Man wird deshalb zum Beispiel bei der Verwendung von Bergmanschen Integraloperatoren bemüht sein, möglichst einfache Erzeugende (Kerne) zu bestimmen. In diesem Zusammenhang sind Erzeugende, die Polynome in einer Hilfsvariablen darstellen, von besonderem Interesse. In der vorliegenden Arbeit werden zunächst für eine Klasse von Differentialgleichungen des Typs

$$w_{z_1 z_2} + b(z_1, z_2) w_{z_2} + c(z_1, z_2) w = 0$$

Polynomerzeugende ermittelt. Mit Hilfe eines von M. Kracht und E. Kreyszig bewiesenen Satzes werden die gewonnenen Lösungen in eine integralfreie Form überführt. Eine solche Form der Darstellung ist für die Untersuchung der funktionentheoretischen Eigenschaften der Lösungen besonders geeignet. Die Verwendung geeigneter Differentialoperatoren gestattet sodann eine besonders einfache Darstellung (Satz 1). Für einen Spezialfall der betrachteten Differentialgleichung lassen sich darüber hinaus Lösungen ermitteln, die zwei beliebige holomorphe Funktionen enthalten (Satz 2). Abschliessend wird auf einige wichtige Differentialgleichungen hingewiesen, die in der hier behandelten Differentialgleichung (1) enthalten sind.

1. EINLEITUNG

Im Anschluss an die grundlegenden Ergebnisse von St. Bergman [2] und I.N. Vekua [11] über die Darstellung von Lösungen partieller Differentialgleichungen durch Integraloperatoren wurden in zahlreichen Arbeiten spezielle Typen von Operatoren untersucht.

Im Zusammenhang mit der Anwendung funktionentheoretischer Methoden bei partiellen Differentialgleichungen ist zum Beispiel die Bestimmung von Bergman Operatoren mit besonders einfachen Erzeugenden von grossem Interesse. So haben under anderem H. Florian [3] und E. Kreyszig [7] Operatoren untersucht, deren Erzeugende einer gewöhnlichen Differentialgleichung genügen. In jüngster Zeit wurden ausserdem von M. Kracht und E. Kreyszig [6,8] sowie H. Florian und G. Jank [4] Bergman Operatoren mit Polynomen als Erzeugenden betrachtet. Hier gilt generell, dass die Darstellung der Lösung auch in eine integralfreie Form gebracht werden kann. Diese Form der Darstellung ist für die Untersuchung der funktionentheoretischen Eigenschaften der Lösungen besonders geeignet. In [9] wurde von E. Kreyszig eine Klasse von Differentialgleichungen explizit charakterisiert, für die Polynomerzeugende von besonders einfacher Form existieren. In der vorliegenden Arbeit werden in Zylindergebieten $G_1 \times G_2$ definierte Lösungen der Differentialgleichung

$$\omega^2 RSw + \lambda\omega Sw = n(n+1-\lambda)w = 0, \quad n \in \mathbb{N}^*), \quad \lambda \in \mathbb{C}, \tag{1}$$

unter Verwendung von Bergman Operatoren mit Polynomen als Erzeugenden ermittelt. Dabei bezeichnen G_1 und G_2 einfach zusammenhängende Gebiete der komplexen Zahlenebene; ausserdem gilt

$$\omega = \varphi(z_1) + \psi(z_2), \quad R = \frac{1}{\varphi'}\frac{\partial}{\partial z_1}, \quad S = \frac{1}{\psi'}\frac{\partial}{\partial z_2}, \tag{2}$$

wobei die Funktionen φ und ψ den folgenden Bedingungen genügen:

(1) $\varphi(z_1)$ und $\psi(z_2)$ holomorph in G_1 bzw. G_2,

(2) $\omega\varphi'\psi' \neq 0$ in $G_1 \times G_2$.

---

* Mit $\mathbb{N}$ bzw. $\mathbb{C}$ wird die Menge der natürlichen bzw. komplexen Zahlen bezeichnet.

Mit Hilfe eines in [6] bewiesenen Satzes werden die Lösungen in eine integralfreie Form überführt und unter Verwendung der in (2) definierten Differentialoperatoren R und S in einer besonders einfachen Weise dargestellt. Sodann wird ein Spezialfall der Differentialgleichung (1) diskutiert, bei dem es möglich ist, mit dem hier behandelten Verfahren Lösungen zu gewinnen, die zwei beliebige in G_1 bzw. G_2 holomorphe Funktionen $g(z_1)$ und $h(z_2)$ enthalten. Abschliessend wird auf einige wichtige Differentialgleichungen hingewiesen, die in (1) enthalten sind.

2 BESTIMMUNG DER POLYNOMERZEUGENDEN

Es bezeichne L den Differentialoperator

$$L := \frac{\partial^2}{\partial z_1 \partial z_2} + \frac{\lambda\varphi'}{\omega}\frac{\partial}{\partial z_2} + \frac{\sigma\varphi'\psi'}{\omega^2}, \quad \lambda, \sigma \in \mathbb{C}. \tag{3}$$

Dann erhält man mit Hilfe des Bergman Operators B gemäss

$$(Bf)(z_1) = \int_{\mathscr{L}} E(z_1,z_2,t)\, f\left(\frac{z_1}{2}(1-t^2)\right) \frac{dt}{(1-t^2)^{1/2}} \tag{4}$$

eine Lösung der Differentialgleichung

$$Lw = 0, \tag{5}$$

falls die Erzeugende $E(z_1,z_2,t)$ in $G_1 \times G_2 \times \mathscr{L}$ eine Lösung der Differentialgleichung

$$(1-t^2)E_{z_2 t} - \frac{1}{t}E_{z_2} + 2tz_1 LE = 0 \tag{6}$$

darstellt, die gewissen Stetigkeitsbedingungen genügt (vgl. [2] bzw. [5]). Dabei ist f eine beliebige in G_1 holomorphe Funktion, während $\mathscr{L}$ eine geeignete rektifizierbare Kurve von -1 bis $+1$ in der t-Ebene bezeichnet.

Setzt man die Erzeugende E als Polynom in t an, so genügt es,

$$E = \sum_{\mu=0}^{n} z_1^{\mu} p_{\mu}(z_1,z_2)\, t^{2\mu} \tag{7}$$

zu betrachten, da die ungeraden Potenzen von t keinen Beitrag zum Integral in (4) liefern. (Der Faktor z_1^μ wurde in Hinblick auf eine Vereinfachung der Rechnung abgespalten.) Setzt man E gemäss (7) in (6) ein, so erhält man durch Koeffizientenvergleich

$$\frac{2\mu+1}{2} p_{\mu+1,z_2} + p_{\mu,z_1 z_2} + \frac{\lambda\varphi'}{\omega} p_{\mu,z_2} + \frac{\sigma\varphi'\psi'}{\omega^2} p_\mu = 0 \tag{8}$$

$$\mu = 0,1,\ldots,n, \quad p_{0,z_2} = 0.$$

Verwendet man speziell

$$p_\mu = \sum_{k=0}^{\mu} \frac{H_k^\mu(z_1)}{\omega^k}, \quad H_k^\mu(z_1) \text{ holomorph in } G_1, \tag{9}$$

so folgt

$$-\frac{2\mu+1}{2}(k+1) H_{k+1}^{\mu+1} - (k+1) H_{k+1}^{\mu'} + \{k(k+1) - \lambda k + \sigma\} \varphi' H_k^\mu = 0, \tag{10}$$

$$\mu = 0,1,\ldots,n, \quad k = 0,1,\ldots,\mu, \quad H_m^j = 0 \text{ für } j > n \text{ und } j < m.$$

Als Abbruchbedingung erhält man mit $\mu = k = n$

$$n(n+1) - \lambda n + \sigma = 0. \tag{11}$$

Damit folgt, dass der Koeffizient σ die Form

$$\sigma = -n(n+1-\lambda), \quad n \in \mathbb{N}, \tag{12}$$

haben muss (der triviale Fall $n = 0$ wurde weggelassen). Nach Wahl von $H_n^n(z_1)$ sind sodann die Funktionen H_k^μ eindeutig gemäss (10) bestimmt.

3. INTEGRALFREIE DARSTELLUNG DER LÖSUNGEN

Unter Verwendung einer in [6], Satz 1 bewiesenen Aussage erhält man bei Berücksichtigung von (10) mit

$$w = \sum_{\mu=0}^{n} \frac{(2\mu)!}{\mu!2^{2\mu}} f_1^{(n-\mu)}(z_1) \sum_{k=0}^{\mu} \frac{H_k^{\mu}(z_1)}{\omega^k} \tag{13}$$

eine integralfreie Darstellung. Dabei ist $f_1(z_1)$ die holomorphe Fortsetzung

$$f_2(z_1) \sum_{m=0}^{\infty} \frac{m!}{2^m(m+n)!} B\left(\frac{1}{2}, m+\frac{1}{2}\right) a_m z_1^{m+n}$$

in G_1, wobei $B(p,q)$ die Betafunktion bezeichnet und

$$f(z_1) = \sum_{m=0}^{\infty} a_m z_1^m .$$

gesetzt wurde. Unter Verwendung der Differentialoperatoren R und S und

$$(a)_\mu = a(a+1)\ldots(a+\mu-1), \quad a \in \mathbb{C},$$

lässt sich eine erheblich einfachere Lösungsdarstellung erzielen, und zwar gilt

Satz 1: Sei $g(z_1)$ ein beliebige in G_1 holomorphe Funktion; dann stellt

$$w = \sum_{k=0}^{n} (-1)^{n-k} \binom{n}{k} (n-\lambda+1)_{n-k} \frac{R^k g}{\omega^{n-k}} \tag{14}$$

eine Lösung von (1) in $G_1 \times G_2$ dar.

Beweis: Ordnet man in (13) nach Potenzen von ω, so folgt

$$w = \sum_{k=0}^{n} \frac{1}{\omega^{n-k}} \sum_{r=0}^{k} \frac{(2n-2r)!}{(n-r)!2^{2n-2r}} H_{n-k}^{n-r} f_1^{(r)}(z_1). \tag{15}$$

Mit

$$g(z_1) = \frac{(2n)!(-1)^n H_n^n(z_1)}{n!2^{2n}(n-\lambda+1)_n} f_1(z_1)$$

erhält (14) die Form

$$w = \sum_{k=0}^{n} \frac{(-1)^k(2n)!(n-\lambda+1)_{n-k}}{k!(n-k)!(n-\lambda+1)_n 2^{2n}} \frac{R^k(H_n^n f_1)}{\omega^{n-k}} . \tag{16}$$

Durch vollständige Induktion über k zeigt man sodann unter Verwendung von (10), dass

$$\sum_{r=0}^{k} \frac{(2n-2r)!2^{2r}}{(n-r)!} H_{n-k}^{n-r} f_1^{(r)}(z_1) = \frac{-1^k (2n)!}{k!(n-k)!(2n-\lambda-k+1)_k} R^k (H_n^n f_1)$$

für $k = 0,1,\ldots,n$ gilt.

Transformiert man die Differentialgleichung (1) gemäss $w = \omega^{-\lambda} W$, so erhält man

$$\omega^2 RSW + \lambda^* \omega RW + \sigma^* W = 0 \tag{17}$$

mit

$$\lambda^* = -\lambda, \ \sigma^* = \lambda - n(n+1) - \lambda) \ . \tag{18}$$

Man erhält mit einer entsprechenden Überlegung wie oben Lösungen von (17) mit einer beliebigen holomorphen Funktion $h(z_2)$, falls die Koeffizienten λ^* und σ^* der zu (11) analogen Abbruchbedingung

$$n^*(n^*+1) - \lambda^* n^* + \sigma^* = 0, \quad n^* \in \mathbb{N},$$

genügen. Setzt man λ^* und σ^* gemäss (18) ein, so folgt $\lambda = n - n^*$, d.h. die Differentialgleichungen (1) und (17) erhalten die Form

$$\omega^2 RSw + (n-n^*)\omega Sw - n(n^*+1)w = 0 , \tag{19}$$

$$\omega^2 RSW + (n^*-n)\omega RW - n^*(n+1)W = 0 . \tag{20}$$

Damit gilt nach geeigneter Normierung der folgende

Satz 2: Seien $g(z_1)$ und $h(z_2)$ beliebige in G_1 bzw. G_2 holomorphe Funktionen; dann stellt

$$w = \sum_{k=0}^{n} \frac{(-1)^k (n+n^*-k)!}{k!(n-k)!} \frac{R^k g}{\omega^{n-k}} + \sum_{k=0}^{n^*} \frac{(-1)(n+n^*-k)!}{k!(n^*-k)!} \frac{S^k h}{\omega^{n-k}} \tag{21}$$

eine Lösung von (19) in $G_1 \times G_2$ dar.

Die hier behandelte Klasse von Differentialgleichungen (1) enthält als Spezialfälle einige Differentialgleichungen, die in verschiedenen Arbeiten untersucht wurden. Setzt man zum Beispiel in (1) $\lambda = 0$, so erhält man die in [4] betrachtete Differentialgleichung. Verwenden man ausserdem $z_1 = z$ und $z_2 = \overline{z}$, so liegt eine in [1] behandelte Differentialgleichung vor, für deren Lösungen ohne Verwendung von Integraloperatoren rekursionsfreie allgemeine Darstellungen ermittelt wurden. Auf Differentialgleichungen dieses Typs wird man zum Beispiel bei der mathematischen Behandlung von zirkulationslosen Unterschallströmungen geführt [10]. Setzt man in (1)

$$\lambda = \frac{\tau}{K}, \quad \varphi(z_1) = Kz_1 ,$$

so liegt eine in [9] behandelte Differentialgleichung vor, der im Zusammenhang mit durch E. Kreyszig untersuchten Klasse P_0 eine besondere Bedeutung zukommt. Verwendet man in (1) $\varphi(z_1) = z_1$ und $\psi(z_2) = -z_2$ und transformiert gemäss $\omega^n w = v$, so erhält man eine Eulersche Differentialgleichung.

References

1 K.W. Bauer und G. Jank, Differentialoperatoren bei einer inhomogenen elliptischen Differentialgleichung. Rend. Ist. Mat. Univ. Triests, Heft II, (1971).

2 St. Bergman, Integral operators in the theory of linear partial differential equations. Ergebn. Math. Grenzgeb. Bd. 23, Berlin Springer (1961).

3 H. Florian, Normale Integraloperatoren. Monatsh. Math. 69 (1965) 18-29.

4 H. Florian und G. Jank, Polynomerzeugende bei einer Klasse von Differentialgleichungen mit zwei unabhängigen Variablen. Monatsh. Math. 75 (1971) 31-37.

5 R.P. Gilbert, Function theoretic methods in partial differential equations. Math. in Science and Eng., New York - London, Academic Press (1969).

6 M. Kracht und E. Kreyszig, Bergman-Operatoren mit Polynomen als Erzeugenden. Manuscripta Math. 1 (1969) 369-376.

7 E. Kreyszig, Über zwei Klassen Bergmanscher Operatoren. Math.Nachr. 37 (1968) 197-202.

8 E. Kreyszig, Bergman-Operatoren der Klasse P. Monatsh.Math. 74 (1970) 437-444.

9 E. Kreyszig, On Bergmanoperators for partial differential equations in two variables. Pac.Journ.Math. 36 (1971) 201-208.

10 E. Lanckau, Eine Anwendung der Bergmanschen Operatorenmethode auf Profilströmungen im Ultraschall. Will. Z. Techn. Hochsch. Dresden 8 (1958/59).

11 I.N. Vekua, Novye metody resenija elliptceskikh (New methods for solving elliptic equations). OGIZ, Moscow and Leningrad (1948): Wiley, New York (1967).

Prof. Dr. Karl Wilhelm Bauer
1. Lehrkanzel und Institut für Mathematik
Technische Hochschule in Graz
Kopernikusgasse 24
A-8010 Graz

Prof. Dr. Helmut Florian
Lehrkanzel und Institut für Angewandte Mathematik
und Informationsverarbeitung
Technische Hochschule in Graz
Steyrergasse 17
A-8010 Graz

J W DETTMAN

8 Related semi-groups and the abstract Hilbert transform

1. INTRODUCTION

This paper will be concerned with three related initial value problems:

$$P_1 \quad \begin{cases} u'' = Au, \ t > 0 \\ u(0) = f \\ u'(0) = 0 \end{cases}$$

$$P_2 \quad \begin{cases} v' = Av, \ t > 0 \\ v(0) = f \end{cases}$$

$$P_3 \quad \begin{cases} w'' = -Aw, \ t > 0 \\ w(0) = f \end{cases}$$

where the prime refers to d/dt and the functions involved are all in some complete Hausdorff barreled locally convex linear topological space X defined over the field R of real numbers. The topology of X is determined by a family of continuous semi-norms $\mathfrak{F} = \{p, q, r, \ldots\}$. The closed linear operator A, which does not depend on t, is defined on $D(A)$, which is dense in X. We shall put conditions on A which will make each of the above problems uniformly well posed in $\overline{R}_+ = \{t \mid 0 \leq t < \infty\}$ so that we can exploit the intimate relationships between them. By uniformly well posed in $\overline{R}_+$ we shall mean (following Fattorini [10]) that in each case there is a dense subspace D such that for every $f \in D$ there exists a unique solution (twice continuously differentiable in $\overline{R}_+$ for P_1 and P_3 and once continuously differentiable for P_2) belonging to $D(A)$, satisfying the differential equation, assuming the initial values in the strong sense, and continuously dependent on the data in the following sense: if $\{f_\alpha\}$ is a generalized sequence in D converging to zero, then the corresponding sequence of solutions converges to zero uniformly on compact subsets of $\overline{R}_+$. In connection with P_3, we shall introduce two operators which are analogous to the real and imaginary parts of an analytic function. In an abstract sense they satisfy generalized Cauchy-Riemann equations. One of these operators yields the solution to P_3 and the other (in the limit) leads to an abstract Hilbert transform. The classical case for these problems is the case where $A = d^2/dx^2$ and X is say $L^p(-\infty,\infty)$. Associated with this operator is the classical Hilbert transform, which can be written

$$Hf = \lim_{y \to 0} \frac{1}{\pi} \int_{-\infty}^{\infty} \frac{t\,f(x-t)}{t^2+y^2}\,dt.$$

The theory of this transform in the case where $f \in L^p(-\infty,\infty)$, $1 < p < \infty$, is well known, and it contains the inversion formula $H^2 f = -f$. Our abstract Hilbert transform will be written as

$$\tilde{H}f = \lim_{y \to 0} \frac{1}{\pi} \int_{-\infty}^{\infty} \frac{t\,U(t)f}{t^2+y^2}\,dt.$$

where $U(t)$ is an equicontinuous group of operators whose infinitesmal generator is $B = A^{1/2}$. We shall prove the inversion formula $\tilde{H}^2 f = -f$. and shall show that the infinitesmal generator of the semi-group associated with P_3 is $B\tilde{H}$. The idea of replacing the translation $f(x-t)$ by $U(t)f$ has been used in an analogous case by Westphal [17] in characterizing fractional powers of infinitesmal group generators.

2. ORIENTATION

Fattorini [10] has shown (see also Goldstein [13]) that if P_1 is uniformly well-posed in $\overline{R}_+$ and the solution is of no more than exponential growth in t, then P_2 is also uniformly well-posed in $\overline{R}_+$ and hence that A must be the infinitesmal generator of a strongly continuous semi-group. The principal tool in this result is the transformation

$$v(t) = \frac{1}{\sqrt{(\pi t)}} \int_0^{\infty} e^{-s^2/4t}\, u(s)\,ds$$

which takes a solution $u(t)$ of P_1 into a solution $v(t)$ of P_2 (see also [3],

[4], [5], [9]). Balakrishnan [1] and others (see [18]) have shown that if A is the infinitesmal generator of an equicontinuous semi-group $\Omega(t)$, then $-(-A)^{1/2}$ is the generator of an equicontinuous semi-group $\hat{\Omega}(t)$ given by

$$\hat{\Omega}(t)f = \frac{t}{\sqrt{(4\pi)}} \int_0^\infty e^{-t^2/4s} s^{-3/2}\Omega(s) f\, ds.$$

If $v(t) = \Omega(t)f$, $f \in D(A)$, is the unique solution of P_2, then $w(t) = \hat{\Omega}(t)f$ is the solution of P_3 (see also [6], [8], [9]). We can guarantee that each of the problems P_1, P_2 and P_3 are uniformly well-posed in $\overline{R}_+$, if we start with the assumption that P_1 is uniformly well-posed in $\overline{R}_+$ and of zero type, i.e. if $S(t)$ is the operator which maps f into $u(t)$, the solution of P_1, then $S(t)$ is equicontinuous. This will assure us that A has a square root $B = -i(-A)^{1/2}$. Fattorini has shown [10] that if an additional assumption (assumption 6.4 of [10]) is made then B is the infinitesimal generator of a strongly continuous group. In particular, this assumption is automatically satisfied if $X = L^p$, $1 < p < \infty$, (see [11]). To avoid having to make Fattorini's additional assumption or specialize to L^p spaces, we shall assume that B is the infinitesimal generator of an equicontinuous group $U(t)$ and hence that

$$S(t) = \frac{1}{2}[U(t) + U(-t)].$$

We shall also need that the associated problem

$$P_1' \begin{cases} u'' = Au, \ t > 0 \\ u(0) = 0 \\ u'(0) = g \end{cases}$$

is uniformly well-posed in $\overline{R}_+$ and of zero type. If $T(t)$ is the operator which maps g into the solution of P_1', then it is easy to show that

$$T(t)g = \int_0^t S(s)g\, ds.$$

Since $S(t)$ is equicontinuous, for every semi-norm p there exists a positive number M and a semi-norm q such that $p(S(t)g) \leq M\, q(g)$ for all $t \geq 0$ and $g \in X$. Hence,

$$p(T(t)g) \leq \int_0^t p(S(s)g)ds \leq t\, Mq(g).$$

This, however, does not make $T(t)$ equicontinuous. Therefore, we make one additional assumption, namely that zero is in the resolvent set* of B. We can then show that

Theorem 2.1: If zero is in the resolvent set of B, where $B^2 = A$, then P_1' is of zero type.

Proof: B^{-1} is closed, continuous, and has a domain dense in X. Let g be in the domain of B^{-1}. Then $B^{-1}g = f$, $g = Bf$, and

$$T(t)g = \int_0^t S(s)g\,ds = \frac{1}{2}\int_0^t [U(s) + U(-s)]Bf\,ds$$

$$= \frac{1}{2}[U(t) - U(-t)]f = \frac{1}{2}[U(t) - U(-t)]B^{-1}g$$

Given a semi-norm p, there exists a positive number K and a semi-norm q such that $p(U(t)f) \leq K\, q(f)$ for all t and all $f \in X$, and on account of the continuity of B^{-1} a semi-norm r such that $q(B^{-1}g) \leq N r(g)$ for all $g \in D(B^{-1})$. Hence, if $g \in D(B^{-1})$,

$$p(T(t)g) = p([\frac{1}{2} U(t) - U(-t)]B^{-1}g) \leq Kq(B^{-1}g) \leq KNr(g)$$

Since the inequality holds in a dense subspace it holds for all $f \in X$. This establishes the equicontinuity of $T(t)$.

3. GENERALIZED CAUCHY-RIEMANN EQUATIONS

In this section we introduce two new operators and show that they are related by a set of generalized Cauchy-Riemann equations. These operators are defined for $y > 0$ by

---

* Goldstein [12] has shown that if X is a Banach space and zero is in the resolvent set of B then B generates a group. One can then show that Fattorini's assumption 6.4 is satisfied.

$$\Phi(y)f = \frac{y}{\pi}\int_{-\infty}^{\infty} \frac{U(t)f}{t^2+y^2}\,dt$$

$$\psi(y)f = \frac{1}{\pi}\int_{-\infty}^{\infty} \frac{tU(t)f}{t^2+y^2}\,dt.$$

The first is clearly defined for all $f \in X$ by the equicontinuity of $U(t)$ and the integrability of $(t^2+y^2)^{-1}$. The second is not so obvious. We shall show that it exists in the sense of a Cauchy principal value. We first prove the following:

Lemma 3.1: The operator $\int_0^t [U(s) - U(-s)]ds$ is equicontinuous for $t \geq 0$.

Proof: We show that

$$\int_0^t [U(s) - U(-s)]ds = 2[U(\tfrac{t}{2}) - U(-\tfrac{t}{2})]T(\tfrac{t}{2})$$

and hence is equicontinuous since both $U(t)$ and $T(t)$ are equicontinuous. The operators on both sides of this equation are clearly continuous for each $t \geq 0$. Therefore, we need only show that the equality holds on a dense subspace of X. Suppose f belongs to the domain of B^{-1}. Then $g = B^{-1}f$, $f = Bg$, and

$$\int_0^t [U(s) - U(-s)]f\,ds = \int_0^t [U(s) - U(-s)]Bg\,ds = [U(t) + U(-t)]g - 2g$$

$$T(\tfrac{t}{2})f = \frac{1}{2}\int_0^{t/2} [U(s) + U(-s)]Bg = \frac{1}{2}[U(\tfrac{t}{2}) - U(-\tfrac{t}{2})]g$$

$$2[U(\tfrac{t}{2}) - U(-\tfrac{t}{2})]\;T(\tfrac{t}{2})f = [U(\tfrac{t}{2}) - U(-\tfrac{t}{2})]^2 g = [U(t) + U(-t)]g - 2g$$

This completes the proof.

Now consider

$$\int_0^N \frac{t[U(t) - U(-t)]f}{t^2+y^2}\,dt = \frac{t}{t^2+y^2}\,\alpha(t)\Big|_0^N + \int_0^N \frac{t^2-y^2}{(t^2+y^2)^2}\,\alpha(t)\,dt$$

where $\alpha(t) = \int_0^t [U(s) - U(-s)]f\,ds$. By Lemma 3.1 and the integrability of $(t^2-y^2)/(t^2+y^2)^2$, we can pass to the limit as $N \to \infty$. This shows that $\psi(y)f$ is defined for each $y > 0$ and for all $f \in X$ as a Cauchy principal value, i.e.

$$\psi(y)f = \text{P.V.} \int_{-\infty}^{\infty} \frac{t\,U(t)f}{\pi(t^2+y^2)}\,dt = \lim_{N\to\infty} \frac{1}{\pi} \int_{-N}^{N} \frac{t\,U(t)f}{t^2+y^2}\,dt$$

$$= \lim_{N\to\infty} \frac{1}{\pi} \int_{0}^{N} \frac{t\,[U(t)-U(-t)]f}{t^2+y^2}\,dt = \frac{1}{\pi} \int_{0}^{\infty} \frac{t^2-y^2}{(t^2+y^2)^2}\,\alpha(t)\,dt$$

Theorem 3.1: If $f \in D(B)$ then $(\Phi(y)f)' = B\psi(y)f$ and $(\psi(y)f)' = -B\Phi(y)f$, for $y > 0$.

Proof: By the integrability of $(t^2-y^2)/(t^2+y^2)^2$ and $2ty/(t^2+y^2)^2$, we can differentiate under the integral signs and

$$(\Phi(y)f)' = \frac{1}{\pi} \int_{-\infty}^{\infty} \frac{t^2-y^2}{(t^2+y^2)^2}\,U(t)f\,dt$$

$$(\psi(y)f)' = \frac{-2y}{\pi} \int_{-\infty}^{\infty} \frac{t\,U(t)f}{(t^2+y^2)^2}$$

If $f \in D(B)$ then $(U(t)f)' = BU(t)f = U(t)Bf$ and $U(t)Bf$ is strongly continuous. Therefore,

$$B\Phi(y)f = \frac{y}{\pi} \int_{-\infty}^{\infty} \frac{B\,U(t)f}{t^2+y^2}\,dt = \frac{y}{\pi} \int_{-\infty}^{\infty} \frac{(U(t)f)'}{t^2+y^2}\,dt$$

$$= \frac{y\,U(t)f}{\pi(t^2+y^2)}\Bigg|_{-\infty}^{\infty} + \frac{2y}{\pi} \int_{-\infty}^{\infty} \frac{t\,U(t)f}{(t^2+y^2)^2}\,dt = -(\psi(y)g)'$$

where we have used integration by parts and the equicontinuity of $U(t)$. An analogous computation leads to $B\psi(y)f = (\Phi(y)f)'$. In the classical case, where $B = -d/dx$ and $X = L^p(-\infty,\infty)$, $1 \le p < \infty$, these equations are the standard Cauchy-Riemann equations.

Now let $f \in D(B^2)$. Then $(\Phi f)' = B\psi f = \psi Bf$ and $(\psi f)' = -B\Phi f = -\Phi Bf$. Furthermore,

$$(\Phi f)'' = (\psi Bf)' = B\Phi Bf = -B^2\Phi f$$

$$(\psi f)'' = -(\Phi Bf)' = -B\psi Bf = -B^2\psi f$$

which shows that both $\Phi(y)f$ and $\psi(t)f$ satisfy the differential equation of P_3. We shall show that if $f \in D(B^2)$ then $w(y) = \Phi(y)f$ is the solution of P_3 showing that

$$\lim_{y \to 0^+} \Phi(y)f = f.$$

We can write

$$\Phi(y)f - f = \frac{y}{\pi} \int_{-\infty}^{\infty} \frac{U(t)f - f}{t^2 + y^2} dt.$$

Given any semi-norm $p \in \mathfrak{F}$

$$p(\Phi(y)f - f) \leq \frac{y}{\pi} \int_{-\infty}^{\infty} \frac{p(U(t)f - f)}{t^2 + y^2} dt.$$

Clearly, (see [16] p.35) the integral on the right can be made arbitrarily small by taking y sufficiently small using the fact that $p(U(t)f - f) \to 0$ as $t \to 0$ and the equicontinuity of $U(t)$. This shows that $p(\Phi(y)f - f) \to 0$ as $y \to 0^+$ for every semi-norm $p \in \mathfrak{F}$, which shows that $\lim_{y \to 0^+} \Phi(y)f = f$.

By a simple change of variable we can write

$$\Phi(y)f = \frac{2y}{\pi} \int_{0}^{\infty} \frac{[U(t) + U(-t)]f}{2(t^2 + y^2)} dt.$$

If $f \in D(A)$

$$u(t) = \frac{1}{2}[U(t) + U(-t)]f$$

is the solution of P_1 and $\Phi(y)f$ is the solution of P_3. Hence, we have established the relation between the solutions of P_1 and P_3 previously cited in [8] and [9].

4. THE ABSTRACT HILBERT TRANSFORM

We now come to the question of the definition of $\lim_{y \to 0^+} \psi(y)f$. When we write

$$\psi(y)f = \frac{1}{\pi}\int_0^\infty \frac{t[U(t) - U(-t)]f}{t^2 + y^2}\,dt$$

the existence of $\lim_{y\to 0^+} \psi(y)f$ clearly becomes a question of the behaviour of $[U(t) - U(-t)]f$ near $t = 0$. Since $[U(t) - U(-t)]f \to 0$ as $t \to 0$ it is to be expected that the limit exists for a large class of $f \in X$. However, it is not clear that the limit exists for all $f \in X$. We can, nevertheless, prove the following:

Theorem 4.1: If $f \in X$ such that for every continuous semi-norm p in $\mathfrak{F}$, $p([U(t) - U(-t)]f)$ is $0(t^\alpha)$, as $t \to 0$, $0 < \alpha \leq 1$, then $\lim_{y\to 0^+} \psi(y)f$ exists.

Proof: Clearly from Lemma 3.1 and the given order condition $\int_0^\infty t^{-1}[U(t) - U(-t)]f\,dt$ exists. Now consider

$$\psi(y)f - \frac{1}{\pi}\int_0^\infty t^{-1}[U(t) - U(-t)]f\,dt$$

$$= \frac{-y^2}{\pi}\int_0^1 \frac{[U(t) - U(-t)]f}{t(t^2 + y^2)}\,dt + \frac{-y^2}{\pi}\int_1^\infty \frac{[U(t) - U(-t)]}{t(t^2 + y^2)}\,dt.$$

By the equicontinuity of $U(t)$ the second of these integrals goes to zero as $y \to 0^+$. Let p be any continuous semi-norm from $\mathfrak{F}$. Then

$$p([U(t) - U(-t)]f \leq M(p,f)\,t^\alpha$$

and

$$p\left(\frac{-y^2}{\pi}\int_0^1 \frac{[U(t) - U(-t)]f}{t(t^2 + y^2)}\,dt\right)$$

$$\leq \frac{y^2}{\pi} M(p,f) \int_0^1 \frac{t^{\alpha-1}}{t^2 + y^2}\,dt$$

$$\leq \frac{y^\alpha}{\pi} M(p,f) \int_0^\infty \frac{\tau^{\alpha-1}}{1 + \tau^2}\,d\tau \to 0$$

as $y \to 0^+$. Therefore,

$$\lim_{y\to 0^+} \psi(y)f = \frac{1}{\pi}\int_0^\infty \frac{[U(t) - U(-t)]f}{t}\,dt.$$

When $\lim_{y \to 0^+} \psi(y) f$ exists we call it the abstract Hilbert Transform of f and write

$$\tilde{H} f = \lim_{y \to 0^+} \frac{1}{\pi} \int_{-\infty}^{\infty} \frac{t\, U(t) f}{t^2 + y^2} \, dt$$

where the integral is taken as a Cauchy principal value both at the origin and at infinity.

Theorem 4.2: If $f \in D(B)$ or $D(A)$ then

$$\tilde{H} f = \frac{1}{\pi} \int_0^{\infty} \frac{[U(t) - U(-t)] f}{t} \, dt .$$

Proof: Since $D(A) \subset D(B)$ we need only prove the result for $D(B)$. If $f \in D(B)$ then

$$[U(t) - U(-t)] f = \int_0^t [U(s) + U(-s)] Bf ds .$$

Since $U(t)$ is equicontinuous, to every semi-norm p there exists a semi-norm q such that $p([U(t) + U(-t)] Bf \leq K q(Bf)$. Hence

$$p([U(t) - U(-t)] f) \leq \int_0^t p([U(s) + U(-s)] Bf) ds \leq K q(Bf) t .$$

The result follows from Theorem 4.1 with $\alpha = 1$.

There are no doubt many other subsets covered by Theorem 4.1. This is the case which Butzer and Berens [7] refer to as the case of non-optimal approximation. For example, if X is a Banach space and g is of the form

$$g = \int_0^{\beta} (\beta - t)^{\alpha - 1} U(t) f dt$$

$0 < \beta$, $0 < \alpha < 1$, $f \in X$, then $\| [U(t) - U(-t)] g \| = 0(t^{\alpha})$ (see Hille and Phillips [15] p. 327).

In general, it is not to be expected that the hypotheses of Theorem 4.1 will be met for all $f \in X$. There are cases, however, where we can prove that $\tilde{H}$ is a continuous operator defined on X.

Theorem 4.3: If B is a continuous operator, then $\tilde{H}f$ exists for all $f \in X$ and $\tilde{H}$ is continuous.

Proof: We shall show that $\psi(y)$ is equicontinuous and, since $\lim_{y \to 0^+} \psi(y) f = \tilde{H} f$ exists for $f \in D(B)$, by Theorem 4.2, the continuity of $\tilde{H}$ will follow from the Banach-Steinhaus theorem.

Given a semi-norm p, we have

$$p(\psi(y) f) \leq \frac{1}{\pi} \left(\int_0^1 + \int_1^\infty \right) \frac{|t^2 - y^2|}{(t^2 + y^2)^2} \, p(\alpha(t)) dt$$

$$\leq \frac{1}{\pi} \int_0^1 \frac{p(\alpha(t))}{t^2 + y^2} dt + \frac{1}{\pi} \int_1^\infty \frac{p(\alpha(t))}{t^2} dt$$

By Lemma 3.1, the continuity of B, and an inequality already obtained in the proof of Theorem 4.2, there are semi-norms q and r such that $p(\alpha(t)) \leq K q(f)$ and $p(\alpha(t)) \leq M r(f) t^2$. Hence

$$p(\psi(y) f) \leq \frac{M r(f)}{\pi} \int_0^1 \frac{t^2}{t^2 + y^2} dt + \frac{K q(f)}{\pi} \int_1^\infty t^{-2} dt \leq \frac{M r(f) + K q(f)}{\pi}$$

Let $q^*(f) = \sup [q(f), r(f)]$. Then $q^* \in \mathfrak{F}$ and $p(\psi(y) f) \leq \frac{K + M}{\pi} q^*(f)$.

This shows that $\psi(y)$ is equicontinuous and the theorem follows.

Now we come to the representation of the infinitesimal generator associated with P_3. It is well known that $\Phi(y)$ is a C_0 semi-group with infinitesimal generator $-(-A)^{1/2}$. We now prove that

Theorem 4.4: If $f \in D(A)$ then $-(A)^{1/2} f = B\tilde{H}f$.

Proof: By Theorem 4.2 $\tilde{H}f$ exists. If $f \in D(A) = D(B^2)$ $\lim_{y \to 0^+} (\Phi(Y(f)' =$ $= -(A)^{1/2} f$. By Theorem 3.1,

$$\lim_{y\to 0^+} (\Phi(y)f)' = \lim_{y\to 0^+} B\psi(y)f = B \lim_{y\to 0^+} \psi(y)f = B\tilde{H}f$$

where we have used, in addition, the fact that B is closed.

If $X = L^p(-\infty,\infty)$, $1 < p < \infty$, and $B = -d/dx$, then $U(t)f = f(x-t)$ and the transform reduces to the classical Hilbert transform. However, the classical theory does not follow as a special case because zero is not in the resolvent set of B. The classical theory depends on the special structure of the underlying space, where of course the integral is the Lebesgue integral.

5. THE INVERSION FORMULA

In this section, we deal with the inversion of the abstract Hilbert transform. We prepare for this result by proving

Lemma 5.1: If $f \in X$ and F is an arbitrary continuous linear functional defined on X then if we take the integrals in the Cauchy principal value sense

$$\lim_{y\to\infty} \frac{y}{\pi} \int_{-\infty}^{\infty} \frac{<F,U(-t)f>}{(x-t)^2+y^2}\, dt = 0$$

$$\lim_{y\to\infty} \frac{1}{\pi} \int_{-\infty}^{\infty} \frac{(x-t)<F,U(-t)f>}{(x-t)^2+y^2}\, dt = 0$$

Proof: $<F,U(-t)f>$ is a real-valued, bounded, continuous function of t. The integrals exist since they are respectively $<F,U(-x)\Phi(y)f>$ and $<F,U(-x)\psi(y)f>$. In fact,

$$<F,U(-x)\frac{y}{\pi}\int_{-\infty}^{\infty} \frac{U(t)f}{t^2+y^2}\, dt> = <F,\frac{y}{\pi}\int_{-\infty}^{\infty}\frac{U(t-x)f}{t^2+y^2}\, dt>$$

$$= \frac{y}{\pi}\int_{-\infty}^{\infty}\frac{<F,U(-t)f>}{(x-t)^2+y^2}\, dt$$

$$<F,U(-x)\frac{1}{\pi}\int_{-\infty}^{\infty}\frac{tU(t)f}{t^2+y^2}\, dt> = <F,\frac{1}{\pi}\int_{-\infty}^{\infty}\frac{tU(t-x)f}{t^2+y^2}\, dt>$$

$$= \frac{1}{\pi}\int_{-\infty}^{\infty}\frac{(x-t)<F,U(-t)f>}{(x-t)^2+y^2}\, dt$$

Now, integrating by parts, we have

$$\frac{y}{\pi}\int_{-\infty}^{\infty}\frac{U(t)f}{t^2+y^2}\,dt = \frac{2y}{\pi}\int_{-\infty}^{\infty}\frac{S(t)f}{t^2+y^2}\,dt$$

$$= \frac{4y}{\pi}\int_{-\infty}^{\infty}\frac{tT(t)f}{(t^2+y^2)^2}\,dt$$

$$= \frac{4}{\pi y}\int_{-\infty}^{\infty}\frac{\eta T(y\eta)g}{(1+\eta^2)^2}\,d\eta$$

There exists a semi-norm p such that $|<F,g>| \le p(g)$. Also there exist semi-norms q and r and positive numbers K and M such that $p(U(-x)f) \le Kq(f)$ and $q(T(y\eta)f) \le M\,r(f)$. Hence

$$|<F,U(-x)\Phi(y)f>| \le p(U(-x)\Phi(y)f) \le K\,q(\Phi(y)f)$$

$$\le \frac{4K}{\pi y}\int_{0}^{\infty}\frac{\eta q(T(y\eta)f)}{(1+\eta^2)^2}\,d\eta$$

$$\le \frac{4KM}{\pi y}\,r(f)\int_{0}^{\infty}\frac{\eta}{(1+\eta^2)^2}\,d\eta\,.$$

From this it follows that $\lim_{y\to\infty} <F,U(-x)\Phi(y)f> = 0$. The other integral can be handled similarly.

Theorem 5.1: If $\tilde{H}f$ has a Hilbert transform then $\tilde{H}^2 f = -f$.

Proof: Let F be an arbitrary continuous linear functional defined on X. Then since $\psi(y)f \to \tilde{H}f$, $<F,\psi(y)f> \to <F,\tilde{H}f>$ and $<F,U(-x)\psi(y)f> \to <F,U(-x)\tilde{H}f>$. Now let $\tilde{H}f = g$. Then $<F,U(-x)\psi(y)g> \to <F,U(-x)\tilde{H}^2 f>$. If we can show that $<F,U(-x)\tilde{H}^2 f> = <F,-U(-x)f>$, then we will have shown that $U(-x)\tilde{H}^2 f = -U(-x)f$ and hence that $\tilde{H}^2 f = -f$. The problem then reduces to showing that $\lim_{y\to 0^+} <F,U(-x)\psi(y)g> = -<F,U(-x)f>$. Consider

$$\phi(x,y) = \frac{y}{\pi}\int_{-\infty}^{\infty}\frac{<F,U(-t)f>}{(x-t)^2+y^2}\,dt = <F,U(-x)\Phi(y)f>.$$

Since $<F,U(-t)f>$ is a bounded continuous function of t, $\phi(x,y)$ is a bounded real-valued harmonic function in the upper half-plane $y > 0$, and $\lim_{y\to 0^+} \phi(x,y) = <F,U(-x)f>$. Also consider

$$\psi(x,y) = \frac{1}{\pi}\int_{-\infty}^{\infty} \frac{(x-t)<F,U(-t)f>}{(x-t)^2+y^2}\,dt = <F,U(x)\psi(y)f>$$

which is also a bounded real-valued harmonic function in the upper half-plane, and $\lim_{y\to 0^+} \psi(x,y) = <F,U(-x)\tilde{H}f>$. We define

$$\Omega(z) = \phi(x,y) + i\psi(x,y) = \frac{i}{\pi}\int_{-\infty}^{\infty} \frac{<F,U(-t)f>}{z-t}\,dt$$

which is a bounded analytic function in the upper half-plane. According to Hille [14], p.445, $\Omega(z)$ has a Poisson integral representation

$$\Omega(z) = \frac{y}{\pi}\int_{-\infty}^{\infty} \frac{<F,U(-t)f> + i<F,U(-t)\tilde{H}f>}{(x-t)^2+y^2}\,dt.$$

Now consider, $g = \tilde{H}f$

$$\phi^*(x,y) = \frac{y}{\pi}\int_{-\infty}^{\infty} \frac{<F,U(-t)g>}{(x-t)^2+y^2}\,dt$$

$$\psi^*(x,y) = \frac{1}{\pi}\int_{-\infty}^{\infty} \frac{(x-t)<F,U(-t)g>}{(x-t)^2+y^2}\,dt$$

$$\Omega^*(z) = \phi^*(x,y) + i\psi^*(x,y)$$

$$= \frac{y}{\pi}\int_{-\infty}^{\infty} \frac{<F,U(-t)g> + i<F,U(-t)\tilde{H}g>}{(x-t)^2+y^2}\,dt.$$

Then $\mathrm{Re}(\Omega^*) = \mathrm{Im}(\Omega)$. Therefore, we have $\mathrm{Im}(\Omega^*) = -\mathrm{Re}(\Omega)$ to within a constant. We can conclude that the unknown constant is zero by using Lemma 5.1. Therefore, $\psi^*(x,y) = -\phi(x,y)$ and

$$\lim_{y\to 0^+} \psi^*(x,y) = <F,U(-x)\tilde{H}g> = -\lim_{y\to 0^+} \phi(x,y) = -<F,U(-x)f>.$$

This completes the proof.

(This research was supported partly by N.S.F. Grant GP-17867.)

References

1 A.V. Balakrishnan, Fractional powers of closed operators and the semigroups generated by them. Pacific J.Math. 10 (1960) 419-437.

2 L.R. Bragg, Hypergeometric operator series and related partial differential equations. Trans.Amer.Math.Soc. 143 (1969) 319-336.

3 L.R. Bragg and J.W. Dettman, Related problems in partial differential equations. Bull.Amer.Math.Soc. 74 (1968) 375-378.

4 L.R. Bragg and J.W. Dettman, Related partial differential equations and their applications. SIAM J.Appl.Math. 16 (1968) 459-467.

5 L.R. Bragg and J.W. Dettman, An operator calculus for related partial differential equations. J.Math.Anal. and Appl. 22 (1969) 261-271.

6 L.R. Bragg and J.W. Dettman, A class of related Dirichlet and initial value problems. Proc.Amer.Math.Soc. 21 (1969) 50-56.

7 B. L. Butzer and H. Berens, Semi-groups of operators and approximation. Springer-Verlag New York Inc. (1967).

8 J.W. Dettman, Initial-boundary value problems related through the Stieltjes transform. J.Math.Anal. and Appl. 25 (1969) 341-349.

9 J.W. Dettman, The wave, Laplace, and heat equations and related transforms. Glasgow Math.J. 11 (1970) 117-125.

10 H.O. Fattorini, Ordinary differential equations in linear topological spaces I. J.Diff.Eq. 5 (1968) 72-105.

11 H.O. Fattorini, Ordinary differential equations in linear topological spaces II. J.Diff.Eq. 6 (1969) 50-70.

12 J.A. Goldstein, Semi-groups and second order differential equations. J.Funct.Anal. 4 (1969) 50-70.

13 J.A. Goldstein, On a connection between first and second order differential equations in Banach spaces. J.Math.Anal. and Appl. 30 (1970) 246-251.

14 E. Hille, Analytic Function Theory, Vol.II. Ginn and Company, Boston (1962).

15 E. Hille and R.S. Phillips, Functional analysis and semi-groups. Amer.Math. Soc.Coll.Publ. 31, Providence, R.I. (1957).

16 I.C. Titchmarsh, Introduction to the theory of Fourier Integrals. Oxford University Press (1937).

17 U. Westphal, Ein Kalkül für gebrochene Potenzen infinitesmaler Erzeuger von Halbgruppen und Gruppen von Operatoren. Compositio Math. 22 (1970) 67-136.

18 K. Yosida, Functional Analysis. Springer-Verlag New York Inc. (1968).

John W. Dettman
Department of Mathematics
Oakland University
Rochester
Michigan 48063
U S A

R V DUDUCHAVA

9 On singular integral operators on piecewise smooth lines

In N. Muskhelishvili's well-known monograph [15] Chapter V is devoted to singular integral equations of the type

$$a(t)\varphi(t) + \frac{b(t)}{\pi i}\int_\Gamma \frac{\varphi(\tau)\,d\tau}{\tau - t} + \int_\Gamma k(t,\tau)\varphi(\tau)\,d\tau = f(t) \tag{1}$$

where Γ is a piecewise smooth line and $a(t)$, $b(t)$, $f(t)$ and $k(t,\tau)$ are piecewise Hölder continuous functions on Γ. In particular if the condition $\inf|a^2(t) - b^2(t)| > 0$ $(t \in \Gamma)$ holds and solutions of (1) are sought in a certain subclass $h(c_1, \ldots, c_n)$ of the class H^* Noether's three theorems are valid; then the solution of the equation (1) is reduced to the solution of a more simple Fredholm integral equation.

Systems of the above-mentioned equations on a smooth line Γ were studied by N.P. Vekua [16].

In [1] a Banach space of Hölder functions with weight $H^0_\mu(\Gamma, \rho)$ is introduced where $H^0_\mu(\Gamma, \rho)$ is the subclass of H^*, the necessary and sufficient conditions for (1) to be the Noether equation in † $H^0_\mu(\Gamma, \rho)$ were found and the index of the equation was calculated (cf. [1-4]). Then in [4,5] it was indicated how to derive from these results Noether's classic theorems concerning equations (1) and systems of such equations.

In the present paper similar results are obtained for a piecewise smooth line Γ.

The studies here, as in the earlier papers [1-4], are carried out through the use of functional analysis and some of the recent propositions of I.C. Gohberg and N.Ja. Krupnik from [8-14] where they studied the same equations in the spaces L_p

† Equation $Ax = y$ in the Banach space $\mathfrak{B}$ is called the Noether equation if the image ImA of the operator A is closed and the homogeneous equation $Ax = 0$ and its conjugate equation $A^*f = 0$ have a finite number dim ker A and dim coker A of linearly independent solutions respectively; the integer Ind A = dim ker A - dim coker A is called the index of the equation (cf. [7]).

and L_p with weight.

A few words on the content of the paper.

The paper consists of six sections. In Section 1, using some estimations of the Cauchy type integrals from [15], we prove the boundedness of the singular integral operator

$$(S\varphi)(t) = \frac{1}{\pi i} \int_\Gamma \frac{\varphi(\tau)\,d\tau}{\tau - t}$$

on a piecewise smooth line Γ in the space $\overset{0}{H}_\mu(\Gamma,\rho)$ under some assumptions on the weight function $\rho(t)$. Section 2 is an auxiliary one. In Section 3 and 4 we study equation (1) with some additional restrictions on the coefficients $a(t)$ and $b(t)$ and the line Γ. In Section 5 we prove the basic theorem about equation (1). In the last section the relation of our results with the classical theory of the equation (1) are explained. In particular, on the basis of the general theorems from [5], Noether's theorems are recovered in their classical formulations (cf. [15,16]) and some well-known results of N.I. Muskhelishvili, concerning the behaviour of the solutions of equation (1) in the neighbourhood of knots of the line Γ, are obtained (cf. [15], Sections 97, 98).

1. ON BOUNDEDNESS OF THE SINGULAR INTEGRAL OPERATOR

(1) A line Γ is called piecewise smooth if it consists of a finite number of closed or open arcs $\gamma_1,\dots\gamma_r$; the common points of $\gamma_1,\dots,\gamma_r$ along with the end points of the open arcs are called the knots of Γ (cf. [15]); all other points of Γ will be called simple. (If needed, any simple point of the line Γ can be attached to the knots of Γ.)

Now let $c_1,\dots,c_n$ be knots of Γ; put $\Gamma = \bigcup_{k=1}^{r} \gamma_k$ where γ_k are smooth arcs with the endpoints as knots of Γ. Denote by $H_\mu(\Gamma)$ $(0 < \mu < 1)$ the Banach space of Hölder continuous functions of order μ on each arc γ_k $(k = 1,2,\dots,r)$ and continuous elsewhere on Γ with the norm †

† If two arcs of Γ have no touching point, an equivalent norm can be introduced in $H_\mu(\Gamma)$ by the equality
$$\|\varphi\|_\mu = \max_{t\in\Gamma} |\varphi(t)| + \sup_{\substack{t_1,t_2\in\Gamma \\ t_1\neq t_2}} \frac{|\varphi(t_2)-\varphi(t_1)|}{|t_2 - t_1|^\mu}.$$

$$\|\varphi\|_\mu = \max_{t\in\Gamma} |\varphi(t)| + \max_{1\le k\le r} \sup_{\substack{t_1,t_2\in\gamma_k \\ t_1\ne t_2}} \frac{|\varphi(t_2)-\varphi(t_1)|}{|t_2-t_1|^\mu} .$$

Denote by $\overset{0}{H}_\mu(\Gamma,c_1,\dots,c_n)$ the subspace of functions of $H_\mu(\Gamma)$ which vanish at the knots $c_1,\dots,c_n$ of Γ.

Put $\rho(t) = \prod_{k=1}^{n} |t-c_k|^{\alpha_k}$, where $\alpha_k > 0$ $(k = 1,2,\dots,n)$ and let $\overset{0}{H}_\mu(\Gamma,\rho)$ be the Banach space of functions $\varphi(t)$ for which $\rho(t)\varphi(t) \in \overset{0}{H}_\mu(\Gamma,c_1,\dots,c_n)$ and $\|\varphi\|_{\rho,\mu} = \|\rho\varphi\|_\mu$.

(2) Now we will prove the following theorem.

<u>Theorem 1.1</u>: Let Γ be a piecewise smooth line with knots $c_1,\dots,c_n$ and $\rho(t) = \prod_{k=1}^{n} |t-c_k|^{\alpha_k}$.

If the conditions

$$0 < \mu < 1, \quad \mu < \alpha_k < \mu + 1 \quad (k = 1,2,\dots,n) \tag{1.1}$$

hold, then the singular integral operator

$$(S\varphi)(t) = \frac{1}{\pi i} \int_\Gamma \frac{\varphi(\tau)\,d\tau}{\tau - t} \tag{1.2}$$

where the integral is understood in the sense of Cauchy principal value, is a linear bounded transformation in the space† $\overset{0}{H}_\mu(\Gamma,\rho)$.

<u>Proof</u>: Obviously it is sufficient to prove the boundedness of the operator $A = \rho S \rho^{-1} I$ in the space $\overset{0}{H}_\mu(\Gamma,c_1,\dots,c_n)$. For this purpose let $\Gamma = \bigcup_{k=1}^{p} \gamma'_k$ where γ'_k is an open smooth arc if it contains two knots which coincide with its ends or γ'_k is a closed smooth contour containing no more than one knot.

Under these assumptions the function $(A\varphi)(t)$, where $\varphi(t) \in \overset{0}{H}_\mu(\Gamma,c_1,\dots,c_n)$, can be represented in the form

---

† By the method offered in [10] it is easy to prove that conditions (1.1) are also necessary for the boundedness of the operator S in $\overset{0}{H}_\mu(\Gamma,\rho)$.

$$(A\varphi)(t) = \sum_{k=1}^{p} (A_k\varphi)(t), \quad (A_k\varphi)(t) = \frac{\rho(t)}{\pi i} \int_{\gamma_k'} \frac{\varphi(\tau)\,d\tau}{\rho(\tau)(\tau-t)} .$$

By $M_1, M_2, \ldots$ we denote constants depending only on the space $\overset{0}{H}_\mu(\Gamma, c_1, \ldots, c_n)$.

Taking into account that $\varphi(c_j) = 0$ $(j = 1,2,\ldots,n)$ we easily get

$$\max_{t\in\Gamma} |(A\varphi)(t)| \le \max_{t\in\Gamma} \sum_{k=1}^{n} \left| \frac{1}{\pi i} \int_{\gamma_k'} \frac{[\rho(t)-\rho(\tau)]\varphi(\tau)d\tau}{\rho(\tau)(\tau-t)} + \frac{1}{\pi i} \int_{\gamma_k'} \frac{\varphi(\tau)-\varphi(t)}{\tau - t} d\tau + \right.$$

$$\left. + \frac{\varphi(t)}{\pi i} \int_{\gamma_k'} \frac{d\tau}{\tau-t} \right| \le M_1 \|\varphi\|_\mu . \tag{1.3}$$

Assume now that

$$\|\psi\|_\gamma = \sup_{\substack{t_1,t_2\in\gamma \\ t_1 \ne t_2}} \frac{|\psi(t_2) - \psi(t_1)|}{|t_2 - t_1|^\mu} \qquad (\gamma \subset \Gamma) .$$

If $\Gamma - \bigcup_{k=1}^{r} \gamma_k$ is the partition of Γ, involved in the definition of the space $\overset{0}{H}_\mu(\Gamma,\rho)$, for the completion of the proof it is sufficient to show that inequalities

$$\|A_k\varphi\|_{\gamma_j} \le M_2 \|\varphi\|_\mu \tag{1.4}$$

hold for $k = 1,2,\ldots,p$, $j = 1,2,\ldots,r$. In fact: if $\rho_1(t) = \prod_{k=1}^{n} |t-c_k|^{-\mu/2}$ and $\varphi_1(t) = \rho_1(t)\varphi(t)$, then (cf. [15], Section 5) $\varphi_1(t) \in \overset{0}{H}_{\mu/2}(\Gamma, c_1, \ldots, c_n)$; it is easy to prove (cf. (1.2)) that if

$$\psi_0(t) = \frac{\rho_2(t)}{\pi i} \int_\Gamma \frac{\varphi_1(\tau)\,d\tau}{\rho_2(\tau)(\tau-t)}$$

where $\rho_2(t) = \rho_1(t)\rho(t)$, then $\max_{t\in\Gamma} |\psi_0(t)\psi < \infty$; hence $(A\varphi)(c_j) = \rho_1^{-1}(c_j)\psi_0(c_j) = 0$ $(j = 1,2,\ldots,n)$.

These equalities together with (1.3 and (1.4) yield

$$\|A\varphi\|_\mu \leq M_3 \|\varphi\|_\mu .$$

Now let us prove the inequalities (1.4).

If $\gamma_j \subset \gamma_k'$ and γ_k' is a closed smooth contour or an open arc, the proof of (1.4) is given in [2]; it is easy to prove (1.4) in the case when γ_j and γ_k' have no common points (cf. [2]).

If now $\tilde{\rho}(t) = \prod_{k=1}^{n} (t-c_k)^{\alpha_k} = \rho(t)\,[i \sum_{k=1}^{n} \arg(t-c_k)]$, then instead of (1.4) it is sufficient to prove the inequalities (cf. [15], Section 6)

$$\|\tilde{A}_k \varphi\|_{\gamma_j} \leq M_4 \|\varphi\|_\mu \quad (k = 1,2,\ldots,p; \quad j = 1,2,\ldots,r) \tag{1.5}$$

where

$$(\tilde{A}_k \varphi)(t) = \frac{\tilde{\rho}(t)}{\pi i} \int_{\gamma_k'} \frac{\varphi(\tau)\,d\tau}{\rho(\tau)(\tau - t)} .$$

Now assume that the knot c_q is the common point of γ_k' and γ_j; then by $D(c_q)$ we denote a closed disk with the center in c_q containing no other knots as well as the arcs γ_k' and γ_j entirely; let a be a point on γ_k' and $a \notin \gamma_j \cup D(c_q)$. Evidently it is sufficient to prove the following inequalities

$$\|\psi_m\|_\gamma \leq M_5 \|\varphi\|_\mu \tag{1.6}$$

$$\|\psi_m\|_{\gamma_j \setminus \gamma} \leq M_6 \|\varphi\|_\mu \qquad (m = 1,2) \tag{1.7}$$

where $\gamma = \gamma_j \cap D(c_q)$ and

$$\psi_1(t) = \frac{\tilde{\rho}(t)}{\pi i} \int_{\smile{c_q a}} \frac{\varphi(\tau)\,d\tau}{\rho(\tau)(\tau - t)}, \qquad \psi_2(t) = (A_k \varphi)(t) - \psi_1(t) .$$

When $\gamma \subset \overset{\smile}{c_q a}$ $(\gamma \subset \gamma_k \setminus \overset{\smile}{c_q a})$ and $m = 1$ $(m = 2)$ the inequality (1.6) is proved in [2]. If the arcs $\overset{\smile}{c_q a}$ and γ have the common knot c_q, then we continue the arc $\overset{\smile}{c_q a}$ past the end c_q to the boundary of the disk $D(c_q)$ so that the obtained arc γ' is smooth and has with γ only one common point c_q.

Consider now the function

$$\psi_1'(t) = \frac{\tilde{\rho}(t)}{\pi i} \int_{\gamma'} \frac{\varphi_1(\tau)d\tau}{\rho(\tau)(\tau - t)},$$

where $\varphi_1(t) = \varphi(t)$ if $t \in \overset{\smile}{c_q a}$ and $\varphi_1(t) = 0$ if $t \in \gamma' \setminus \overset{\smile}{c_q a}$. It is easy to verify that $\psi_1'(t) = \psi(t)$ if $t \in \overset{\smile}{c_q a} \cup \gamma$ and $\varphi_2(t) \in H_\mu(\gamma')$ because $\varphi(c_q) = 0$; hence we obtain (cf. [2])

$$\| \psi_1' \|_{\gamma''} \leq M_7 \| \varphi \|_\mu ,$$

where $\gamma'' = \gamma' \cap D(c_q)$. Since the function

$$\Psi(z) = \frac{\tilde{\rho}(z)}{\pi i} \int_{\gamma'} \frac{\varphi_1(\tau)d\tau}{\rho(\tau)(\tau - z)}$$

is analytic in the disk $D(c_q)$ cut along the arc γ'' and its boundary values

$$\Psi^\pm(t) = \pm \frac{\tilde{\rho}(t)}{\rho(t)} \varphi_1(t) + \psi_1'(t) \qquad (t \in \gamma'')$$

on γ'' satisfy the condition $\| \Psi^\pm \|_{\gamma''} \leq M_8 \| \varphi \|_\mu$ (cf. [15] Section 21), inequality (1.6) holds when $m = 1$. The proof of (1.6) when $m = 2$ and the arcs γ and $\gamma' \setminus \overset{\smile}{c_q a}$ have only one common knot c_q, is the same as in the previous case.

(1.7) is easily proved in the case $m = 1$ (cf. [2]) if we take into account that $\overset{\smile}{c_q a}$ and $\gamma_j \setminus \gamma$ have no common points or $\gamma_j \setminus \gamma \subset \overset{\smile}{c_q a}$. Now we note that either the arcs $\gamma_k' \setminus \overset{\smile}{c_q a}$ and $\gamma_j \setminus \gamma$ i) have no common points; ii) $\gamma_k' \setminus \overset{\smile}{c_q a} \supset \gamma_j \setminus \gamma$; iii) may have a common knot. In the first two cases inequality (1.7), in which $m = 2$, is not difficult to prove (cf. [2]) and in the last case the proof is similar to that of (1.6). Q.E.D.

2. AUXILIARY SUGGESTIONS

(1) The following lemma is valid (cf. [15] p.443).

Lemma 2.1: Let Γ be a piecewise smooth line and $a(t) \in H_\nu(\Gamma)$ $(0 < \mu < \nu \leq 1)$. There exists a sequence of functions $\{a_k(t)\}_1^\infty$ with properties i) $da_k(t)/dt \in H_1(\Gamma)$; ii) $a_k(t)$ is constant in a small neighbourhood of the knots $c_1, \ldots, c_n$ of Γ; iii) $\lim_k \| a - a_k \|_\mu = 0$.

With the help of preceding lemma we can easily prove the following lemma (cf. [15] Section 49)

Lemma 2.2: Let Γ be a piecewise smooth line with knots $c_1, \ldots, c_n$ and $\rho(t) = \prod_{k=1}^{n} |t - c_k|^{\alpha_k}$, where the numbers $\mu, \alpha_1, \ldots, \alpha_n$ satisfy (1.1).

If $a(t) \in H_\nu(\Gamma)$ and $\nu > \mu$ then the operator

$$(T\varphi)(t) = \frac{1}{\pi i} \int_\Gamma \frac{[a(\tau) - a(t)]\varphi(\tau)\, d\tau}{\tau - t}$$

is compact in the space† $\overset{0}{H}_\mu(\Gamma, \rho)$.

(2) Let Γ be a piecewise smooth line with knots $c_1, \ldots, c_n$. Denote by $H_\mu(\Gamma, c_1, \ldots, c_n)$ the set of Hölder continuous functions of order μ on each smooth arc from Γ with possible discontinuity at the knots.

The piecewise smooth line Γ is called star-shaped if it consists of closed contours $\Gamma_1, \ldots, \Gamma_s$ oriented counterclockwise having only one common point c_1 and without any common interior point. Without loss of generality we suppose that the origin is placed in the area bounded by the contour Γ_1.

If Γ is a star-shaped line $\Gamma = \bigcup_{k=1}^{s} \Gamma_k$ and $a(t) \in H_\mu(\Gamma, c_1)$, then by a_k^+ and a_k^- $(k = 1, 2, \ldots, s)$ denote the limits of $a(t)$ when $t \to c_1$ along Γ_k counterclockwise and clockwise respectively.

Denote by $H_\mu^+(\Gamma, c_1)$ $(H_\mu^-(\Gamma, c_1))$ the subclasses of $H_\mu(\Gamma, c_1)$ consisting of functions $a(t)$ with properties $a_k^- = a_k^+$ $(a_k^+ = a_{k-1}^-, a_s^+ = a_s^-;\ k = 1, 2, \ldots, s)$.

Denote by $\mathfrak{S}(\overset{0}{H}_\mu(\Gamma, \rho))$ the set of compact operators in the space $\overset{0}{H}_\mu(\Gamma, \rho)$.

If Γ is the star-shaped line and the operators P and Q are defined as follows

$$P = \frac{1}{2}(I + S), \qquad Q = \frac{1}{2}(I - S)$$

where I is the identity transformations and S is defined by (1.2), then P and Q are projections in $\overset{0}{H}_\mu(\Gamma, \rho)$: $P^2 = P$, $Q^2 = Q$ (cf. [15] Section 90); they

---

† If $\inf |t_2 - t_1| / s(t_1, t_2) > 0$, where $s(t_1, t_2)$ is the length of the arc $t_1 t_2$, then we can suppose $\mu = \nu$.

project $H_\mu^0(\Gamma,\rho)$ onto subspaces of functions having an analytic continuation in the interior F^+ and exterior F^- of the line Γ respectively.

Using Lemma 2.2 we can easily prove the following lemma (cf. [14]).

Lemma 2.3: Let Γ be the star-shaped line $\rho(t) = \prod_{k=1}^{n} |t-c_k|^{\alpha_k}$ and let the numbers $\mu, \alpha_1, \ldots, \alpha_n$ satisfy (1.2).

If $a(t) \in H_\nu^+ \in (\Gamma, c_1)$, $b(t) \in H_\nu^-(\Gamma, c_1)$ and $\nu > \mu$ (see the footnote on p.115) then $PaP - aP$, $QbQ - bQ \in \mathfrak{S}(H_\mu^0(\Gamma,\rho)$.

Corollary: If $a(t), b(t) \in H_\nu^+(\Gamma, c_1)$ or $a(t), b(t) \in H_\nu^-(\Gamma, c_1)$ then $PaPbP - PabP$, $QaQbQ - QabQ \in \mathfrak{S}(H_\mu^0(\Gamma,\rho))$.

3. SINGULAR INTEGRAL OPERATORS WITH COEFFICIENTS FROM $H_\nu^\pm(\Gamma, c_1)$

Throughout this section we suppose that $\Gamma = \bigcup_{k=1}^{s} \Gamma_k$ is the star-shaped line $\rho(t) = \prod_{k=1}^{n} |t-c_k|^{\alpha_k}$ and the numbers satisfy the relations (1.1).

(1) Now we prove the theorem.

Theorem 3.1: Let $a(t), b(t) \in H_\nu^+(\Gamma, c_1)$ or $a(t), b(t) \in H_\nu^-(\Gamma, c_1)$ and $\nu > \mu$. The operator $A = aP + bQ$ $(A = PaI + QbI)$ is a Φ_+- or a Φ_-- operator in the space† $H_\mu^0(\Gamma,\rho)$ if and only if the condition

$$\inf |a(t) \cdot b(t)| > 0 \quad (t \in \Gamma) \tag{3.1}$$

holds. If the condition is fulfilled, then the invertibility of the operator A depends on the integer†† κ where ‡

† A linear bounded operator A in a Banach space $\mathfrak{B}$ is called a Φ_+ (Φ_-) -operator if its image ImA is closed in $\mathfrak{B}$ and dimker $A < \infty$ (dimcoker $A < \infty$); if A is a Φ_+- and a Φ_--operator it is called Φ-operator (or Noether operator) and the integer Ind A = dimker A - -dimcoker A is called the index of the operator.

†† It is said (cf. [6]) that invertibility of an operator A depends on an integer κ if A is invertible, left or right invertible according to κ being zero, positive or negative respectively.

‡ By the symbol $[g(t)]_\Gamma$ we denote the increase of the function $g(t)$ when t runs Γ counterclockwise.

$$\kappa = -\operatorname{ind}_\Gamma (a^{-1} b) = -\frac{1}{2\pi} \sum_{k=1}^{s} [\arg \frac{b(t)}{a(t)}]_{\Gamma_k} \tag{3.2}$$

and

$$\operatorname{Ind} A = \operatorname{ind}_\Gamma (a^{-1} b) .$$

Proof: First we prove sufficiency of the condition (3.1) for the operator $A = aP + bQ$, when $a(t), b(t) \in H_\nu^+(\Gamma, c_1)$. If $g(t) = b(t)/a(t)$ then $g(t) \in H_\nu^+(\Gamma, c_1)$ and hence $g_k = g_k^+ = g_k^-$ $(k = 1, 2, \dots, s)$. If $u(t) = \sum_{k=1}^{s} g_k \chi_k(t)$ where $\chi_k(t)$ is the characteristic function of Γ_k then $Pu^{\pm 1} P = u^{\pm 1} P$ and $h(t) = g(t)/u(t) = b(t)/a(t) u(t) \in H_\nu(\Gamma)$. Consider the factorization

$$h(t) = h_-(t) t^- \; h_+(t) \tag{3.3}$$

where κ is defined by (3.2) and $h_+(t) = \exp[P(\ln h(t))]$, $h_-(t) = \exp[Q(\ln h(t))]$; evidently $h_\pm^{\pm 1}(t) \in H_\nu(\Gamma, c_1, \dots, c_n)$ and

$$Ph_+^{\pm 1} P\varphi = h_+^{\pm 1} P\varphi, \qquad Qh_-^{\pm 1} Q\varphi = h_-^{\pm 1} Q\varphi \qquad (\varphi(t) \in \overset{0}{H}_\mu(\Gamma, \rho)) .$$

Taking into account the above equalities we easily get $BA\varphi = \varphi$ if $\kappa \geq 0$ and $AB\varphi = \varphi$ if $\kappa \leq 0$ $(\varphi(t) \in \overset{0}{H}_\mu(\Gamma, \rho))$, where

$$B = (uP + Q)(h_+ P + h_-^{-1} Q)(P + t^\kappa Q) a^{-1} h_+^{-1} u^{-1} ,$$

$$A = aP + bQ = ah + u(P + t^{-\kappa} Q)(h_+^{-1} P + h_- Q)(u^{-1} P + Q) .$$

From the last equality it follows also that $\operatorname{Ind} A = \operatorname{Ind}(P + t^{-\kappa} Q) = \operatorname{ind}_\Gamma t^{-\kappa} = -\kappa$, because the operators $A_1 = h_+^{-1} P + h_- Q$ and $A_2 = u^{-1} P + Q$ have the inverses $A_1^{-1} = h_+ P + h_-^{-1} Q$ and $A_2^{-1} = uP + Q$. Now let $a(t), b(t) \in H_\nu^-(\Gamma, c_1)$.

Denote by z_k an arbitrary point from the interior F_k^+ of contour Γ_k $(k = 1, 2, \dots, s)$. The function $v_j(t) = \beta_j [(t - z_j)/(t - z_{j+1})]^{\delta_j}$ $(z_{s+1} = z_1$; $j = 1, 2, \dots, s)$ is analytic in the complex plane cut along any simple continuous arc, connecting the points z_j, c_1 and z_{j+1} and lying entirely in $F_j^+ \cup F_{j+1}^+ \cup \{c_1\}$ $(F_{s+1}^+ = F_1^+$; $j = 1, 2, \dots, s)$. Choosing the numbers β_j and δ_j in a special way we can suppose that $(v_j)_k^+ = g_k^+$ and $(v_j)_k^- = 1$ $(j,k = 1,2,\ldots,s)$ where $g(t) = b(t)/a(t)$ (for the definition of $(v_j)_k^\pm$ cf. Section 2, point (2)). Obviously the function $v^{\pm 1}(t) = \prod_{k=1}^{s} v_k^{\pm 1}(t)$ is analytic outside the area $\Gamma \cup F_1^+ \cup \ldots \cup F_s^+$; hence $Qv^{\pm 1}Q = v^{\pm 1}Q$.

It is easy to verify that the function $h(t) = g(t)/v(t) = b(t)/a(t)\,v(t)$ is continuous on Γ and $h(c_1) = 1$; hence $h(t) \in H_\nu(\Gamma)$.

Taking the factorization of the function $h(t)$ (cf. (3.3)) we obtain $BA\varphi = \varphi$ if $\kappa \geq 0$ and $AB\varphi = \varphi$ if $\kappa \leq 0$, where $\varphi(t) \in \overset{0}{H}_\mu(\Gamma,\rho)$ and

$$B = (P + v^{-1}Q)(h_+ P + h_-^{-1}Q)(P + t^\kappa Q)a^{-1}h_+^{-1}I ,$$

$$A = aP + bQ = ah_+(P + t^{-\kappa}Q)(h_+^{-1}P + h_-Q)(P + vQ) .$$

We have also

$$\operatorname{Ind} A = \operatorname{Ind}(P + t^{-\kappa}A) = -\kappa .$$

For the operator $A = PaI + QbI$ the sufficiency of the conditions of the theorem follows now from the equality $A = bB_1(dP + Q)B_2$ where the operators $B_1 = I + PdQ$ and $B_2 = I - QdP$ $(d(t) = a(t)/b(t))$ are invertible $B_1^{-1} = I - PdQ$, $B_2^{-1} = I + QdP$.

By means of Lemma 2.1 the necessity of the conditions of the theorem can be proved by the method proposed by M.G.Krein (cf. [8]). Q.E.D.

(3) Denote by $\overset{0}{H}_\mu(\Gamma,\rho)$ the direct sum of m copies of the spaces $\overset{0}{H}_\mu(\Gamma,\rho)$ and by $H_\mu^{(m\times m)\pm}(\Gamma,c_1)$ - the set of matrix-functions $a(t) = \|a_{jk}(t)\|_{j,k=1}^m$ with components from $H_\mu^\pm(\Gamma,c_1)$. Let $S_m = \|\delta_{jk}S\|_{j,k=1}^m$, $P_m = (1/2)(I + S_m)$ and $Q_m = I - P_m$ where S is the operator defined by (1.2).

Theorem 3.2: Let $a(t), b(t) \in H_\nu^{(m\times m)+}(\Gamma,c_1)$ or $a(t), b(t) \in H_\nu^{(m\times m)}(\Gamma,c_1)$ where $\nu > \mu$.

The operator $A = aP_m + bQ_m$ $(A = P_m aI + Q_m bI)$ is a Φ_+- or a Φ_--operator in the space $H_\mu^{(m)}(\Gamma,\rho)$ if and only if the condition

$$\inf|\det[a(t)\cdot b(t)]| > 0 \quad (t \in \Gamma) \tag{3.5}$$

holds. If the condition is fulfilled then

$$\operatorname{Ind}A = \frac{1}{2\pi}\sum_{k=1}^{s} [\arg\det(a^{-1}(t)\,b(t))]_{\Gamma_k}\,. \tag{3.6}$$

Proof: For definiteness we will suppose that $A = aP_m + bQ_m$ where $a(t), b(t) \in H_\nu^{(m\times m)+}(\Gamma, c_1)$. If the condition (3.5) holds, then by Lemma 2.3 the operator A can be rewritten as

$$A = P_m a P_m + Q_m b Q_m + P_m b Q_m + T$$

where $T \in \mathfrak{S}(H_\mu^{(m)}(\Gamma,\rho))$; evidently the operator

$$B = P_m a^{-1} P_m + Q_m b^{-1} Q_m - P_m a^{-1} P_m b Q_m b^{-1} Q_m$$

is the regularizer of the operator A (i.e. $AB - I,\ BA - I \in \mathfrak{S}(H_\mu^{(m)}(\Gamma,\rho))$), hence (cf. [7]) A is a Φ-operator in $H_\mu^{(m)}(\Gamma,\rho)$. Applying then the homotopy described in [15], application VI, we easily get the diagonal operator $\widetilde{A} = \widetilde{a}P_m + \widetilde{b}Q_m$ where $\widetilde{a}(t) = \|\delta_{jk}a_k(t)\|_{j,k=1}^m$, $\widetilde{b}(t) = \|\delta_{jk}b_k(t)\|_{j,k=1}^m \in H_\nu^{(m\times m)+}(\Gamma,c_1)$ is a Φ-operator and $\operatorname{Ind}\widetilde{A} = \operatorname{Ind}A$, $\det\widetilde{a}(t) \equiv \det a(t)$, $\det\widetilde{b}(t) \equiv \det b(t)$. It is easy to prove then that

$$\operatorname{Ind}\widetilde{A} = \sum_{k=1}^{n} \operatorname{Ind}(a_k P + b_k Q)$$

and hence by Theorem 3.1 we get

$$\operatorname{Ind}A = \operatorname{Ind}\widetilde{A} = \frac{1}{2\pi}\sum_{k=1}^{m}\sum_{j=1}^{s} \left[\arg\frac{b_k(t)}{a_k(t)}\right]_{\Gamma_j} = \frac{1}{2\pi}\sum_{j=1}^{s} [\arg\det(a^{-1}(t)b(t))]_{\Gamma_j}\,.$$

By means of the theorem about stability of $\Phi_\pm$-operators (cf. [7]) and the Lemma 2.1 we easily prove that if $A = aP_m + bQ_m$ is a $\Phi_\pm$-operator then it is a Φ-operator in $H_\mu^{(m)}(\Gamma,\rho)$.

Let now A be a Φ-operator and $\inf|\det[a(t)\cdot b(t)]| = 0$ $(t \in \Gamma)$. Without loss of generality we can suppose that $\inf|\det b(t)| = 0$ $(t \in \Gamma)$; using again, if needed, Lemma 2.1 and the theorem about stability of a Φ-operator (cf. [7]), we can suppose without loss of generality that $\inf|\det a(t)| > 0 \quad (t \in \Gamma)$. But then

$$A = (P_m a P_m + Q_m b Q_m) D + T_1 ,$$

where $T_1 \in \mathfrak{S}(H_\mu^{(m)}(\Gamma,\rho))$ and the operator $D = I - P_m a^{-1} P_m b Q_m$ is invertible $D^{-1} = I + P_m a^{-1} P_m b Q_m$. Hence (cf. [7]) the operator $A' = P_m a P_m + Q_m b Q_m$ is a Φ-operator in $H_\mu^{(m)}(\Gamma,\rho)$ and by the lemma of N. Ja. Krupnik (cf. [6]. p.269) the operator $\det A' = P(\det a)P + Q(\det b)Q + T_2$ where $T_2 \in \mathfrak{S}(H_\mu^0(\Gamma,\rho))$ (cf. the corollary of Lemma 2.3) is also a Φ-operator in $H_\mu^0(\Gamma,\rho)$. But $\inf|\det a(t)| > 0$ and from the equality

$$\det A' = [(\det a)P + (\det b)Q] D_1 + T_3 ,$$

where $T_3 \in \mathfrak{S}(H_\mu^0(\Gamma,\rho))$, $D_1 = I - P(\det a^{-1})P(\det b)Q$ we conclude that the operator $A' = (\det a)P + (\det b)Q$ is a Φ-operator in $H_\mu^0(\Gamma,\rho)$. This contradicts Theorem 3.1 because $\inf|\det b(t)| = 0 \quad (t \in \Gamma)$. Q.E.D.

4. SINGULAR INTEGRAL OPERATORS WITH COEFFICIENTS FROM $H_\nu^{(m)}(\Gamma, c_1)$

Throughout this section we suppose that Γ is the star-shaped line $\rho(T) = \prod_{k=1}^{n} |t - c_k|^{\alpha_k}$ and the numbers $\mu, \alpha_1, \ldots, \alpha_n$ satisfy the relations (1.1).

(1) Denote by $\xi(x)$ the function, defined as follows

$$\xi(x) = \begin{cases} \dfrac{\sin x\theta}{\sin\theta} \exp[i(x-1)\theta] & \text{if } \theta \neq 0 \\ x & \text{if } \theta = 0 \end{cases} \qquad (4.1)$$

where $\theta = \pi - 2\pi(\alpha_1 - \mu)$. If $\theta \neq 0$, then the set of the values of the function $\xi(x) \ (0 \leq x \leq 1)$ form the arc of the circle on the complex plane joining the origin with the point 1; the angle between the lines joining each point of this arc with the points 0 and 1 is equal to $\tilde{\theta} = 2\pi(\alpha_1 - \mu)$ and $\operatorname{Im}\xi(x) \geq 0 \ (\operatorname{Im}\xi(x) \leq 0)$ if $\alpha_1 - \mu > \frac{1}{2} \ (\alpha_1 - \mu < \frac{1}{2})$.

<u>Theorem 4.1</u>: Let $a(t), b(t) \in H_\nu^{(m\times m)}(\Gamma, c_1)$ and $\nu > \mu$. The operator $A = aP_m + bQ_m \ (A = P_m aI + Q_m bI)$ is a Φ_+- or a Φ_--operator in the space

$H_\mu^{(m)}(\Gamma,\rho)$ if and only if the following conditions hold:

$$\inf |\det(a(t) - b(t)| > 0 \quad (t \in \Gamma), \tag{4.2}$$

$$\det[\xi(x) h_0 + (1 - \xi(x) E_m] \neq 0 \quad (0 \leq x \leq 1), \tag{4.3}$$

where $h_0 = (a_1^-)^{-1} b_1^- b_1^+)^{-1} a_1^+ \dots (b_s^+)^{-1} a_s^+ / h_0 = b_1^- (a_1^-)^{-1} a_1^+ (b_1^+)^{-1} \dots a_s^+ (b_s^+)^{-1} /$ (cf. Section 2) and E_m is the unit matrix of order m.

If the conditions (4.2) and (4.3) hold, then A is a Φ-operator in $H_\mu^{(m)}(\Gamma,\rho)$ and

$$\operatorname{Ind} A = \frac{1}{2\pi} \sum_{k=1}^{s} \{\arg \det [a^{-1}(t) b(t)]\}_{\tau_\kappa}$$

$$+ \frac{1}{2\pi} \{\arg \det [\xi(x) h_0 + (1 - \xi(x)) E_m]\}_{x \in [0,1]} .$$

If $m = 1$ and the conditions (4.2) and (4.3) are fulfilled, then the invertibility of the operator A depends on the integer $\kappa = -\operatorname{Ind} A$.

Proof: We prove only the last part of the theorem since the other parts of the theorem can be proved as in [14], Section 4 by means of the Lemma 2.1 and theorems from† [1,3,4].

Let us suppose $m = 1$ and the conditions (4.2) and (4.3) are fulfilled. We consider only the operator $A = aP + bQ$ (cf. the proof of Theorem 3.1).

The function $b(t)/a(t)\ (\in H_\nu(\Gamma,c_1))$ can be represented in the form (cf. [14])

$$\frac{b(t)}{a(t)} = d^{(1)}(t) h(t) d^{(2)}(t)$$

where $d^{(1)}(t) \in H_\nu^-(\Gamma,c_1)$, $d^{(2)}(t) \in H_\nu^+(\Gamma,c_1)$ and the function $h(t)(\in H_\nu(\Gamma,c_1))$ have the following properties: i) $h(t) = 1$ on $\Gamma \setminus \Gamma_1$; ii) $h_1^+ = 1$ and

$$h_1^- = \prod_{k=1}^{s} a_k^+ b_k^- \left(\prod_{k=1}^{s} a_k^- b_k^+\right)^{-1} .$$

The functions $d^{(1)}(t)$ and $d^{(2)}(t)$ can be expressed in the form

---

† In [1-4] it was supposed that Γ is the smooth line, but all suggestions are valid for a closed piecewise smooth contour Γ.

$d^{(1)}(t) = v(t)\tilde{v}(t)$, $d^{(2)}(t) = u(t)\tilde{u}(t)$ where $\tilde{u}(t), \tilde{v}(t) \in H_\nu(\Gamma)$, $v^{\pm 1}(t) \in H_\nu^-(\Gamma, c_1)$, $u^{\pm 1}(t) \in H_\nu^+(\Gamma, c_1)$ and $Qv^{\pm 1}Q = v^{\pm 1}Q$, $Pu^{\pm 1}P = u^{\pm 1}P$ (cf. the proof of Theorem 3.1). If now $\tilde{v}(t)\,\tilde{u}(t) = d_-(t)\,t^\kappa\,d_+(t)$ is the factorization of this function (cf. (3.3)) then

$$\frac{b(t)}{a(t)} = g_-(t)\tilde{h}(t)g_+(t) \ , \tag{4.4}$$

where $g_-(t) = v(t)d_-(t)$ and

$$g_+(t) = \begin{cases} g(t)\,d_+(t)\,t^{\kappa_1} & \text{if } t \in \Gamma \setminus \Gamma_1 \ , \\ g(t)\,d_+(t) & \text{if } t \in \Gamma_1 \ , \end{cases}$$

$$\tilde{h}(t) = \begin{cases} 1 & \text{if } t \in \Gamma \setminus \Gamma_1 \ , \\ t^{\kappa_1} h(t), & \text{if } t \in \Gamma . \end{cases} \tag{4.5}$$

It is easy to prove the equalities

$$Qg_-^{\pm 1}Q = g_-^{\pm 1}Q, \qquad Pg_+^{\pm 1}P = g_+^{\pm 1}P \tag{4.6}$$

From (4.4) and (4.5) it follows that

$$A = A_1(P + \tilde{h}Q)A_2 \ ,$$

where $A_1 = g_+ aI$ and $A_2 = g_+^{-1}P + g_-Q$ are the invertible operators $A^{-1} = g_+^{-1}a^{-1}I$, $A_2^{-1} = g_+P + g_-^{-1}Q$. Hence we can consider only the operator $B = P + \tilde{h}Q$.

In the space $H_\mu^0(\Gamma_1, \rho)$ consider the operator $B^{(1)} = P^{(1)} + \tilde{h}Q^{(1)}$, where $P^{(1)} = (1/2)(I + S^{(1)})$, $Q^{(1)} = I - P^{(1)}$ and

$$(S\varphi)(t) = \frac{1}{\pi i}\int_\Gamma \frac{\varphi(\tau)\,d\tau}{\tau - t} .$$

By means of (4.5) it is obvious that it suffices to consider the operator $B^{(1)}$ in the space† $H^0_\mu(\Gamma,\rho)$. The proof is completed now by means of theorems proved in [3] and the equality

$$\operatorname{Ind} B^{(1)} = \kappa_1 + \frac{1}{2\pi}[\arg h(t)]_{\Gamma_1} + \frac{1}{2\pi}\{\arg[h_0\xi(x) + 1 - \xi(x)]\}_{x\in[0,1]}$$

$$= \frac{1}{2\pi}\sum_{k=1}^{s}[\arg\frac{b(t)}{a(t)h(t)}]_{\Gamma_k} + \frac{1}{2\pi}[\arg h(t)]_{\Gamma_1}$$

$$+ \frac{1}{2\pi}\{\arg[h_0\xi(x) + 1 - \xi(x)]\}_{x\in[0,1]} = \operatorname{Ind} A .$$

5. SINGULAR INTEGRAL OPERATORS ON PIECEWISE SMOOTH LINES AND THEIR SYMBOLS

Throughout this section we will suppose that Γ is a piecewise smooth line with knots at $c_1,\dots,c_n$, $\rho(t) = \prod_{k=1}^{n}|t-c_k|^{\alpha_k}$ and the numbers $\mu,\alpha_1,\dots,\alpha_n$ satisfy the relations (1.1).

Let $\Gamma = \bigcup_{k=1}^{r}\gamma_k$ where γ_k are the smooth arcs with the end-points at knots of Γ. Denote by $\gamma_1^{(k)},\dots,\gamma_{r_k}^{(k)}$ the smooth arcs from $\gamma_1,\dots,\gamma_r$ having a common end-point at the knot c_k $(k = 1,2,\dots,n)$ and numerated in the order in which we meet them moving counterclockwise around the knot; let $\epsilon(k,j) = 1$ if $\gamma_j^{(k)}$ is outgoing and $\epsilon(k,j) = -1$ if $\gamma_j^{(k)}$ is entering the knot c_k. For a matrix-function $a(t)$ in $H_\mu^{(m\times m)}(\Gamma,d_1,\dots,c_n)$ we denote the limit of $a(t)$ when $t\to c_k$ along $\gamma_j^{(k)}$ $(j = 1,2,\dots,r;\ k = 1,2,\dots,n)$ by $a_j^{(k)}$; if $\inf|\det a(t)| > 0$ then we denote $a^{(k)} = (a_1^{(k)})^{\epsilon(k,1)}\dots(a_{r_k}^{(k)})^{\epsilon(k,r_k)}$.

Denote by $\xi_k(x)$ $(0 \le x \le 1)$ the function defined by (4.1) in which $\theta = \theta_k = \pi - 2\pi(\alpha_k - \mu)$ $(k = 1,2,\dots,n)$.

(1) Now we prove the theorem.

<u>Theorem 5.1</u>: Let $a(t), b(t) \in H_\nu(\Gamma,c_1,\dots,c_n)$ and $\nu > \mu$. The operator $A = aP + bQ$ $(a = PaI + QbI)$ is a Φ_+- or a Φ_--operator in the space $H^0_\mu(\Gamma,\rho)$ if and only if the following conditions hold:

---

† Operators B and $B^{(1)}$ are simultaneously invertible (or not invertible) from the left and right in the spaces $H^0_\mu(\Gamma,\rho)$ and $H^0_\mu(\Gamma_1,\rho)$ respectively and Ind B = Ind $B^{(1)}$ (cf. [9]).

I. $\inf |a(t) \cdot b(t)| > 0 \quad (t \in \Gamma)$;

II. $g^{(k)} \xi_k(x) + 1 - \xi_k(x) \neq 0 \quad (k = 1,2,\dots,n;\ 0 \leq x \leq 1)$ where $g(t) = b(t)/a(t)$.

If the conditions I and II hold then the invertibility of an operator A depends on the integer κ

$$\kappa = -\frac{1}{2\pi} \sum_{j=1}^{r} [\arg g(t)]_{\gamma_j} - \frac{1}{2\pi} \sum_{k=1}^{n} \{\arg [g^{(k)} \xi_k(x) + 1 - \xi_k(x)]\}_{x \in [0,1]}$$

and $\operatorname{Ind} A = -\kappa$.

We give only the short plan of the proof, which is borrowed from [14], and leave the details to the reader.

First let Γ consist of a finite number of open arcs with only one common end-point at the knot c_1; let $A = P + bQ$, where $b(t) \in H_\nu(\Gamma, c_1)$ and $b(c_2) = \dots = b(c_n) = 1$. Complementing the line Γ to the star-shaped line $\tilde{\Gamma}$ and supposing $b(t) = 1$ when $t \in \tilde{\Gamma} \setminus \Gamma$, we easily reduce (cf. [9]) this case to Theorem 4.1.

In the general case we represent the function $b(t)$ in such a way $b(t) = \prod_{k=0}^{n} b_k(t)$ where: i) the function $b_k(t)$ differs from 1 only in a neighbourhood $u_k (\subset \Gamma)$ of the knot c_k and these neighbourhoods do not intersect† $(k = 1,2,\dots,n)$ ii) $b_0(t) \in H_\nu(\Gamma)$ and $b_0(t) = 1$ when t is in small neighbourhoods of the knots $c_1,\dots,c_n$.

By the Lemma 2.2 we get then ††

$$P + bQ = (P + b_1 Q)\dots(P + b_n Q)(P + b_0 Q) + T ,$$

where $T \in \mathfrak{S}(\overset{0}{H}_\mu(\Gamma,\rho))$. The above equality enables us to reduce the proof of the theorem to the first case.

The operator $A = PaI + QbI$ can be treated as in the Theorem 3.1. Q.E.D.

(2) According to the plan given for the proof of the Theorem 5.1 we can prove also the theorem

---

† From this condition it follows that $b(t) = b_0(t)\, b_k(t)$ when $t \in u_k$ $(k = 1,2,\dots,n)$.

†† We easily reduce the investigation of the operator $aP + bQ$ to the investigation of the operator $P + gQ$.

Theorem 5.2: Let $a(t), b(t) \in H_\nu^{(m\times m)}(\Gamma, c_1, \ldots, c_n)$ and $\nu > \mu$. The operator $A = aP_m + bQ_m$ $(A = P_m aI + Q_m bI)$ is a Φ_+- or a Φ_--operator in the space $H_\mu^{(m)}(\Gamma, \rho)$ if and only if the following two conditions are fulfilled:

I. $\inf |\det [a(t) \cdot b(t)]| > 0 \quad (t \in \Gamma)$;

II. $\det\{g^{(k)} \xi_k(x) + [1 - \xi_k(x)] E_m\} \neq 0 \quad (0 \le x \le 1 \; ; \; k = 1,2,\ldots,n)$ where $g(t) = a^{-1}(t) b(t)$ $(g(t) = b(t) a^{-1}(t))$ and E_m is an identity matrix of order m.

If the conditions I and II hold then A is a Φ-operator in the space $H_\mu^{(m)}(\Gamma, \rho)$ and

$$\operatorname{Ind} A = \frac{1}{2\pi} \sum_{j=1}^{r} [\operatorname{argdet} g(t)]_{\gamma_j}$$

$$+ \frac{1}{2\pi} \sum_{k=1}^{n} \{\operatorname{argdet} [g^{(k)} \xi_k(x) + (1 - \xi_k(x)) E_m]\}_{x \in [0,1]} .$$

(3) The symbol of an operator $A = aP_m + bQ_m$ where $a(t), b(t) \in H^{(m\times m)}(\Gamma, c_1, \ldots, c_n)$ in the space $H_\mu^{(m)}(\Gamma, \rho)$ will be called the matrix-function $A(t,x)$ $(t \in \Gamma \; ; \; 0 \le x \le 1)$ defined as follows.

If $t \neq c_1, \ldots, c_n$ $(t \in \Gamma)$ then

$$A(t,x) = \begin{pmatrix} a(t) & 0 \\ 0 & b(t) \end{pmatrix} \qquad (0 \le x \le 1) .$$

Let $\gamma_1^{(k)}, \ldots, \gamma_{r_k}^{(k)}$ be the arcs defined in the beginning of this section; but we suppose in addition that $r_k = 2p_k$ where p_k is an integer and the arcs $\gamma_{2j-1}^{(k)}$ are outgoing and $\gamma_{2j}^{(k)}$ are entering at † c_k $(j = 1,2,\ldots,r_k \; ; \; k = 1,2,\ldots,n)$. Now we assume

$$A(c_k, x) = \| a_{j\ell}^{(k)}(x) \|_{j,k=1}^{r_k} \qquad (0 \le x \le 1)$$

where

---

† If that is not so we can add smooth arcs to $\gamma_1^{(k)}, \gamma_2^{(k)}, \ldots, \gamma_{r_k}^{(k)}$ and suppose on them $a(t) = b(t) = 1$.

$$a_{j\ell}^{(k)}(x) = \begin{cases} \xi_k(x)\, a_\ell^{(k)} + [1 - \xi_k(x)]\, a_{\ell+1}^{(k)} & \text{if } j = \ell \text{ and } \ell \text{ is odd,} \\ \xi_k(x)\, b_\ell^{(k)} + [1 - \xi_k(x)]\, b_{\ell+1}^{(k)} & \text{if } j = \ell \text{ and } \ell \text{ is even,} \\ (-1)^{j+1} g_\ell^{(k)} [\xi_k(x)]^s [1 - \xi_k(x)]^{1-s} & \text{if } j < \ell, \\ (-1)^j g_\ell^{(k)} [\xi_k(x)]^{1-s} [1 - \xi_k(x)] & \text{if } j > \ell \end{cases}$$

and

$$g_\ell^{(k)} = \begin{cases} b_\ell^{(k)} - b_{\ell+1}^{(k)} & \text{if } \ell \text{ is even,} \\ a_\ell^{(k)} - a_{\ell+1}^{(k)} & \text{if } \ell \text{ is odd.} \end{cases}$$

In these equalities we supposed that $b_{r_k+1}^{(k)} = b_1^{(k)}$, $a_{r_{k+1}}^{(k)} = a_1^{(k)}$ and the function $\xi_k(x)$ is defined by (4.1 in which $\theta = \theta_k = \pi - 2\pi(\alpha_k - \mu)$ $(k = 1,2,\ldots,n.)$

Denote by $\mathfrak{K}_m$ the non-closed algebra of operators

$$A = \sum_{j=1}^{k} A_{j1} A_{j2} \ldots A_{js} \tag{5.1}$$

where $A_{j\ell} = a_{j\ell} P_m + b_{j\ell} Q_m$ and $a_{j\ell}(t), b_{j\ell}(t) \in H_\nu^{(m\times m)}(\Gamma, c_1, \ldots, c_n)(\nu > \mu)$. With the operator A we associate the symbol

$$A(t,x) = \sum_{j=1}^{k} A_{j1}(t,x) A_{j2}(t,x) \ldots A_{js}(t,x)$$

where $A_{j\ell}(t,x)$ is the symbol of the operator $A_{j\ell}$. It can be proved that the symbol $A(t,x)$ does not depend on the representation of the operator A in the form (5.1) (cf. [11, 13, 14]).

Theorem 5.3: The operator A from the algebra $\mathfrak{K}_m$ is a Φ_+- or a Φ_--operator in the space $H_\mu^{(m)}(\Gamma, \rho)$ if and only if the condition $\inf |\det A(t,x)| > 0$ $(t \in \Gamma;\ 0 \leq x \leq 1)$ holds. If the condition if fulfilled then A is a Φ-operator in $H_\mu^{(m)}(\Gamma, \rho)$ and

$$\text{Ind } A = \frac{1}{2\pi} \sum_{j=1}^{r} \{\arg\det [a^{-1}(t)\, b(t)]\}_{\gamma_j} + \frac{1}{2\pi} \sum_{j=1}^{n} [\arg\det A(c_k, x)]_{x \in [0,1]} \tag{5.2}$$

where $a(t) = \sum_{j=1}^{k} a_{j1}(t) \ldots a_{js}(t)$, $b(t) = \sum_{j=1}^{k} b_{j1}(t) \ldots b_{js}(t)$.

By means of Theorem 5.2 the above theorem can be proved like the same suggestion from [11, 13, 14].

6. RELATIONS WITH THE CLASSIC THEORY

Throughout this section we adhere to the definitions from the preceding section.

(1) Let us consider the operator

$$A = \sum_{j=1}^{k} (a_{j1} I + b_{j1} S_m) \ldots (a_{js} I + b_{js} S_m) + K \tag{6.1}$$

where $a_{j\ell}(t)$, $b_{j\ell}(t) \in H_\nu^{(m\times m)}(\Gamma, c_1, \ldots, c_n)$ $(\nu > \mu)$; the operator

$$(K\varphi)(t) = \int_\Gamma k(t, \tau)\, \varphi(\tau)\, d\tau$$

is a compact operator in the space † $H_\mu^0(\Gamma, \rho)$.

If for the symbol $A(t,x)$ of the operator A the condition $\det A(t,x) \neq 0$ $(t \in \Gamma\ ;\ 0 \leq x \leq 1)$ holds then by the Theorem 5.3 A is a Φ-operator in the space $H_\mu^{(m)}(\Gamma, \rho)$. But in the definition of a Φ-operator the conjugate space and the conjugate operator participate †† and in the present case they are indeterminate; so it is desirable to replace them by a simple adjoint space and adjoint operator.

Let us introduce some definitions.

The space $H_\mu^{(m)}(\Gamma, \rho')$ where $\rho'(t) = \prod_{k=1}^{n} |t - c_k|^{1+2\mu-\alpha k}$ will be called

---

† We can suppose, for example, that $k(t, \tau) = [a(t) - a(\tau)](\tau - t)^{-1}$, where $a(t) \in H_\nu^{(m\times m)}(\Gamma)$, or $k(t, \tau)$ belongs to $H_\nu^{(m\times m)}(\Gamma, c_1, \ldots, c_n)$ with respect to both variables (cf. Lemma 2.1).

†† A linear bounded operator A in the Banach space $\mathfrak{B}$ is a Φ-operator if (cf. [7]) : (a) the equation $Ax = y$ $(y \in \mathfrak{B})$ has a solution if and only if $f(y) = 0$ for all the solutions of the conjugate homogeneous equation $A^* f = 0$; (b) the homogeneous equations $Ax = 0$ and $A^* f = 0$ have only a finite number of linearly independent solutions.

adjoint to the space $H_\mu^{(m)}(\Gamma,\rho)$.

Denote by $a' = \|a_{kj}\|_{j,k=1}^{m}$ the transposed matrix of $a = \|a_{jk}\|_{j,k=1}^{m}$.

The operator

$$A' = \sum_{j=1}^{k} (a'_{js} I - S_m b'_{js} I) \dots (a'_{j1} I - S_m b'_{j1} I) + K' \ ;$$

where

$$(K'\varphi)(t) = \int_\Gamma K'(\tau,t)\,\varphi(\tau)\,d\tau$$

is called adjoint to the operator A defined by (6.1).

If the symbol $A(t,x)$ of the operator A satisfies the condition $\det A(t,x) \neq 0$ $(t \in \Gamma \ ; \ 0 \leq x \leq 1)$ then by means of the theorems proved in [5] we get the following theorems of Noether † :

I. The adjoint homogeneous equations $A\varphi = 0$ and $A'\psi = 0$ have only a finite number dimker A and dimker A' of linearly independent solutions from spaces $H_\mu^{(m)}(\Gamma,\rho)$ and $H_\mu^{(m)}(\Gamma,\rho')$ respectively.

II. The equation $A\varphi = f\ (f(t) \in H_\mu^{(m)}(\Gamma,\rho))$ has the solution $\varphi(t)(\in H_\mu^{(m)}(\Gamma,\rho))$ if and only if the condition

$$\int_\Gamma f(t)\,\psi(t)\,dt = 0$$

holds for all the functions $\psi(t) \in H_\mu^{(m)}(\Gamma,\rho')$ which are the solutions of the adjoint homogeneous equation $A'\psi = 0$.

III. dimker A - dimker A' = Ind A (cf. the formula (5.2)).

(2) Now we consider the behaviour of the solutions of the singular integral equation

$$(A\varphi)(t) \equiv a(t)\varphi(t) + \frac{b(t)}{\pi i} \int_\Gamma \frac{\varphi(\tau)\,d\tau}{\tau - t} = \chi(t) \qquad (6.2)$$

where $a(t),\ b(t) \in H_\nu(\Gamma,c_1,\dots,c_n)\ (\nu > \mu)$, $\chi(t) \in H_\mu(\Gamma,c_1,\dots,c_n)$ and the solutions are sought in the space $\overset{0}{H}_\mu(\Gamma,\rho)$. We assume that $\inf|a^2(t) - b^2(t)| > 0 \ (t \in \Gamma)$.

† It is easy to prove (cf. [5]) that the condition $\det A(t,x) \neq 0$ $(t \in \Gamma \ ; \ 0 \leq x \leq 1)$ is not only sufficient but also necessary for Noether's theorems I and II to be valid.

The solutions of equation (6.2) depend on the weight function $\rho(t) = \prod_{k=1}^{n} |t-c_k|^{\alpha_k}$. Taking certain numbers $\alpha_1,\ldots,\alpha_n$ the bounded solutions in preassigned knots $c_1,\ldots,c_n$ can be founded.

Consider the numbers

$$\beta_k = \frac{1}{2\pi} \ell_n g^{(k)} , \quad -1 < \text{Re } \beta_k < 1 \quad (k = 1,2,\ldots,n) \tag{6.3}$$

where $g(t) = [a(t) + b(t)]/[a(t) - b(t)]$ and the numbers $g^{(k)}$ are defined in Section 5. If $\text{Re } \beta_k = 0$ then the knot c_k is called a special one (cf. [15]).

Without loss of generality we can suppose that only the knots $c_{p+1},\ldots,c_n$ are special $(0 \le p \le n)$.

If we are interested in the solutions of the equation (6.2), which are bounded in non-special knots $c_1,\ldots,c_q (q \le p)$, i.e. in the solutions of the class $h(c_1,\ldots,c_q)$ (cf. [15], Section 82), then we assume $\text{Re}\beta_k > 0$ $(k=1,2,\ldots,q)$, $\text{Re}\beta_k < 0$ $(k= q+1,\ldots,p)$ and select the numbers $\alpha_1,\ldots,\alpha_n$ satisfying the relations

$$\mu < \alpha_k < \mu + 1 , \quad \alpha_k - \mu - 1 < \text{Re}\beta_k < \alpha_k - \mu \quad (k = 1,2,\ldots,n) \tag{6.4}$$

It turns out that if the relations (6.3) and (6.4) hold, then all the solutions of the equation (6.2) of the space $\overset{0}{H}_\mu(\Gamma,\rho)$ (if such solutions exist) belong to the class $h(c_1,\ldots,c_q)$. This suggestion is valid even if $\chi(t) = \chi_1(t) \prod_{k=q+1}^{n} |t - c_k|^{-\delta_k}$ where $\mu < \delta_k < \mu + 1$ and $\chi_1(t) \in H_\mu(\Gamma,c_1,\ldots,c_n)$, $\chi_1(c_k \pm 0) = 0$ $(k = q+1,\ldots,n)$.

The proof of the same suggestion for the smooth line Γ was provided in [4].

It is not difficult to prove the following results (cf. [4]):

(i) If the solutions of equation (6.2) in the space $\overset{0}{H}_\mu(\Gamma,\rho)$ belong to the class $h(c_1,\ldots,c_q)$, then the solutions of the adjoint equation

$$(A'\psi)(t) \equiv a(t)\psi(t) - \frac{1}{\pi i} \int_\Gamma \frac{b(\tau)\psi(\tau)\,d\tau}{\tau - t} = \chi(t)$$

in the space $\overset{0}{H}_\mu(\Gamma,\rho')$ belong to the adjoint class $h(c_{q+1},\ldots,c_p)$, i.e. are bounded in the non-special knots $c_{q+1},\ldots,c_p$;

(ii) The index of the equation (6.2) depends on the choice of the numbers $\alpha_1,\ldots,\alpha_n$ i.e. on the class $h(c_1,\ldots,c_q)$ in which the solutions are sought; if the solutions of the equation (6.2) are sought in classes $h(c_1,\ldots,c_q)$ and $h(c_1,\ldots,c_{q+1})$ respectively, then the indices κ_1 and κ_2 of the equation satisfy the equality $\kappa_1 = \kappa_2 + 1 \quad (0 \leq q \leq p - 1)$.

References

1 R.V. Duduchava, On singular integral operators in a Hölder space with weight. Dokl.Akad.Nauk SSSR, t.191 (1970) 16-19; Soviet Math.Dokl. v.11, N.2 (1970) 304-308.

2 R.V. Duduchava, On boundedness of an operator of singular integration in Hölder spaces with weight. Matem,Issledov., Kishinev, v.5, N.1 (1970) 56-76 (Russian).

3 R.V. Duduchava, Singular integral equations in Hölder spaces with weight, I. Hölder coefficients. Matem.Issledov., Kishinev, v.5, N.2 (1970) 104-124 (Russian).

4 R.V. Duduchava, Singular integral equations in Hölder spaces with weight, II. Piecewise Hölder coefficients. Matem.Issledov., Kishinev, v.5, N.3 (1970) 58-82 (Russian).

5 R.V. Duduchava, On the Noether's theorems for singular integral operators in Hölder spaces with weight. Proceedings of the Symposium on Continuum Mechanics and Related Problems of Analysis, "Metsniereba", Tbilisi, v.I (1973) (Russian).

6 I.C. Gohberg and I.A. Feldman, Equations in convolution and projective methods of their solution. "Nauka", Moscow (1971) (Russian).

7 I.C. Gohberg and M.G. Krein, Principal statements on defective numbers, root numbers and indices of linear operators. Uspekhi Matem. Nauk 12, N.2 (1957) 43-118 (Russian).

8 I.C. Gohberg and N.Ja. Krupnik, On spectrum of singular integral operators in L_p. Studia Mathem. 31 (1968) 347-362 (Russian).

9 I.C. Gohberg and N.Ja. Krupnik, On spectrum of singular integral operators in

space L_p with weight. Dokl. Akad. Nauk SSSR, t.185, N.4 (1969) 745-748 (Russian).

10 I.C. Gohberg and N. Ja. Krupnik, On singular integral operators with unbounded coefficients. Matem. Issledov., Kishinev, t.V, N.3 (1970) 46-57 (Russian).

11 I.C. Gohberg and N. Ja. Krupnik, Banach algebras generated by singular integral operators. Colloquia mathem. soc. Janos Bolyai 5. Hilbert space operators, Tihany (Hungary)(1970) 239-264.

12 I.C. Gohberg and N. Ja. Krupnik, On singular integral operators on a complex contour. Bulletin of the Academy of Sciences of the Georgian SSR, v.64, N.1 (1971) 21-24 (Russian, English summary).

13 I.C. Gohberg and N. Ja. Krupnik, Singular integral operators with piecewise continuous coefficients and their symbols. Izvestia Akad. Nauk SSSR, ser. matem. t.35, N.4 (1971) 940-964 (Russian).

14 I.C. Gohberg and N. Ja. Krupnik, On the symbol of singular integral operators on complex contour. Proceedings of the Symposium on Continuum Mechanics and Related Problems of Analysis, "Metsniereba", Tbilisi, v.I (1973) (Russian).

15 N.I. Muskhelishvili, Singular integral equations. 3rd ed. "Nauka", Moscow (1968); English translation of the 1st ed. Nordhoff, Groningen (1953).

16 N.P. Vekua, Systems of singular integral equations. 2nd ed. "Nauka", Moscow (1971).

R.V. Duduchava
Tbilisi Mathematical Institute
Academy of Sciences of the Georgian SSR
USSR

E G GORDADZE and B V KHVEDELIDZE

10 On singular integral operators

The paper deals with the integral operators (3) and (5) in the case when the integration line Γ belongs to some general class. The results obtained are applied to solve the problem of regularization of the operator (6) in the space $L_p(\Gamma)$, $p > 1$.

1. DEFINITIONS. STATEMENT OF THE PROBLEM

Let Γ be either a closed or an open line in the complex plane. In what follows, without stating it explicitly, Γ is assumed to be a rectifiable line. By an arc we mean a simple open continuous line.

Denote by $\underset{\sim}{C}(\Gamma)$ a set of complex functions, continuous on Γ (in the case when Γ is open, continuity is to be implied also at end-points). Further, denote by $\underset{\sim}{C}(\Gamma; c_1, \ldots, c_m)$ a set of bounded piecewise continuous functions with first order discontinuities at points $c_1, \ldots, c_m$. This class will sometimes be denoted by $\underset{\sim}{C}_0(\Gamma)$.

Let $\rho(t)$ be some measurable, non-negative function, almost everywhere different from zero. Denote by $L_p(\Gamma;\rho)$ a set of functions, measurable on Γ, for which the product $\rho(t)\,|\varphi(t)|^p$ is summable on Γ. We write $L_p(\Gamma)$ instead of $L_p(\Gamma;1)$, and $L(\Gamma)$ instead of $L_1(\Gamma)$. Later $L_p(\Gamma;\rho)$ and $L_p(\Gamma)$ will also denote Banach spaces in which the norm is defined as follows

$$\|\varphi\|_{p,\rho} = \{\int_\Gamma \rho(t)\,|\varphi(t)|^p\,ds\}^{1/p}, \tag{1}$$

$$\|\varphi\|_p = \{\int_\Gamma |\varphi(t)|^p\,ds\}^{1/p}, \tag{2}$$

where $ds = |dt|$.

Denote by $\underset{\sim}{S}$ a singular integral operator with the Cauchy kernel, i.e.

$$(\underset{\sim}{S}\varphi)(t) = \frac{1}{\pi i}\int_\Gamma \frac{\varphi(\tau)\,d\tau}{\tau - t}, \quad t \in \Gamma_1 . \tag{3}$$

where the integral is understood in the sense of Cauchy principal value.

The operator of multiplication by some function, defined on Γ, will be denoted by $\underset{\sim}{U}_\omega$, i.e.

$$\underset{\sim}{U}_\omega = \omega I , \tag{4}$$

where I is the identiy operator.

Then we shall consider the operators $T_\omega = \underset{\sim}{S}\underset{\sim}{U}_\omega - \underset{\sim}{U}_\omega \underset{\sim}{S}$, i.e.

$$(T_\omega \varphi)(t) = \frac{1}{\pi i} \int_\Gamma \frac{\omega(t) - \omega(t)}{\tau - t} \varphi(\tau)\, d\tau , \tag{5}$$

and $A = \underset{\sim}{U}_a + \underset{\sim}{U}_b \underset{\sim}{S} + \underset{\sim}{V}$, i.e.

$$(A\varphi)(t) = a(t)\varphi(t) + b(t)(\underset{\sim}{S}\varphi)(t) + (\underset{\sim}{V}\varphi)(t) , \tag{6}$$

where $a(t)$, $b(t)$ are functions, defined on Γ, and $\underset{\sim}{V}$ is a compact operator in $L_p(\Gamma)$.

The equation

$$(A\varphi)(t) = f(t) , \quad t \in \Gamma , \tag{7}$$

where $f(t)$ is a function, defined on Γ, A is an operator, defined by the equality (6), φ is an unknown function in the space where A is defined, is usually called the singular integral equation with the Cauchy kernel.

In the monograph by N.I. Muskhelishvili [10] the complete theory of equations is given with the following restrictions:

Γ is a piecewise-smooth line, the functions $a(t)$ and $b(t)$ satisfy a Hölder condition except possibly at a finite number of points, where they have first order discontinuities, and $a(t)$ and $b(t)$ satisfy the condition

$$\inf_{t \in \Gamma} | a^2(t) - b^2(t) | > 0 . \tag{8}$$

The functions $f(t)$, $\varphi(t)$ also satisfy a Hölder condition with the exception of a finite number of points where they can have an infinite limit of order less than 1.

The generalization of this theory to the case when Γ is a line of continuous curvature, and $a(t)$, $b(t)$ are continuous functions, was given by S.G. Mikhlin [9] and for the case when Γ is a Ljapunov line †, $f, \varphi \in L_p(\Gamma)$, $p > 1$, was considered by B.V. Khvedelidze. Later the complete theory of equations (7) in $L_p(\Gamma)$, $p > 1$, when Γ is a Ljapunov line, $a(t), b(t) \in \underset{\sim}{C}_0(\Gamma)$ was constructed in the works of I.C. Gohberg and N.Ja. Krupnik (see [1]).

In constructing the theory of equations (7) in L_p, one has to bear in mind the following:

1^0. If $\varphi \in L(\Gamma)$, then the integral (3) exists almost everywhere.

2^0. The operator $\underset{\sim}{S}$ is bounded in $L_p(\Gamma)$, $p > 1$, i.e.

$$\forall \varphi \in L_p(\Gamma), \quad \|\underset{\sim}{S}\varphi\|_p \leq \text{const } \|\varphi\|_p . \tag{9}$$

3^0. If $\omega(t) \in \underset{\sim}{C}(\Gamma)$, then the operator (5) is compact in $L_p(\Gamma)$, $p > 1$.

It is known [12], [13] that the operator (3) has the properties 1^0 and 2^0 in case when Γ is a circle. Those results can be easily generalized to the case when Γ is a Ljapunov line (see [6]). There naturally arises the problem: to prove 1^0 and 2^0 for more general lines than the Ljapunov line. There are a number of works in this direction. It is known (see [2]) that 1^0 and 2^0 are retained for the lines which are piecewise-smooth in Ljapunov's sense, i.e. for those lines which form the union of a finite number of Ljapunov arcs, are mutually devoid of common interior points as well as for some other classes of lines which have been considered in the papers of Khavin, Daniljuk and Shelepov, Dzhvarsheishvili, Khuskivadze and others (for these works the reader is referred to [2]).

However, it is not so far known whether the properties 1^0 and 2^0 are retained for smooth Γ.

Thus it was hitherto impossible to transfer the results of the monograph [10] to the case $f, \varphi \in L_p$, $p > 1$ so that the conditions, imposed on Γ, include the smooth lines. Therefore in constructing the theory of equations (7) for $f, \varphi \in L_p$, $p > 1$, it is advisable to assume that the operator (3) has all the necessary properties without indicating any particular class of lines. The theory thus constructed

† The angle between the tangential of the line Γ and some fixed direction satisfies the Hölder condition as a function of point of contact.

will automatically hold true for those classes of lines in which it is possible to show these properties.

It will be said that Γ belongs: i) to the class $\mathfrak{R}$, if the operator $\underset{\sim}{S}$ is bounded† in $L_p(\Gamma)$, $p > 1$; ii) to the class $\mathfrak{K}$, if there exists a positive number k, such that

$$\forall t_1 t_2 \in \Gamma : \quad \frac{|t_1 - t_2|}{s(t_1, t_2)} \geq k, \tag{10}$$

where $S(t_1, t_2)$ denotes the length of the smallest arc on Γ bounded by the points t_1 and t_2; iii) to the class $\mathfrak{G}_0$, if Γ is piecewise smooth, i.e. is the union of a finite number of smooth arcs which are mutually devoid of common interior points.

In the present paper the operators (3) and (5) will be considered under the assumption that $\Gamma \in (\mathfrak{R} \cap \mathfrak{K}) \cup (\mathfrak{R} \cap \mathfrak{G}_0)$. Later on the results which have been obtained will be used for solving the problem of regularization. We can formulate the problem as follows:

Let E and E_1 be Banach spaces. Denote by $[E \to E_1]$ a set of linear bounded operators, acting from E into E_1.

Let $A \in [E \to E]$ and $B \in [E \to E]$. The operator B will be called the regularizer for the operator A, if $AB = I + \underset{\sim}{V}_1$, $BA = I + \underset{\sim}{V}_2$, where $\underset{\sim}{V}_1$ and $\underset{\sim}{V}_2$ are the compact operators in E.

The problem of finding the operator B will be called the problem of regularization of the operator A.

In what follows, without stating the opposite, it will be assumed that $p > 1$, and

$$q = p(p-1)^{-1}.$$

2. SOME PROPERTIES OF THE OPERATORS $\underset{\sim}{S}$ AND T_ω

Lemma 1: If $\Gamma \in \mathfrak{K} \cup \mathfrak{G}_0$ then the relation

$$\int_\Gamma \frac{|d\tau|}{|\tau - c|^\alpha |\tau - t|^\beta} = 0(|t-c|^{1-\alpha-\beta}), \tag{11}$$

† If the operator $\underset{\sim}{S}$ is bounded in $L_p(\Gamma)$, where $p_0 > 1$ is a fixed number, then it will be bounded also in $L_p(\Gamma)$, for any $p > 1$ (see, e.g. [11]).

holds, where $c \in \Gamma$, $0 < \alpha, \beta < 1$ and $\alpha + \beta > 1$.

Proof: Let $t = t(s)$, $0 \leq s \leq \gamma$ be the equation of the line Γ in arc coordinates (γ is the length of the arc Γ, and S is the arc coordinate).

If $\Gamma \in \mathfrak{K}$, then the relation (10) immediately follows from the condition (11) and the estimate

$$\int_0^\gamma \frac{d\sigma}{|\sigma - s_0|^\alpha |\sigma - s|^\beta} = 0(|s - s_0|^{1-\alpha-\beta}),$$

where $c = t(s_0)$, $\tau = t(\sigma)$.

Assume now that $\Gamma \in \mathfrak{G}_0$ i.e. $\Gamma = \bigcup_{j=1}^n \Gamma_j$, where Γ_j are smooth arcs. Clearly, it suffices to estimate the integral

$$\psi(t) = \int_{\Gamma_j} \frac{d\sigma}{|\tau - c|^\alpha |\tau - t|^\beta}, \quad t \in \Gamma$$

where c is one of the ends of Γ_k. If $t \in \Gamma_k$ and $\Gamma_k \cap \Gamma_j = \phi$, or Γ_k and Γ_j have an end in common and $\Gamma_k \cup \Gamma_j \in \mathfrak{K}$ then evidently

$$\psi(t) = 0(|t - c|^{1-\alpha-\beta}).$$

If, on the other hand, the common end of lines Γ_k and Γ_j (which will be denoted by c) is the cusp of the line $\Gamma_k \cup \Gamma_j$, then one can apply here the reasoning, similar to that of [2].

It will be assumed that the arcs Γ_k and Γ_j are standard (see the definition of a standard arc in [10], p. 10) and α_0 is the corresponding angle.

Draw from the point $t \in \Gamma_k$ the perpendicular to the tangent of the line Γ at the point c and denote the point of intersection of this perpendicular with the arc Γ_j by t_0. Then for $t \in \Gamma_j$ we can write

$$t = t_0 \pm i|t - t_0|e^{i\alpha_1}$$

where α_1 is the angle between the tangent to Γ at the point c and the real axis. Since

$$|\tau - t| = |(\tau - t_0) \pm i|t - t_0| e^{i\alpha_1}| = |(\tau - t_0) e^{-i\alpha_1} \pm i|t - t_0|| \geq$$

$$|\mathrm{Re}\,(\tau - t_0) e^{-i\alpha_1}| = |\mathrm{Re}\,|\tau - t_0| e^{i[\arg(\tau - t_0) - \alpha_1]}| \geq |\tau - t_0| \cos \alpha_0 ,$$

we obtain

$$\psi(t) \leq \frac{1}{\cos \alpha_0} \int_{\Gamma_j} \frac{d\sigma}{|t - c|^{\alpha} |\tau - t_0|^{\beta}} .$$

Now, since $t_0 \in \Gamma_j$, we get

$$\psi(t) = 0(|t_0 - c|^{1-\alpha-\beta}) = 0(|t - c|^{1-\alpha-\beta}) .$$

Theorem 1: If $\Gamma \in (\mathfrak{R} \cap \mathfrak{K}) \cup (\mathfrak{R} \cap \mathfrak{G})$ then the operator $\underset{\sim}{S}$ is bounded in $L_p(\Gamma_{i\rho})_1\ p > 1$, where

$$\rho(t) = \prod_{k=1}^{m_1} |t - c_k|^{\alpha_k(p-1)} \prod_{k=m_1+1}^{m} |t - c_k|^{-\alpha_k} , \tag{12}$$

$$0 < \alpha_k < 1 , \quad c_k \in \Gamma , \quad k = 1, \ldots, m , \quad 0 \leq m_1 \leq m .$$

Proof: This theorem for the Ljapunov line was proved in [6]. Now it will be shown that the similar reasoning may be also applied to the present case.

Let $\varphi(t) \in L_p(\Gamma_{i\rho})$. Then

$$\varphi(t) = \rho^{-1/p}(t)\, f(t) , \tag{13}$$

where $f \in L_p(\Gamma)$. It is easily verified that

$$\int_{\Gamma} \rho(t)\, |(\underset{\sim}{S}\varphi)(t)|^p\, ds \leq 2^{p-1} \int_{\Gamma} |\psi(t)|^p ds + 2^{p-1} \int_{\Gamma} |(\underset{\sim}{S}f)(t)|^p\, ds , \tag{14}$$

where

$$\psi(t) = \frac{1}{\pi i} \int_{\Gamma} \frac{\rho^{1/p}(t) - \rho^{1/p}(\tau)}{\tau - t}\, d\tau . \tag{15}$$

In virtue of the conditions of the theorem $\Gamma \in \mathfrak{R}$, i.e. for $\forall f \in L_p(\Gamma)$

$$\int_\Gamma |(\underset{\sim}{S}f)(t)|^p ds \leq \text{const} \int_\Gamma |f(t)|^p ds ,$$

or due to (13)

$$\int_\Gamma |(\underset{\sim}{S}f)(t)|^p ds \leq \text{const} \int_\Gamma \rho(t) |\varphi(t)|^p ds . \tag{16}$$

Assume now that

$$\rho(t) = |t-c|^{\alpha(p-1)}, \qquad 0 < \alpha < 1 . \tag{17}$$

In this case (15) will take the form

$$\psi(t) = \frac{1}{\pi i} \int_\Gamma \frac{|t-c|^{\alpha/q} - |\tau - c|^{\alpha/q}}{\tau - t} \varphi(\tau) d\tau . \tag{18}$$

Introduce the notations: $\nu = 1 - \frac{\alpha}{q}$, $\alpha_1 = \alpha + \epsilon$ where ϵ is an arbitrary number such that $0 < \epsilon < 1 - \alpha$. Using the Hölder inequality, we shall have

$$|\psi(t)|^p \leq \left(\frac{1}{\pi} \int_\Gamma \frac{|\varphi(\tau)|}{|\tau - t|^\nu} d\sigma\right)^p = \frac{1}{\pi}\left(\int_\Gamma \frac{|\varphi(\tau)|\,|\tau - c|^{\alpha_1/q}}{|\tau - t|^{\nu/p}} \cdot \frac{1}{|\tau - t|^{\nu/q} |\tau - c|^{\alpha_1/q}} d\sigma\right)^p$$

$$\leq \frac{1}{\pi^p} \int_\Gamma \frac{|\varphi(\tau)|^p |\tau - c|^{\alpha_1(p-1)}}{|\tau - t|^\nu} d\sigma \cdot \left(\int_\Gamma \frac{d\sigma}{|\tau - t|^\nu |\tau - c|^{\alpha_1}}\right)^{p-1} \tag{19}$$

Hence, according to Lemma 1,

$$\int_\Gamma \frac{d\sigma}{|\tau - t|^\nu |\tau - c|^{\alpha_1}} = 0\left(|t-c|^{-\frac{\alpha}{p} - \epsilon}\right) . \tag{20}$$

Let ϵ be chosen so small that $\epsilon(p-1) + \frac{\alpha}{q} < 1$. By applying first the Fubini theorem and then Lemma 1, we obtain

$$\int_\Gamma |\psi(t)|^p ds \leq \text{const} \int_\Gamma |\tau - c|^{\alpha(p-1)} |\varphi(\tau)|^p d\sigma . \tag{21}$$

From (14), (16) and (21) the validity of the theorem follows for the case when is defined by the equality (17).

ıppose now that

$$(t) = |t-c|^{-\alpha} \qquad 0 < \alpha < 1 , \qquad c \in \Gamma . \tag{22}$$

s case, the equality (15) will take the form

$$(t) = |t-c|^{-\alpha/p} \psi_1(t)_1 \tag{23}$$

where

$$\psi_1(t) = \frac{1}{\pi i} \int_\Gamma \frac{|\tau-c|^{\alpha/p} - |t-c|^{\alpha/p}}{\tau - t} f(\tau)\, d\tau , \tag{24}$$

and $f(t) \in L_p(\Gamma)$.

Let $\nu = 1 - \frac{\alpha}{p}$, and let α_1 be a number such that

$$\frac{\alpha}{q} < \alpha_1 < \min(\nu, p-1) .$$

Applying the Hölder inequality and Lemma 1, one has

$$\psi_1(t) \leq \text{const}\, |t-c|^{\alpha/q-\alpha_1} \int_\Gamma \frac{|\tau-c|^{\alpha_1} |f(\tau)|^p}{|\tau-t|^\nu}\, d\sigma .$$

From this inequality and equality (23) one obtains

$$\int_\Gamma |\Psi(t)|^p ds \leq \text{const} \int_\Gamma |\tau-c|^{\alpha_1} |f(\tau)|^p d\sigma \int_\Gamma \frac{ds}{|t-c|^{\alpha_1+\alpha/p} |\tau-t|^\nu} . \tag{25}$$

Since $\alpha_1 + \frac{\alpha}{p} < 1$, $\alpha_1 + \frac{\alpha}{p} + \nu = 1 + \alpha_1$, in virtue of Lemma 1, the relations (25) will take the form

$$\int_\Gamma |\Psi(t)|^p ds \leq \text{const} \int_\Gamma |f(t)|^p ds .$$

Hence, due to (13),

$$\int_\Gamma |\Psi(t)|^p ds \leq \text{const} \int_\Gamma |t-c|^{-\alpha} |\varphi(t)|^p ds . \tag{26}$$

From (14), (16) and (26) the validity of the theorem follows in the case when $\rho(t)$ is defined by the equality (22).

If now we divide the line Γ into parts Γ_k, $k = 1, \ldots, m$, so that on each of these parts there is no more than one point c_k, then it is easily verified that the theorem holds true also in the case when $\rho(t)$ is defined by the equality (12).

Theorem 2: Let $\omega(t) \in \underset{\sim}{C}(\Gamma)$. If $\Gamma \in \mathfrak{R}$ then the operator T_ω is compact in $L_p(\Gamma)$, $p > 1$; if $\Gamma \in (\mathfrak{R} \cap \mathfrak{K}) \cup (\mathfrak{R} \cap \mathfrak{G}_0)$, then T_ω is compact also in $L_p(\Gamma; \rho)$, $p > 1$ where $\rho(t)$ is the function defined by (12).

Proof: As it is known (see [8], p. 417), the function $\omega(t)$ which is continuous on Γ can be uniformly approximated by means of rational functions $\omega_n(t)$ whose poles do not lie on Γ. Thus

$$\lim_{n \to \infty} \max_{t \in \Gamma} |\omega_n(t) - \omega(t)| = 0 . \tag{27}$$

It is easily observed that the operator T_{ω_n} is finite-dimensional in the spaces $L_p(\Gamma)$, $p > 1$ and $L_p(\Gamma; \rho)$, $p > 1$. Besides,

$$\| T_\omega - T_{\omega_n} \| \leq \text{const} \max_{t \in \Gamma} |\omega(t) - \omega_n(t)| . \tag{28}$$

From the relations (27) and (28) there follows the theorem which was to be proved.

Theorem 2 for the space $L_2(\Gamma)$ when Γ is the line with continuous curvature was first proved by S. G. Mikhlin [9]; later for the Ljapunov line and spaces $L_p(\Gamma; \rho)$ it was shown in [6].

There arises the following question: Does the above theorem remain valid if the function $\omega(t) \in \underset{\sim}{C}_0(\Gamma)$?

In [1] I. C. Gohberg and N. Ja. Krupnik have proved as a consequence of one general theorem that if $\omega(t) \in \underset{\sim}{C}(\Gamma) \cup \underset{\sim}{C}_0(\Gamma)$, where Γ is the Ljapunov line then the operator (5) will be compact if and only if $\omega(t)$ is a continuous function.

A function $\omega(t) \in \underset{\sim}{C}_0(\Gamma)$ will now be given for which the operator (5) is not compact in $L_p(\Gamma)$.

Let Γ_{ab} be an open line of the class $\mathfrak{R} \cap \mathfrak{K}$, a and b its finite points. Let c be an interior point of the line Γ_{ab}. Consider the function

$$\omega_0(t) = \begin{cases} 0 , & \text{when} \quad t \in \Gamma_{ac} \\ 1 , & \text{when} \quad t \in \Gamma_{cb} \end{cases}$$

where

$$\Gamma_{ab} = \Gamma_{ac} \cup \Gamma_{cb} .$$

It will be shown that the operator $T\omega_0$ is not compact in $L_p(\Gamma_{ab})$.

Consider in $L_p(\Gamma_{ab})$ as the set of functions

$$\{\varphi_\epsilon(t)\} \qquad 0 < \epsilon < 1 , \tag{29}$$

where

$$\varphi_\epsilon(t) = \begin{cases} 0 , & \text{when} \quad t \in \Gamma_{ac} \cup \Gamma_{c_1 b} , \\ \epsilon^{-1/p}, & \text{when} \quad t \in \Gamma_{cc_1} , \end{cases}$$

and c_1 is a point on Γ_{cb} such that the length of the arc Γ_{cc_1} is equal to ϵ.

It is clear that

$$\| \varphi_\epsilon \|_p = 1 .$$

Find the image of the set (29) in case of the transformation T_{ω_0}:

$$(T_{\omega_0} \varphi_\epsilon)(t) = \int_{\Gamma_{cc_1}} \frac{\omega_0(\tau) - \omega_0(t)}{\tau - t} \varphi(\tau)\, d\tau = \epsilon^{-1/p} \psi_t(\Gamma_{ac}) \ln \frac{c_1 - t}{c - t} ,$$

where $\psi_t(\Gamma_{ac})$ are the characteristic functions of the set Γ_{ac}.

Let γ be the length of the arc Γ_{ac} and

$$\Phi_\epsilon(t) = (T_\omega \varphi_\epsilon)(t) .$$

Then for $\forall h > 0$ we obtain

$$\int_{\Gamma_{ab}} |\Phi_\epsilon(t(s+h)) - \Phi_\epsilon(t(s))|^p ds \geq \int_{\gamma - \frac{h}{2}}^{\gamma} |\Phi_\epsilon(t(s))|^p ds \geq \frac{1}{\epsilon} \int_{\gamma - \frac{h}{2}}^{\gamma} \left| \ln \frac{|c_1 - t|}{|c - t|} \right|^p ds. \tag{30}$$

Denote the length of the arcs Γ_{tc_1} and Γ_{tc}, respectively, by γ_{tc_1} and γ_{tc}, then since $\Gamma_{ab} \in \mathfrak{K}$, we have

$$\frac{|c_1 - t|}{|c - t|} \geq k\left(1 + \frac{\gamma_{cc_1}}{\gamma_{tc}}\right) \geq k\left(1 + \frac{2\epsilon}{h}\right).$$

If now ϵ and h are connected by means of the relation $h = 2\epsilon k$, then

$$\frac{|c_1 - t|}{|c - t|} \geq 1 + k .$$

From the relation (30) we have

$$\int_{\Gamma_{ab}} |\Phi_\epsilon(t(s+h)) - \Phi_\epsilon(t(s))|^p ds > k\,|\ell n(1+k)|^p .$$

Consequently, in virtue of the theorem of M. Riesz on the compactness of a set of functions in the space L_p (see [4], p.273), the set of functions $\{\Phi_\epsilon(t)\}$ is not compact in $L_p(\Gamma_{ab})$. Hence, the operator T_{ω_0} is not compact in $L_p(\Gamma_{ab})$.

The example which has just been constructed now enables us to indicate a simple method of proving the above result of I.C. Gohberg and N.Ja. Krupnik in the space $L_p(\Gamma)$.

Theorem 3: If $\omega(t) \in \underset{\sim}{C}(\Gamma) \cup \underset{\sim}{C}_0(\Gamma)$, $\Gamma \in \mathfrak{R} \cap \mathfrak{K}$, then the operator (5) will be compact in $L_p(\Gamma)$ if and only if $\omega(t) \in \underset{\sim}{C}(\Gamma)$.

Proof: If $\omega(t) \in \underset{\sim}{C}(\Gamma)$, then due to Theorem 2, the operator (5) will be compact in $L_p(\Gamma_{ab})$.

Suppose now that the operator T_ω is compact. Assume for simplicity that $\omega(t) \in \underset{\sim}{C}(\Gamma_{ab}; c)$, where c is an interior point of the line Γ_{ab}. Let $\alpha = \omega(c+0) - \omega(c-0)$ and consider the function

$$\omega_1(t) = \omega(t) - \alpha\,\omega_0(t) .$$

It is clear that $\omega_1(t) \in \underset{\sim}{C}(\Gamma)$ and therefore, in virtue of Theorem 1, T_{ω_1} will be compact in $L_p(\Gamma_{ab})$. Now from the evident equality

$$T_\omega = \alpha\, T_{\omega_0} + T_{\omega_1}$$

it follows that the operator T_ω is not compact in $L_p(\Gamma_{ab})$.

Let Γ be an arbitrary line of the class $\mathfrak{R} \cap \mathfrak{K}$, $\omega(t) \in \underset{\sim}{C}_0(\Gamma)$ and c one of the points of discontinuity of the function $\omega(t)$. Denote by $\Gamma_{ab} = \Gamma_{ac} \cup \Gamma_{cb}$ the neighbourhood of the point c on the line Γ such that there are no other points of discontinuity on it except the point c. Due to the above theorem the operator T_ω will not be compact in $L_p(\Gamma)$, but to within an isometry $L_p(\Gamma_{ab}) \subset L_p(\Gamma)$. Hence T_ω will not be compact in $L_p(\Gamma)$.

Note: The above theorem remains valid also for the space $L_p(\Gamma;\rho)$, $p > 1$. The proof can be carried out in a similar way. Consider now the operator (5) when it acts from one functional Lebesgue space into another.

Theorem 4: If $\omega(t) \in \underset{\sim}{C}(\Gamma;c_1,\dots,c_m)$, $\sigma(t) \in \underset{\sim}{C}(\Gamma)$ and vanishes at points $c_1,\dots,c_m$ then the operator $U_\sigma T_\omega$ will be compact in the space $L_p(\Gamma)$, $p > 1$.

Proof: The functions $\sigma(t)$, $\omega(t)$ are bounded, therefore $\underset{\sim}{U}_\sigma T_\omega \in [L_p(\Gamma) \to L_p(\Gamma)]$.

On account of the equality

$$\omega(\tau) - \omega(t) = \frac{\omega(\tau)\sigma(\tau) - \omega(t)\sigma(t)}{\sigma(t)} - \frac{\sigma(\tau) - \sigma(t)}{\sigma(t)}\,\omega(t),$$

we shall have

$$(T_\omega\varphi)(t) = \frac{1}{\sigma(t)}\int_\Gamma \frac{\omega(\tau)\sigma(\tau) - \omega(t)\sigma(t)}{\tau - t}\,\varphi(\tau)\,d\tau - \frac{1}{\sigma(t)}\int_\Gamma \frac{\sigma(\tau) - \sigma(t)}{\tau - t}\,\omega(\tau)\varphi(\tau)\,d\tau.$$

This equality can also be written as

$$\underset{\sim}{U}_\sigma T_\omega = T_{\omega\sigma} - T_\sigma \underset{\sim}{U}_\omega\,. \tag{31}$$

From the Theorem 2 and equality (31) there follows the validity of the theorem if we take into account that the functions $\omega(t)\sigma(t)$ and $\sigma(t)$ are continuous on Γ.

It is clear that in the above theorem the space $L_p(\Gamma)$ may be changed to the space $L_p(\Gamma;\rho)$, where $\rho(t)$ is defined by (12).

Let

$$\rho(t) = \prod_{k=1}^{m} |t - c_k|^{\nu_k}, \tag{32}$$

where $\nu_1, \dots, \nu_m$ are arbitrarily fixed positive numbers. Then we shall have

Theorem 5: If $\omega(t) \in \underset{\sim}{C}(\Gamma; c_1, \dots, c_m)$ then the operator Γ_ω is compact as an operator from $L_p(\Gamma)$ into $L_p(\Gamma;\rho)$, and also as an operator from $L_p(\Gamma)$ into $L_{p-\epsilon}(\Gamma)$, where $\epsilon > 0$ is an arbitrarily small fixed number.

Proof: Consider the function

$$\sigma(t) = \prod_{k=1}^{m} |t - c_k|^{\alpha_k}, \tag{33}$$

where

$$0 < \alpha_k \leq \frac{\nu_k}{p}, \qquad k = 1, \dots, m.$$

It is easily verified that $\underset{\sim}{U}_{1/\sigma} \in [L_p(\Gamma) \rightarrow L_p(\Gamma;\rho)]$. Therefore from the equality

$$T_\omega = \underset{\sim}{U}_{1/\sigma}(\underset{\sim}{U}_\sigma T_\omega)$$

and the preceding theorem there follows the compactness of T_ω as an operator from $L_p(\Gamma)$ into $L_p(\Gamma;\rho)$.

In order to prove the second part of the theorem we have to consider any fixed number p_1, such that

$$1 < p_1 < p$$

and let

$$\max\{\nu_1, \nu_2, \dots, \nu_m\} < \frac{p - p_1}{p_1}. \tag{34}$$

Introduce the notation

$$r = \frac{p}{p_1}$$

Then we shall have

$$\|\varphi\|_{p_1}^{p_1} = \int_\Gamma |\varphi(t)|^{p_1} \rho^{\frac{1}{2}}(t) \rho^{-\frac{1}{2}}(t) ds \leq \left\{\int_\Gamma |\varphi(t)|^{rp_1} \rho(t) ds\right\}^{\frac{1}{2}} \cdot \left\{\int_\Gamma \rho^{-\frac{r'}{r}}(t) ds\right\}^{\frac{1}{2}'} = M^{p_1} \|\varphi\|_{p,\rho}^{p}$$

where

$$M^{p_1} = \{\int_\Gamma \rho^{-\frac{r'}{r}}(t)\,ds\}^{\frac{1}{2}'} \quad , \quad r' = r(r-1)^{-1} .$$

Due to the inequality (34) the summability of the function can be provided. Thus, we have the relation

$$\|\varphi\|_{p_1} \leq M\|\varphi\|_{p,\rho} \cdot \tag{35}$$

From the latter inequality and the part of the theorem which has already been proved we can see that the second part of the theorem is true.

In Theorem 6 $\sigma(t)$ will denote the function (33) and

$$\rho(t) = \prod_{k=1}^{m} |t-c_k|^{\nu_k(p-1)}$$

where α_k, ν_k $(k = 1,\ldots,m)$ are positive numbers, when $\nu_k < 1$ $(k = 1,\ldots,m)$ and

$$\max\{\alpha_1, \alpha_2, \ldots, \alpha_m\} < \frac{1}{q} . \tag{36}$$

Theorem 6: If $\sigma(t)\,\omega(t) \in \underset{\sim}{C}(\Gamma) \cup \underset{\sim}{C}(\Gamma; c_1,\ldots,c_m)$, then the numbers ν_k can be chosen so that the operator T_ω will map the space $L_p(\Gamma)$ into $L_p(\Gamma;\rho^2)$ and will be compact.

Proof: Introduce the notation

$$\omega_0(t) = \sigma(t)\,\omega(t). \tag{37}$$

Due to the condition (36), $\alpha_k q < 1$, $k = 1,\ldots,m$.

Let us choose the numbers ν_k so that

$$\alpha_k q < \nu_k < 1 , \quad k = 1,\ldots,m. \tag{38}$$

Then, as it can easily be verified,

$$\underset{\sim}{U}_\omega \in [L_p(\Gamma) \to L_p(\Gamma;\rho)] . \tag{39}$$

Consider first the case when $\omega_0 \in \underset{\sim}{C}(\Gamma)$. Using the equalities (31), we obtain

$$\underset{\sim}{U}_\sigma T_\omega = T_{\omega_0} - T_\sigma \underset{\sim}{U}_\omega \tag{40}$$

From this equality, Theorem 1, relation (39) and conditions $\sigma(t), \omega(t) \in \underset{\sim}{C}(\Gamma)$ it can easily be concluded that the operator $\underset{\sim}{U}_\sigma T_\omega$ is compact as the operator from $L_p(\Gamma)$ into $L_p(\Gamma;\rho)$.

By the condition (38)

$$\underset{\sim}{U}_{1/\sigma} \in [L_p(\Gamma) \to L_p(\Gamma;\rho^2)] . \tag{41}$$

Therefore, from the equality

$$T_\omega = \underset{\sim}{U}_{1/\sigma}(\underset{\sim}{U}_\sigma T_\omega)$$

it follows that the operator T_ω is compact as the operator from $L_p(\Gamma)$ into $L_p(\Gamma;\rho^2)$.

Let $\omega_0(t) \in \underset{\sim}{C}(\Gamma;c_1,\ldots,c_m)$. Consider the function $\sigma_1(t) = \sigma_0(t)\sigma(t)$, where

$$\sigma_0(t) = \prod_{k=1}^{m} |t - c_k|^{\epsilon_k} ,$$

where ϵ_k are positive numbers to be defined in the sequel.

Taking into account that $\underset{\sim}{U}_{\sigma_1} = \underset{\sim}{U}_{\sigma_0} U_\sigma$, from the equality (40) we obtain

$$\underset{\sim}{U}_{\sigma_1} T_\omega = \underset{\sim}{U}_{\sigma_0} T_{\omega_0} - \underset{\sim}{U}_{\sigma_0} T_\sigma \underset{\sim}{U}_\omega . \tag{42}$$

Assume now that the numbers ν_k are chosen according to the conditions (38). Then from the equality (42), relation (39), Theorem 4 and condition $\omega_0 \in \underset{\sim}{C}_0(\Gamma)$, $\sigma, \sigma_0 \in \underset{\sim}{C}(\Gamma)$ it is easily concluded that the operator $\underset{\sim}{U}_{\sigma_1} T_\omega$ is compact as an operator from $L_p(\Gamma)$ into $L_p(\Gamma;\rho)$.

As it has been assumed, the numbers ν_k satisfy the conditions

$$\nu_k(p-1) - \alpha_k p > 0 , \qquad k = 1,\ldots,m .$$

Let us choose the positive numbers ϵ_k so small that

$$\nu_k (p-1) - p\alpha_k - p\epsilon_k > 0 , \qquad k = 1,\dots,m$$

Then

$$\underset{\sim}{U}_{1/\sigma_1} \in [L_p(\Gamma;\rho) \to L_p(\Gamma;\rho^2)] .$$

From this relation and the above theorem about the operator $U_{\sigma_1} T_\omega$ and the equality

$$T_\omega = \underset{\sim}{U}_{1/\sigma} (\underset{\sim}{U}_{\sigma_1} T_\omega)$$

we can conclude that the mapping T_ω from $L_p(\Gamma)$ to $L_p(\Gamma;\rho^2)$ is compact.

Theorem 7: If $\sigma(t)\,\omega(t) \in \underset{\sim}{C}(\Gamma) \cap \underset{\sim}{C}(\Gamma;c_1,\dots,c_m)$, then the operator T_ω maps compactly the space $L_p(\Gamma)$ into the space $L_{p-\epsilon}(\Gamma)$, where $\epsilon > 0$ is an arbitrarily small fixed number.

Proof: Let p_1 be a fixed number, satisfying the relations

$$\frac{p}{2p-1} < p_1 < p .$$

Select numbers $\alpha_1,\dots,\alpha_m$ in such a way that

$$\max\{\alpha_1,\dots,\alpha_m\} < \frac{p-p_1}{2pp_1} .$$

From these relations we get

$$\alpha_k q < \frac{p-p_1}{2p_1(p-1)} < 1 , \qquad k = 1,\dots,m .$$

Let ν_k $(k = 1,\dots,m)$ be numbers, satisfying the relations

$$\alpha_k q < \nu_k < \frac{p-p_1}{2p_1(p-1)} , \qquad k = 1,\dots,m . \tag{43}$$

Then by means of Theorem 6 the operator T_ω is compact from $L_p(\Gamma)$ into $L_p(\Gamma;\rho^2)$ where the weight $\rho(t) = \prod_{k=1}^{m} |t-c_k|^{\nu_k(p-1)}$. Now to complete the proof it suffices to note that

$$\|\varphi\|_{p_1} \leq \text{const}\,\|\varphi\|_{p,\rho^2}\,. \tag{44}$$

The last relation follows from the following chain of relations (where we assume $r = \frac{p}{p_1}$):

$$\|\varphi\|_{p_1}^{p_1} = \int_\Gamma |\varphi(t)|^{p_1} ds = \int_\Gamma |\varphi(t)|^{p_1} \rho^{2/r}(t)\rho^{-2/r}(t) ds \leq (M\|\varphi\|_{p,\rho^2}^{p_1})\,,$$

where

$$M = \{\int_\Gamma \rho^{-\frac{2r'}{r}}(t)\,ds\}^{\frac{1}{2}'}\,, \quad r' = r(r-1)^{-1}\,.$$

By means of (43) we get $M < +\infty$.

3. ON THE PROBLEM OF REGULARIZATION

Consider the operator (6), i.e. the operator

$$A = \underset{\sim}{U}_a + \underset{\sim}{U}_b\underset{\sim}{S} + \underset{\sim}{V}\,.$$

In $L_p(\Gamma)$, $p > 1$. Later, under certain assumptions, we shall construct the regularizer of the operator (6). This is equivalent to the fact that for the singular integral equation (7) the theorems of Noether are true (see e.g. [1]).

Assume first that $a(t)$ and $b(t)$ are continuous functions.

Theorem 8: If Γ is a closed line of the class $\mathfrak{R}$, $a(t), b(t) \in C(\Gamma)$ and condition (8) is satisfied, then the operator

$$B = \underset{\sim}{U}_{a_1} - \underset{\sim}{U}_{b_1}\underset{\sim}{S}_1 \tag{45}$$

where

$$a_1(t) = \frac{a(t)}{a^2(t) - b^2(t)}\,, \qquad b(t) = \frac{b(t)}{a^2(t) - b^2(t)}\,, \tag{46}$$

is the regularizer of the operator (6) in $L_p(\Gamma)$, $p > 1$.

In order to prove this theorem we shall need

Lemma 2: If Γ is a closed Jordan line of the class $\mathfrak{R}$, $\varphi(t) \in L_p(\Gamma)$, then almost everywhere on Γ the following equality holds.

$$(\underset{\sim}{S}^2 \varphi)(t) = \varphi(t) \tag{47}$$

Proof: Consider the Cauchy type integral

$$\Phi(z) = \frac{1}{2\pi i} \int_\Gamma \frac{\varphi(\tau)\, d\tau}{\tau - z} .$$

According to Privalov's principal lemma (see [12], p. 184), the following equality

$$\Phi^+(t) = \tfrac{1}{2}\varphi(t) + \tfrac{1}{2}(\underset{\sim}{S}\varphi)(t) \tag{48}$$

holds almost everywhere on Γ.

It is clear that $\Phi^+(t) \in L_p(\Gamma)$ and therefore, as it was shown by V.P. Khavin [5], $\Phi(z) \in E_p$ †. Then from V.I. Smirnov's result (see [12], p. 208) it follows that $\Phi(z)$ can be taken as the Cauchy integral

$$\Phi(z) = \frac{1}{2\pi i} \int_\Gamma \frac{\Phi^+(\tau)\, d\tau}{\tau - z} .$$

Hence, almost everywhere on Γ we have

$$(\underset{\sim}{S}\Phi^+)(t) = \Phi^+(t) .$$

If we introduce $\Phi^+(t)$ from (48) into this equality, we shall obtain the equality (47) or, what is the same thing, the operator equality

$$\underset{\sim}{S}^2 = I . \tag{49}$$

Proof of Theorem 8: Note that if $\alpha(t)$ is the bounded measurable function on Γ, then

$$\underset{\sim}{S}\underset{\sim}{U}_\alpha = \underset{\sim}{U}_\alpha \underset{\sim}{S} + T_\alpha . \tag{50}$$

On account of equalities (49) and (50) we have

† See the definition of the class E_p, (e.g. [12], p. 203).

$$BA = (\underset{\sim}{U}_{a_1} - \underset{\sim}{U}_{b_1}\underset{\sim}{S})(\underset{\sim}{U}_a + \underset{\sim}{U}_b\underset{\sim}{S}) = \underset{\sim}{U}_{a_1}\underset{\sim}{U}_a + \underset{\sim}{U}_{a_1}\underset{\sim}{U}_b\underset{\sim}{S} - \underset{\sim}{U}_{b_1}(\underset{\sim}{U}_a\underset{\sim}{S} + T_a) - \underset{\sim}{U}_{b_1}(\underset{\sim}{U}_b\underset{\sim}{S} + T_b)\underset{\sim}{S}$$

$$= (\underset{\sim}{U}_{a_1}\underset{\sim}{U}_a - \underset{\sim}{U}_{b_1}\underset{\sim}{U}_b) + (\underset{\sim}{U}_{a_1}\underset{\sim}{U}_b - \underset{\sim}{U}_{b_1}\underset{\sim}{U}_a)\underset{\sim}{S} - \underset{\sim}{U}_{b_1}(T_a + T_b\underset{\sim}{S}) = I + \underset{\sim}{V}_1 ,$$

where

$$\underset{\sim}{V}_1 = -\underset{\sim}{U}_{b_1}(T_a + T_b\underset{\sim}{S}) .$$

By the assumption that a and $b \in C(\Gamma)$ and Theorem 1, the operators T_a and T_b will be compact in $L_p(\Gamma)$ and so will the operator $\underset{\sim}{V}_1$.

In a similar way we have

$$AB = I + \underset{\sim}{V}_2 ,$$

where

$$\underset{\sim}{V}_2 = \underset{\sim}{U}_b(T_{a_1} - T_{b_1}\underset{\sim}{S}) .$$

Note: If the functions $a(t)$, $b(t)$ belong to the class $C_0(\Gamma)$ then the regularizer of the operator (6) in a general form will not be written as $B = \underset{\sim}{U}_{a_1} - \underset{\sim}{U}_{b_1}\underset{\sim}{S} + \underset{\sim}{V}$ where $\underset{\sim}{V}$ is compact.

In fact, let $a(t) = 0$, $b(t) \in C_0(\Gamma)$. Then

$$AB = -\underset{\sim}{U}_{b_1}T_b\underset{\sim}{S} + A\underset{\sim}{V} , \quad BA = -\underset{\sim}{U}_b T_{b_1}\underset{\sim}{S} + \underset{\sim}{V}A .$$

But due to Theorem 3, the operators T_b and T_{b_1} are not compact in $L_p(\Gamma)$.

Let us proceed now to construct the regularizer in the case $a(t)$, $b(t) \in \underset{\sim}{C}(\Gamma; c_1, \ldots, c_m)$.

As usual, the operator

$$A_0 = \underset{\sim}{U}_a + \underset{\sim}{U}_b\underset{\sim}{S} \tag{51}$$

will be called the characteristic part of the operator A.

If we introduce the notations (see, e.g. [1])

$$P = \tfrac{1}{2}(I + \underset{\sim}{S}) , \qquad Q = \tfrac{1}{2}(I - \underset{\sim}{S}) \tag{52}$$

and take into account condition (8), then the operator (51) may be written as

$$A_0 = \underset{\sim}{U}_{a+b}(\underset{\sim}{P} + \underset{\sim}{U}_G Q)$$

where

$$G(t) = \frac{a(t) - b(t)}{a(t) + b(t)} .$$

For arbitrary $p > 1$ one can define a piecewise continuous function $\ell n G(t)$ so that

$$-\frac{1}{p} < \operatorname{Re}\gamma_k \leq \frac{1}{q} , \qquad k = 1,\ldots,m , \tag{53}$$

where

$$\gamma_k \equiv \frac{1}{2\pi i} [\ell n\, G(c_k - 0) - \ell n\, G(c_k + 0)] . \tag{54}$$

It is clear that one can always choose the number $p > 1$ so that in (53) strict inequality will hold, i.e.

$$-\frac{1}{p} < \operatorname{Re}\gamma_k < \frac{1}{q} , \qquad k = 1,\ldots,m . \tag{55}$$

Let $\mathscr{D}^+$ be a finite domain, bounded by the line Γ and $\mathscr{D}^-$ - an infinite domain. Define the function

$$\omega(t) = \prod_{k=1}^{m} (t - c_k)^{\gamma_k} \tag{56}$$

where $(t - c_k)^{\gamma_k}$ is the boundary value of some branch of function $(z - c_k)^{\gamma_k}$ which is continuous in $\mathscr{D}^+$.

<u>Theorem 9</u>: If Γ is closed and $\Gamma \in (\mathfrak{R} \cap \mathfrak{K}) \cup (\mathfrak{R} \cup \mathfrak{G}_0)$, $a, b \in \underset{\sim}{C}(\Gamma; c_1, \ldots, c_m)$ the conditions (8), (55) are satisfied, then the operator

$$B = \underset{\sim}{U}_{a_1} - \underset{\sim}{U}_{b_1(a+b)\omega} \underset{\sim}{S}\, \underset{\sim}{U}_{(a+b)^{-1}\omega^{-1}} \tag{57}$$

where a_1 , b_1 and ω are defined by means of formulas (46) and (56), is the regularizer of the operator (6). This theorem in the case when $\Gamma \in \mathfrak{R} \cap \mathfrak{G}_0$

is proved in [3]

In order to prove the theorem a lemma will be reproduced beforehand.

Let

$$\omega_1(t) = \prod_{k=1}^{m} (t-c)^{\gamma_k} , \tag{58}$$

where $c \in D^+$ are numbers satisfying the condition (55), and $(t-c)^{\gamma_k}$ is continuous on Γ, with the exception of the point c_k.

Lemma 3: If $\Gamma \in (\mathfrak{R} \cap \mathfrak{K}) \cup (\mathfrak{R} \cap \mathfrak{G}_0)$ then the operator

$$A_1 = \underset{\sim}{P} + \underset{\sim}{U}_{\omega_1} Q$$

will be transformed into $L_p(\Gamma)$, $p > 1$.

The solution of the equation

$$A_1 \varphi = f$$

in $L_p(\Gamma)$ is equivalent to the solution of the following boundary value problem: to find the function $\Phi(z)$, given by the Cauchy type integral

$$\Phi(z) = \frac{1}{2\pi i} \int_\Gamma \frac{\varphi(t)\,dt}{t-z} , \qquad \varphi \in L_p(\Gamma)$$

whose angular boundary values $\Phi^+(t)$, $\Phi^-(t)$ satisfy the condition

$$\Phi^+(t) = w_1(t)\,\Phi^-(t) + f(t) . \tag{59}$$

The explicit solution of this problem when Γ is the Ljapunov line is given in [6,7]. For the contours considered here the problem is solved in a similar way.

Denote by Γ_1 a simple arc connecting the points c, c_k, ∞ and having no other common points with Γ except c_k. The functions $\prod_{k=1}^{m} (z-c_k)^{\gamma_k}$ and $\prod_{k=1}^{m} (z-c)^{\gamma_k}$ are analytic in the plane, cut along Γ_1. Choose the branch of functions $\prod_{k=1}^{m} (z-c)^{\gamma_k}$ so that its boundary values from $\mathscr{D}^-$ will not coincide with $w_1(t)$ and the branch $(z-c_k)^{\gamma_k}$ so that $\lim_{z\to\infty} (z-c_k)^{\gamma_k} \setminus (z-c)^{\gamma_k} = 1$, $k = 1, \dots, m$.

Then, evidently, the function

$$\underset{\sim}{X}(Z) = \begin{cases} \prod_{k=1}^{m} (z - c_k)^{\gamma_k} , & z \in \mathscr{D}^+ \\ \prod_{k=1}^{m} \left(\frac{z - c_k}{z - c}\right)^{\gamma_k} , & z \in \mathscr{D}^- \end{cases}$$

is piecewise analytic in the plane, cut along Γ and belongs to the class E_p not only in $\mathscr{D}^+$ but also in $\mathscr{D}^-$. Besides, $[\underset{\sim}{X}(z)]^{-1} \in E_q$ in $\mathscr{D}^+$ as well as in $\mathscr{D}^-$ and

$$\omega_1(t) = \frac{\underset{\sim}{X}^+(t)}{\underset{\sim}{X}^-(t)} . \tag{60}$$

Introducing (60) into (59) we obtain

$$\Phi^+(t) = \frac{\underset{\sim}{X}^+(t)}{\underset{\sim}{X}^-(t)} \Phi^-(t) + f(t) ,$$

i.e.

$$\frac{\Phi^+(t)}{\underset{\sim}{X}^+(t)} = \frac{\Phi^-(t)}{\underset{\sim}{X}^-(t)} + \frac{f(t)}{\underset{\sim}{X}^+(t)} . \tag{61}$$

The functions $\Phi(z)$ and $[\underset{\sim}{X}(z)]^{-1}$ belong to the conjugate classes E_p, E_q and therefore $\Psi(z) = \Phi(z)[\underset{\sim}{X}(z)]^{-1} \in E_1$. Obviously we have

$$\Psi^+(t) = \Psi^-(t) + \frac{f(t)}{\underset{\sim}{X}^+(t)}$$

and hence it follows that the solution of problem (59) when it exists is unique and is written as

$$\Phi(z) = \frac{\underset{\sim}{X}(z)}{2\pi i} \int_\Gamma \frac{f(t)\,dt}{X^+(t)(t-z)} . \tag{62}$$

Let $f(t) \in \underset{\sim}{C}(\Gamma)$. Then $\int_\Gamma \underset{\sim}{X}^+(t)(t-z)^{-1} f(t)\,dt \in E_q$ and $\Phi \in E_1$ in $\mathscr{D}^+$ as well as in $\mathscr{D}^-$. Besides, $\Phi^\pm(t) \in L_p(\Gamma)$ and therefore, in case of continuous $f(t)$, (62) is always the solution. Thus, for $f \in \underset{\sim}{C}(\Gamma)$ we have

$$A_1^{-1} f = \Phi^+ - \Phi^- = \underset{\sim}{X}^+ \underset{\sim}{P}\,[(\underset{\sim}{X}^+)^{-1} f] + \underset{\sim}{X}^- Q\,[(\underset{\sim}{X}^+)^{-1} f] \,. \tag{63}$$

In virtue of the Theorem 1 the operator (63) is continuous in $L_p(\Gamma)$ and that is why the formula (63) is true for $f \in L_p(\Gamma)$.

Proof of the theorem: One can reduce the study of the operator A_0 in case of piecewise-continuous coefficients to the case of continuous coefficients by means of the usual method, i.e. representing the function $G(t)$ as the product $G(t) = \omega_1(t)\, G_0(t)$ where $\omega_1(t)$ is the same as in (58), and $G_0(t)$ is continuous.

The two operators N_1 and N_2 will be called similar and we write $N_1 \sim N_2$, if $N_1 - N_2$ is compact.

On the basis of the Theorem 2 it can be easily verified that

$$A \sim A_0 \sim \underset{\sim}{U}_{a+b} A_1 A_2$$

where A_0 is the operator, defined by formula (51),

$$A_1 = \underset{\sim}{P} + \underset{\sim}{U}_{\omega_1} Q \,,$$

$$A_2 = \underset{\sim}{P} + \underset{\sim}{U}_{G_0} Q \,.$$

In fact

$$A_1 A_2 = \underset{\sim}{P}^2 + \underset{\sim}{P}\underset{\sim}{U}_{G_0} Q + \underset{\sim}{U}_{\omega_1} Q \underset{\sim}{P} + \underset{\sim}{U}_{\omega_1} Q U_{G_0} Q \sim \underset{\sim}{P} + \underset{\sim}{U}_{G_0} \underset{\sim}{P} Q + \underset{\sim}{U}_{\omega_1} \underset{\sim}{U}_{G_0} Q^2 =$$

$$= \underset{\sim}{P} + \underset{\sim}{U}_G Q = \underset{\sim}{U}_{(a+b)^{-1}} A_0 \,.$$

Due to the Theorem 8, the regularizer of the operator A_2 can be written as

$$B_2 = \underset{\sim}{P} + \underset{\sim}{U}_{G_0^{-1}} Q \,.$$

Therefore, the operator

$$B = B_2 A_1^{-1} \underset{\sim}{U}_{(a+b)^{-1}} = (\underset{\sim}{P} + \underset{\sim}{U}_{G_0^{-1}} Q)(U_X + \underset{\sim}{P}\underset{\sim}{U}_{(X^+)^{-1}} + \underset{\sim}{U}_X - Q\underset{\sim}{U}_{(X^+)^{-1}})\underset{\sim}{U}_{(a+b)^{-1}} \tag{64}$$

will be the regularizer of the operator A.

Let φ be some bounded function. Since $\Gamma \in \mathfrak{R}$ the function $(\underset{\sim}{P}\underset{\sim}{U}_{(X^+)^{-1}}\varphi)(t)$ will be the boundary value of some function from the class E_q and $(\underset{\sim}{U}_{X^+}\underset{\sim}{P}\underset{\sim}{U}_{(X^+)^{-1}}\varphi)(t)$ be the boundary value of a function from the class E_1. Therefore one has

$$\underset{\sim}{P}\underset{\sim}{U}_{X^+}\underset{\sim}{P}\underset{\sim}{U}_{(X^+)^{-1}}\varphi = \underset{\sim}{U}_{X^+}\underset{\sim}{P}\underset{\sim}{U}_{(X^+)^{-1}}\varphi . \tag{65}$$

In virtue of Theorem 1 the operator $\underset{\sim}{S}$, and therefore $\underset{\sim}{P}$, are continuous in $L_p(\Gamma;\rho)$. That is why (65) is true also for $\varphi \in L_p(\Gamma)$.

In a similar way we can obtain

$$\underset{\sim}{Q}\underset{\sim}{U}_{X^+}\underset{\sim}{P}\underset{\sim}{U}_{(X^+)^{-1}}\varphi = 0 ,$$

$$\underset{\sim}{Q}\underset{\sim}{U}_{X^-}\underset{\sim}{Q}\underset{\sim}{U}_{(X^+)^{-1}}\varphi = \underset{\sim}{U}_{X^-}\underset{\sim}{Q}\underset{\sim}{U}_{(X^+)^{-1}}\varphi , \tag{66}$$

$$\underset{\sim}{P}\underset{\sim}{U}_{X^-}\underset{\sim}{Q}\underset{\sim}{U}_{(X^+)^{-1}}\varphi = 0 .$$

Introducing (65) and (66) into (64)

$$B = (\underset{\sim}{U}_{X^+}\underset{\sim}{P}\underset{\sim}{U}_{(X^+)^{-1}} + U_{G_0^{-1}}\underset{\sim}{U} - \underset{\sim}{Q}\underset{\sim}{U}_{(X^+)^{-1}})\underset{\sim}{U}_{(a+b)^{-1}} .$$

If we note that $\underset{\sim}{X}^-(t) = \underset{\sim}{X}^+(t)/_1(t)$, where $\underset{\sim}{X}^+(t) = \prod_{k=1}^{m}(t-c_k)^{\gamma_k}$, we finally have

$$B = \underset{\sim}{U}_{\omega}\underset{\sim}{P}\underset{\sim}{U}_{\omega^{-1}} + \underset{\sim}{U}_{G_0^{-1}\omega_1^{-1}}U_{\omega}\underset{\sim}{Q}\underset{\sim}{U}_{\omega^{-1}}\underset{\sim}{U}_{(a+b)^{-1}} =$$

$$= (\underset{\sim}{U}_{\omega}\underset{\sim}{P}\underset{\sim}{U}_{\omega^{-1}} + \underset{\sim}{U}_{G^{-1}}\underset{\sim}{U}_{\omega}\underset{\sim}{Q}\underset{\sim}{U}_{\omega^{-1}})\underset{\sim}{U}_{(a+b)^{-1}}$$

or, what is the same

$$B = \underset{\sim}{U}_{a_1} - \underset{\sim}{U}_{b_1}\underset{\sim}{U}_{(a+b)\omega}\underset{\sim}{S}\underset{\sim}{U}_{(a+b)^{-1}\omega^{-1}} .$$

The theorem is proved.

References

1 I.C. Gohberg and N.Ja. Krupnik, Introduction to the theory of one-dimensional singular integral operators. Kishinev (1973) (Russian).

2 E.G. Gordadze, On singular integrals with the Cauchy kernel. Trudy Tbilissk. matem.instit. t.XLII, (1972) (Russian).

3 E.G. Gordadze, On singular integral operators with piecewise continuous coefficients. Bull.of the Acad.of Sciences of the Georgian SSR. 63,N.2 (1971) (Russian).

4 L.V. Kantorovich and G.P. Akimov, Functional analysis in normed spaces. Moscow (1959) (Russian).

5 V.P. Khavin, The boundary properties of Cauchy type integrals and ... Matem.sborn. 68 (1965) 499-517 (Russian).

6 B.V. Khvedelidze, Linear discontinuous boundary value problems of the theory of functions. Trudy Tbilissk.matem.instit. t.XVIII (1957) 3-158 (Russian).

7 B.V. Khvedelidze, The boundary value problem of Riemann-Privalov with the piecewise-continuous coefficients. Trudy Gruz.politekhn. inst. 81 (1962) 11-29 (Russian).

8 A.I. Markushevich, The theory of analytic functions. Moscow-Leningrad (1950) (Russian).

9 S.G. Mikhlin, Singular integral equations. Uspekhi matem.nauk 3 (1948) 29-112 (Russian).

10 N.I. Muskhelishvili, Singular integral equations. Groningen, Noordhoff (1953) English transl. from the 1st Russian edit., Moscow (1946).

11 V.A. Paatashvili, On singular Cauchy integrals. Bull. of the Acad. of Sciences of the Georgian SSR, 53, N.3 (1969) (Russian).

12 I.I. Privalov, Boundary properties of analytic functions. Moscow (1950) (Russian).

13 M. Riesz, Sur les fonctions conjugées. Math. Z., 27 (1927) 218-244.

E.G. Gordadze and B.V. Khvedelidze
Tbilisi Mathematical Institute
Academy of Sciences of the Georgian SSR
USSR

G JANK

11 On integro-differential-operators for partial differential equations

INTRODUCTION

Recently one has succeeded to give representations for solutions of much more general classes of partial differential equations starting from the differential equation

$$(1 + \epsilon z\bar{z})^2 w_{z\bar{z}} + \epsilon n(n+1) w = 0, \qquad \epsilon = \pm 1,\ n \in \mathbb{N}\ ,\ \dagger$$

using differential operators.

In [6] and [8] one proved that these representations may be obtained from a special Bergman operator. These operators are defined on the set of in a certain domain holomorphic or antiholomorphic functions, respectively. St. Ruscheweyh shows in his papers [10,11] that differential operators are easier to be manipulated for function theoretical investigations. So one aims to enlarge the class of differential equations permitting the representation of solutions with differential operators. One such enlargement will be described here. We further discuss the differential equation

$$w_{zz^*} - \frac{n(n+1)}{(z+z^*)^2} w + \psi(z - z^*) w = 0, \qquad n \in \mathbb{N}, \tag{1}$$

in a cylindrical domain $G \times G^* \subset \mathbb{C}^2$ without zeros of $z + z^*$ and ψ being holomorphic in this domain. The domain of definition of the differential operator we are going to introduce here is the set of the solutions (defined in $G \times G^*$) of the differential equation

$$Lw = 0, \tag{2}$$

L being the operator

$$L := \frac{\partial^2}{\partial z \partial z^*} + \psi(z - z^*). \tag{3}$$

† $\mathbb{N}$, $\mathbb{C}$ denotes the set of natural or complex numbers respectively.

For $\psi \equiv 0$ the solutions of (2) are determined as the sum of one in G and one in G^* defined holomorphic function. In this case the differential equation (1) is identical to the differential equation treated in [1].

For

$$\psi = \frac{m(m+1)}{(z-z^*)^2}, \qquad m \in \mathbb{N},$$

the solutions (defined in $G \times G^*$) of the differential equation (2) may be represented by a differential operator. So one can specify a differential operator representing the solutions of (1) and containing derivatives of, in G or G^* respectively, holomorphic functions up to the order $m+n$. In this context it is to be mentioned that in this case if and only if $m = n$, differential equation (1) is a special case of the differential equation treated in [3]. If ψ is not of the form given above the solutions of (2) may only be represented by an appropriate integral operator (comp. [5. 12]). This leads to a certain connection of differential and integral operators in the representation of the solutions.

1. PARTICULAR SOLUTIONS OF DIFFERENTIAL EQUATION (1)

Let us define the differential operator

$$d := \frac{\partial}{\partial z} + \frac{\partial}{\partial z^*}, \tag{4}$$

$$d^0 w = w, \qquad d^\nu w = d d^{\nu-1} w, \qquad \nu \in \mathbb{N},$$

or L according to (3) with

$$L^0 w = w, \qquad L^\nu w = L L^{\nu-1} w, \qquad \nu \in \mathbb{N},$$

then we get particular solutions of (1) by the following

Lemma 1: Let $h(z,z^*)$ be an in $G \times G^*$ defined solution of differential equation (2), then

$$w = \sum_{\mu=0}^{n} A_\mu^n \frac{d^\mu h}{(z+z^*)^{n-\mu}}, \qquad A_\mu^n = \frac{(-1)^{n-\mu}(2n-\mu)!}{\mu!\,(n-\mu)!} \tag{5}$$

is a solution of (1) defined in $G \times G^*$ and it holds

$$L^n [(z + z^*)^n w] = n!\, d^{2n} h . \tag{6}$$

The proof is given by substituting (5) into (1). To prove (6) we use $LH = 0$ and

$$Ldw = dLw \tag{7}$$

to derive the following formula by induction

$$L^k [(z + z^*)^\mu H] = \sum_{\nu=0}^{k} \frac{\mu!}{(\mu - 2k + \nu)!} \binom{k}{\nu} (z + z^*)^{\mu - 2k + \nu} d^\nu H . \tag{8}$$

Let $k = n$ in (8), so, because of $0 \leq \mu$, $\nu \leq n$ and

$$\frac{1}{(\mu + \nu - 2n)!} = \begin{cases} 0, & \text{if } \mu + \nu < 2n \\ 1, & \text{if } \mu + \nu = 2n \end{cases}$$

we get

$$L^n [(z + z^*)^\mu H] = \begin{cases} 0, & \text{if } \mu < n \\ n!\, d^n H, & \text{if } \mu = n. \end{cases} \tag{9}$$

Forming the left-hand side of (6) with w according to (5), we get the assertion made above with $H = d^n h$.

In the following we will give lemmas from which can be concluded that each solution of (1) can be represented in the form (5). These will be combined with Lemma 1 to a general representation theorem.

Lemma 2: Let w be a solution of the differential equation (1), w being defined in $G \times G^*$, then with the abbreviation $\sigma = z + z^*$

$$L^{m+1} (\sigma^m w) = [n(n+1) - m(m+1)]\, L^m (\sigma^{m-1} w) , \quad m \in \mathbb{N} ,^\dagger \tag{10}$$

is valid.

For $m = 0$, (10) becomes (1). To proceed in the proof one derives from

$$L(\sigma f) = df + \sigma Lf$$

† According to a theorem of Picard [9] every solution of (1) is analytic; thus especially yielding the existence of the derivatives of arbitrary order.

by induction according to (7) for a sufficiently often differentiable function f the following formula

$$L^r(\sigma f) = r d L^{r-1} f + \sigma L^r f , \qquad r \in \mathbb{N} . \tag{11}$$

Setting $r = m + 2$, $f = \sigma^m w$, in (11), we get

$$L^{m+2}(\sigma^{m+1} w) = (m+2) d L^{m+1}(\sigma^m w) + L^{m+2}(\sigma^m w) . \tag{12}$$

By repeated application of the assumption of the induction transformed into

$$L^{m+1}(\sigma^m w) = \frac{(n-m)(m+n+1)}{\sigma} L^m(\sigma^{m-1} w) \tag{13}$$

we get furthermore

$$L^{m+1}(\sigma^m w) = [2(m+1)]! \binom{n+m+1}{2(m+1)} \frac{w}{\sigma^{m+2}} , \quad m = 0,1,\ldots,n . \tag{14}$$

Substituting (14) into (12) we get

$$L^{m+2}(\sigma^{m+1} w) = [2(m+1)]! \binom{n+m+1}{2(m+1)} [(m+2)\sigma d(\frac{w}{\sigma^{m+2}}) + \sigma^2 L(\frac{w}{\sigma^{m+2}})] .$$

Using

$$d(\frac{w}{\sigma^{m+2}}) = \frac{dw}{\sigma^{m+1}} - \frac{2(m+2)}{\sigma^{m+2}} w$$

and

$$\sigma^2 L(\frac{w}{\sigma^{m+2}}) = \frac{n(n+1) + (m+2)(m+3)}{\sigma^{m+2}} w - \frac{m+2}{\sigma^{m+1}} dw$$

we get the assertion made above. Setting $n = m$ in Lemma 2 we have proved that each solution of the differential equation (1) must satisfy as well the inhomogeneous differential equation

$$L^n(\sigma^n w) = H(z, z^*), \quad \text{with } LH = 0 \text{ in } G \times G^* . \tag{15}$$

It is now possible to derive an ordinary differential equation from (15) whose set of solutions contains the one of differential equation (1).

2. AN ORDINARY DIFFERENTIAL EQUATION FOR THE SOLUTIONS OF DIFFERENTIAL EQUATION (1)

In the first step we want to derive from (15) a partial differential equation of order n for the solutions of differential equation (1). For $n = 1$ using (15) and

$$L(\sigma w) = dw + \sigma Lw$$

and assuming w to be a solution of (1) we get

$$L(\sigma w) = dw + \frac{2w}{\sigma} = H(z, z^*) .$$

This is a partial differential equation in w of order $n = 1$, H being arbitrary ($LH = 0$). Generally valid is

<u>Lemma 3</u>: Let w be a solution of differential equation (1), w being defined in $G \times G^*$. Then

$$L^n(\sigma^n w) = n! \sum_{j=0}^{n} \frac{(2n-j)!}{(n-j)!j!} d^j\left(\frac{w}{\sigma^{n-j}}\right) . \tag{16}$$

The proof is given with the following formula

$$L^n(\sigma^n w) = \sum_{j=0}^{k-1} \frac{n!(2n-j)!}{(n-j)!j!} d^j\left(\frac{w}{\sigma^{n-j}}\right) + \frac{n!}{(n-k)!} d^k[L^{n-k}(\sigma^{n-k} w)] , \tag{17}$$

(17) is correct for $k = 0$. Abbreviating we write F for the left-hand side and F_k for the sum on the right-hand side of (17). Using (11) and $r = n - k$ and $f = \sigma^{n-k-1} w$ we get

$$F = F_k + \frac{n!}{(n-k)!} d^k [\sigma L^{n-k}(\sigma^{n-k-1} w) + (n-k) dL^{n-k-1}(\sigma^{n-k-1} w)] .$$

Furthermore, using (14)

$$F = F_k + \frac{n!(2n-k)!}{(n-k)!k!} d^k\left(\frac{w}{\sigma^{n-k}}\right) + \frac{n!}{(n-k-1)!} d^{k+1} L^{n-k-1}(\sigma^{n-k-1} w)$$

holds, and finally

$$F = F_{k+1} + \frac{n!}{(n-k-1)!} d^{k+1} L^{n-k-1}(\sigma^{n-k-1} w) .$$

Setting $k = m+1$ in (17) we get the assertion made above. To come to an ordinary differential equation we introduce the new variables

$$\begin{aligned} \sigma &= z + z^* \\ \tau &= z - z^* \end{aligned} \tag{18}$$

Then with $w(z,z^*) = \tilde{w}(\sigma,\tau)$

$$dw = 2\tilde{w}_\sigma \tag{19}$$

and using Lemma 3 and (18) or (19) respectively the differential equation (15) becomes the ordinary differential equation

$$n! \sum_{j=0}^{n} \frac{(2n-j)!2^j}{(n-j)!j!} \frac{\partial^j}{\partial\sigma^j} \left(\frac{\tilde{w}}{\sigma^{n-j}}\right) = \tilde{H}(\sigma,\tau) , \tag{20}$$

with

$$\tilde{H}(\sigma,\tau) = H(z,z^*), \qquad LH = 0 .$$

Firstly we will solve differential equation (20) for $H = 0$. Setting $\tilde{w} = \sigma^\lambda$, $\lambda \in \mathbb{C}$, in (20) with $H = 0$, after some transformations we get

$$\sum_{j=0}^{n} 2^j \binom{2n-j}{n} \binom{\lambda+j-n}{j} = 0 .$$

The roots of this polynomial of order n in λ are given by

$$\sum_{j=0}^{n} (-1)^j 2^j \binom{2n-j}{n}\binom{2k+1}{j} = 0 , \quad k = 0,1,\dots,n-1 , \quad n \in \mathbb{N} , \tag{21}$$

with

$$\binom{\lambda+j-n}{j} = (-1)^j \binom{n-\lambda-1}{j} .$$

We get

$$\lambda_k = n - 2(k+1), \quad k = 0,\dots,n-1 ,$$

or

$$\lambda_\nu = n - 2\nu\ , \quad \nu = 1,\ldots,n\ .$$

The functions $\sigma^{\lambda\nu}$ are a fundamental set of solutions of differential equation (20). To get the general solution of the homogeneous differential equation (20) we formulate

Lemma 4: The solutions $\varphi(\sigma,\tau)$ of the differential equations

$$\sum_{j=0}^{n} \frac{(2n-j)!\,2^j}{(n-j)!\,j!} \frac{\partial^j}{\partial\sigma^j} \left(\frac{\varphi}{\sigma^{n-j}}\right) = 0$$

can be represented in the form

$$\varphi(\sigma,\tau) = \sum_{\nu=1}^{n} a_\nu(\tau)\,\sigma^{n-2\nu}\ , \quad a_\nu(\tau)\ \text{hol.} \tag{22}$$

This representation one can also get using the function

$$h_0 = \sum_{k=0}^{n-1} \alpha_k(\tau)\,\sigma^{2k} \tag{23}$$

with

$$a_{n-k}(\tau) = (-1)^{n-k}\binom{n}{k}\frac{(2k)!\,(2n-2k)!}{n!}\,\alpha_k(\tau)\ , \qquad k = 0,\ldots,n-1\ ,$$

to construct

$$\varphi(\sigma,\tau) = \sum_{j=0}^{n} A_j^n \frac{d^j h_0}{\sigma^{n-j}}\ . \tag{24}$$

The equivalence of the representations (22) and (24) we show by substituting (23) into (24). This gives

$$\varphi(\sigma,\tau) = \sum_{k=0}^{n-1} \alpha_k(\tau)\,\sigma^{2k-n} n! \sum_{j=0}^{n} (-1)^{n-j}\,2^j \binom{2n-j}{n}\binom{2k}{j}$$

and using

$$n! \sum_{j=0}^{n} (-1)^j\,2^j \binom{2n-j}{n}\binom{2k}{j} = (-1)^k \binom{n}{k} \frac{(2k)!\,(2n-2k)!}{n!}\ , \tag{25}$$

$k = 0,\dots,n, \quad n \in \mathbb{N}$,

$$\varphi(\sigma,\tau) = \sigma^n \sum_{k=0}^{n-1} \sigma^{2k-2n} \alpha_k(\tau)(-1)^{n-k} \binom{n}{k} \frac{(2k)!(2n-2k)!}{n!} .$$

Setting $\nu = n-k$, $k = 0,\dots,n-1$, in (22) we get

$$\varphi(\sigma,\tau) = \sigma^n \sum_{k=0}^{n-1} a_{n-k}(\tau)\, \sigma^{2k-2n} ,$$

and comparing the coefficients of σ^{2k-2n} we get the assertion made above.

3. A GENERAL REPRESENTATION THEOREM FOR THE SOLUTIONS OF DIFFERENTIAL EQUATION (1)

We define the differential operator T by

$$Th = \sum_{\mu=0}^{n} A_\mu^n \frac{d^\mu h}{\sigma^{n-\mu}} . \tag{26}$$

Now we can prove the following representation theorem:

<u>Theorem 1</u>: (i) To each solution w (defined in $G \times G^*$) of differential equation (1) there is a solution $h(z,z^*)$ (defined in $G \times G^*$) of the differential equation $Lh = 0$ so that w can be represented according to $w = Th$. This function h will be called the 'generating function' of w.

(ii) Conversely, for each solution h (defined in $G \times G^*$) of the differential equation $Lh = 0$, $w = Th$ represents a solution of differential equation (1), w being defined in $G \times G^*$.

(iii) Given a solution (defined in $G \times G^*$) of the differential equation (1), the function $d^{2n}h$ is determined unambiguously according to (6). The generating function is not determined unambiguously.

(iv) The most general generating function h^* is given by

$$h^*(z,z^*) = h(z,z^*) + \sum_{k=0}^{n-1} c_{2k+1}(z-z^*)(z+z^*)^{2k+1} ,$$

with

$$c''_{2k+1}(z-z^*) - \psi(z-z^*)c_{2k+1}(z-z^*) = (2k+2)(2k+3)c_{2k+3}(z-z^*)$$
$$k = 0, .. , n-1 ,$$
$$c_{2n+1} \equiv 0 .$$

The proof of (ii) and (iii) is given immediately by Lemma 1 and (i). To demonstrate (i) we firstly derive the solutions of the inhomogeneous differential equation (20), because of Lemma 1 and 2 each solution of differential equation (1) being contained in the set of solutions of (20). Lemma 4 gives the general solution of the homogeneous differential equation (20). In addition we need a particular solution of w_p of the inhomogeneous differential equation (20). This solution is given by $w_p = Th$ because of Lemma 1 and 2, if there exists a function h with $Lh = 0$ and

$$d^{2n}h = H . \tag{26}$$

There H denotes a solution of the differential equation $LH = 0$ (comp. (6), (16), (20)). To demonstrate this it is sufficient to show the existence of a function α_0 with $d\alpha_0 = H$ and $L\alpha_0 = 0$. Transforming to the variables σ and τ one can see that α_0 is contained in the set of functions of the form

$$\alpha = \gamma(\sigma, \tau) + a(\tau)$$

with $\gamma_\sigma = \frac{1}{2}H$, $a(\tau)$ hol. The condition $L\alpha = 0$ gives for $a(\tau)$ the inhomogeneous linear differential equation with holomorphic coefficients and holomorphic right-hand side

$$a'' - \psi a = L\gamma .$$

Now we take one of the solutions, being defined in $G \times G^*$ calling it $a_0(\tau)$. This gives the wanted function

$$\alpha_0 = \gamma(\sigma, \tau) + a_0(\tau)$$

with $L\alpha_0 = 0$. By repeated application of this property we get the existence of a suitable generating function h. According to (22) and (24) we can give the general solution of the differential equation (20) in the form

$$w = \sum_{j=0}^{n} A_j^n \frac{d^j(h_0+h)}{\sigma^{n-j}} . \tag{27}$$

h_0 being defined by (23) with arbitrary holomorphic functions α_k. In the set of solutions defined by (27) all solutions of differential equation (1) are contained. We get them by substituting (27) into (1), taking into account that according to Lemma 1 h generates a solution of (1). Transforming differential equation (1) to the variables σ and τ, the following is valid:

$$w_{\sigma\sigma} - w_{\tau\tau} - \frac{n(n+1)}{\sigma^2} w + \psi(\tau) w = 0 .$$

Substituting (27) into this we get the following conditions for the coefficients $\alpha_k(\tau)$

$$\alpha_k'' - \psi(\tau)\alpha_k = (2k+1)(2k+2)\alpha_{k+1}, \qquad k = 0, \ldots, n-1, \tag{28}$$

$$\alpha_n = 0 .$$

It is easily demonstrated that (28) is equivalent to $Lh_0 = 0$, which proves (i).

To (iv): h_1 and h_2 being two generating functions giving the same solution. $h_1 - h_2 = h^0$ is then the generating function of the null-solution $w \equiv 0$. According to (6) it is necessary that $d^{2n} h^0 = 0$ or integrated

$$h^0 = \sum_{r=0}^{2n-1} c_r(\tau)\sigma^r , \qquad c_r(\tau) \text{ hol.} \tag{29}$$

h^0 being generating function induces $Lh^0 = 0$, that is

$$c_r'' - \psi(\tau) c_r = (r+1)(r+2) c_{r+2}, \qquad r = 0, \ldots, 2n-1, \tag{30}$$

$$c_{2n} \equiv c_{2n+1} \equiv 0 .$$

For the null-solution we further get the condition $Th^0 = 0$, i.e.

$$0 = \sum_{r=0}^{2n-1} \sigma^{r-n} c_r(\tau) \sum_{j=0}^{n} A_j^n 2^j \binom{r}{j} j! .$$

By comparing the coefficients of σ^r we get

$$c_r(\tau) \sum_{j=0}^{n} (-1)^j 2^j \binom{2n-j}{n} \binom{r}{j} \equiv 0, \quad r = 0,\dots,2n-1 .$$

The sum over j represents a polynomial in r of order n, having the roots

$$r = 2k+1, \quad k = 0,\dots,n-1, \qquad \text{(comp. (21))}.$$

Therefore

$$c_{2k} \equiv 0$$

$$c_{2k+1}(\tau) \quad \text{according to (30)}$$

$k = 0, 1, \dots, n-1$, which proves (iv). For the real differential equations contained in (1) we finally derive a representation theorem for the real solutions. We deal with the differential equation, defined in the simply connected finite domain G not containing the origin,

$$w_{z\bar{z}} - \frac{n(n+1)}{(z+\bar{z})^2} w + \psi(z-\bar{z}) w = 0 \tag{31}$$

with $\psi = \bar{\psi}$ and analytical.

This differential equation has real solutions with which we will deal in the next section.

4. A REPRESENTATION THEOREM FOR THE REAL SOLUTIONS OF THE DIFFERENTIAL EQUATION (31)

For differential equation (31) we define the analogous operators to (3), (4), (26) by replacing z^* by $\bar{z}$. These operators we will define with $\hat{L}$, $\hat{d}$, $\hat{T}$. So for example $\hat{L}$ becomes according to (3)

$$\hat{L} := \frac{\partial^2}{\partial z \partial \bar{z}} + \psi(z-\bar{z}), \quad \text{with} \quad \psi = \bar{\psi} .$$

Theorem 2: (1) For each real solution w (defined in G) of differential equation (31) there exists a real solution h (defined in G) of the differential equation

$$\hat{L}h = 0 , \tag{32}$$

so that

$$w = \hat{T}h \quad . \tag{33}$$

(ii) Conversely for each real solution h of (32) defined in G, (33) represents a real solution (defined in G) of differential equation (31).

(iii) Given a real solution (defined in G) of the differential equation (31) the function $\hat{d}^{2n}h$ is unambiguous according to

$$\hat{L}^n(\hat{\sigma}w) = n!\,\hat{d}^{2n}h, \quad \hat{\sigma} = z+\overline{z}$$

and is a real function.

(iv) The most general real generating function of a real solution of (31) is given by

$$h^*(z,\overline{z}) = h(z,\overline{z}) + \sum_{k=0}^{n-1} c_{2k+2}(\hat{\tau})\,\hat{\sigma}^{2k+1}, \quad \hat{\tau} = z-\overline{z},$$

the functions $c_{2k+1}(\hat{\tau})$ being real solutions of the system

$$c''_{2k+1} - c_{2k+1} = (2k+2)(2k+3)\,c_{2k+3} \qquad k = 0,\dots,n-1,$$

$$c_{2n+1} = 0$$

and $h(z,\overline{z})$ being a real solution of differential equation (32) (defined in G).

<u>Proof</u>: To (i): Let w be an arbitrary real solution of (31) then using (6)

$$\hat{L}^n[\hat{\sigma}^n(w-\overline{w})] = 0 = n!\,\hat{d}^{2n}(h-\overline{h})$$

h being a generating function of w. $h-\overline{h}$ is a generating function of the null-solution and according to Theorem 1, (iv) is

$$h-\overline{h} = \sum_{k=0}^{n-1} c_{2k+1}(\hat{\tau})\,\hat{\sigma}^{2k+1}.$$

The coefficients $c_{2k+1}(\hat{\tau})$ have to be imaginary quantities and must satisfy the conditions in Theorem 1. $c_{2k+1} \equiv 0$ for $k = 0,\dots,n-1$, satisfies these conditions, which proves the existence of a real generating function.

To (ii): h being a real generating function, then

$$w - \overline{w} = \hat{T}h - \overline{\hat{T}h} = \hat{T}(h-\overline{h}) = 0 .$$

To (iii): if w is a real function and h denotes a generating function of w, then

$$n!\,\overline{\hat{d}^{2n}h} = \overline{\hat{L}^n(\hat{\sigma}^n w)} = \hat{L}^n(\hat{\sigma}^n \overline{w}) = n!\,\hat{d}^{2n}h\ .$$

To (iv): w_1 and w_2 being two identical real solutions of the differential equation (31) h_1 and h_2 being their generating functions, so that

$$w_k = \hat{T}h_k\ , \quad \text{with } \hat{L}h_k = 0 \quad \text{and} \quad h_k = \overline{h_k}\ , \qquad k = 1,2$$

is valid. By application of Theorem 1 we get

$$h_1 - h_2 = \sum_{k=0}^{n-1} c_{2k+1}(\hat{\tau})\,\hat{\sigma}^{2k+1}$$

with the conditions for the coefficient functions c_{2k+1} given under Theorem 1. The most general generating function can be given immediately if the coefficients $c_{2k+1}(\hat{\tau})$ are real solutions for the system mentioned in Theorem 1.

References

1 K.W. Bauer, Über eine partielle Differentialgleichung 2. Ordnung mit zwei unabhängigen komplexen Variablen. Monatsh. Math. 70 (1966) 385-418.

2 K.W. Bauer, Über die Darstellung von Lösungen einer Differentialgleichung mit N komplexen Variablen. Inst. f. Angew. Math. Univ. TH Graz, Bericht Nr 70-4 (1970).

3 K.W. Bauer, Über Differentialgleichungen der Form $F(z,\overline{z})w_{z\overline{z}} - n(n+1)w = 0$. Monatsh. Math. 75 (1971) 1-13.

4 K.W. Bauer and G. Jank, Differentialoperatoren bei einer inhomogenen elliptischen Differentialgleichung, Rend. Ist. Mat. Univ. Trieste 3 (1971) 140-168.

5 St. Bergman, Integral Operators in the Theory of Linear Partial Differential Equations. Erg. Math. Grenzgeb. Bd.23, Berlin: Springer (1971).

6 H. Florian and G. Jank, Polynomerzeugende bei einer Klasse von Differentialgleichungen mit zwei unabhängigen Variablen. Monatsh. Math. 75 (1971) 31-37.

7 R.P. Gilbert, Function Theoretic Methods in Partial Differential Equations. N.Y., London: Academic Press (1969).

8 M. Kracht and E. Kreyszig, Bergman-Operatoren mit Polynomen als Etzeugenden. Manuscripta Math. 1 (1969) 369-376.

9 C. Miranda, Partial Differential Equations of Elliptic Type. Berlin, Heidelberg, New York: Springer (1970).

10 St. Ruscheweyh, Gewisse Klassen verallgemeinerter analytischer Funktionen. Bonner Math. Schr. 39 (1969).

11 St. Ruscheweyh, Operatoren bei partiellen Differentialgleichungen. Teil II: Eigenschaften der Lösungen. Inst. f. Angew. Math. Univ. TH Graz, Vortragsauszug (1970).

12 I.N. Vekua, New Methods for Solving Elliptic Equations. Amsterdam, North. Holl. Publ. Comp. (1967).

Gerhard Jank
1. Lehrkanzel und Institut für Mathematik
Technische Hochschule in Graz
Kopernikusgasse 24
A-8010 Graz
Austria

T L McCOY

12 An application of the Poisson transformation to harmonic functions

1. INTRODUCTION

It is a well-known principle that if $f(z)$ is given in the unit disc by a convergent series, with coefficients a_j which extend to an analytic function $a(j)$, one can expect that $f(z)$ can be continued analytically to a much larger region than the unit disc, (cf. [6], pp 54 ff. for a number of results of this kind).

A rather recent theory, developed by S. Bergman, R.P. Gilbert, and others (cf. [4], Chapter 2 for discussion and an extensive bibliography) enables us to extend harmonic functions in terms of an integral operator applied to an associated holomorphic function of several complex variables. Briefly, if the harmonic function V is given near the origin by a convergent series in a complete system of spherical harmonics, one uses these coefficients to construct an associated function f holomorphic in several variables. There is an integral operator $\underline{I}$, such that $V = \underline{I} f$ locally; however, $\underline{I} f$ may extend to a harmonic function outside the original neighbourhood of definition.

In this note we show that if

$$f(z_1, \ldots, z_N) = \sum a_{j_1} \ldots {}_{j_N} z_1^{j_1} \ldots z_N^{j_N},$$

where

$$a_{j_1} \ldots {}_{j_N} = a(j_1, \ldots, j_N)$$

can be extended to a function holomorphic in the Cartesian product of N copies of the right half-plane and uniformly of exponential type zero there, then f is holomorphic in the Cartesian product of N copies of the plane slit from 1 to ∞ along the positive real axis. The main tools are the Poisson summation formula in N dimensions, and some facts about the Laplace transform.

We discuss the Poisson summation formula, and its application to analytic continuation, in Section 2, in some detail partly for the sake of orientation. Section 3 is devoted to terminology needed for the N-dimensional case, and Section 4 is devoted to a proof of the theorem stated above.

In Section 5, we apply Gilbert's operator, $\underline{G}_3$, to the harmonic function.

$$\operatorname{Re} \sum_{n=0}^{\infty} r^n \sum_{m=0}^{n} a_{nm} P_n^m(\cos\theta)\, e^{im\phi},$$

where $a_{nm}(2m)!/2^m m!$ extends to a biholomorphic function of n,m in the Cartesian product of two copies of the right half-plane, uniformly of exponential type zero. We find that $V(x_1, x_2, x_3)$ extends harmonically everywhere off a certain subset contained in a quadrant of the $x_1 - x_3$-plane.

2. THE POISSON TRANSFORMATION IN THE CASE OF ONE VARIABLE

The Poisson transformation is the following formula, valid if both sides of the equation are absolutely convergent series, $\emptyset$ absolutely integrable over the real line (cf. [2] p. 76);

$$\sum_{j=-\infty}^{\infty} \emptyset(j) = \sum_{k=-\infty}^{\infty} \int_{-\infty}^{\infty} \emptyset(t)^{-2\pi ikt}\, dt .$$

Now let $f(z)$ be given in $|z| < 1$ by the convergent series

$$f(z) = \sum_{j=0}^{\infty} a(j) z^j$$

$$= \sum_{j=0}^{\infty} a(j) e^{j(\log r + i\theta)}, \quad z = re^{i\theta}$$

If we take

$$\emptyset(j) = \begin{cases} a(j) e^{j(\log r+i\theta)}, & j \geq 0 \\ 0, & j < 0 \end{cases}$$

and apply the Poisson transformation, which is justified for $r < 1$, we obtain

$$f(z) = \sum_{k=-\infty}^{\infty} \int_0^{\infty} a(t)\, e^{t(\log r + i\theta - 2\pi i k)}\, dt \tag{2.1}$$

$$= \sum_{k=-\infty}^{\infty} \hat{a}(w + 2\pi i k), \quad w = \log 1/z$$

where $\hat{a}$ is the Laplace transform of a.

The equation (2.1) is valid for $|z| < 1$, or in other words, for w in the half-strip $\operatorname{Re} w > 0$, $|\operatorname{Im} w| \leq \pi$, fixing a particular branch of the logarithm. It is clear that if we can extend the right-hand side of (2.1) analytically outside the half-strip, we obtain an analytic continuation of $f(z)$ outside the disc. We can do this, under certain restrictions on $a(t)$, by means of the following lemma. The proof of the first part of Lemma 2.1 is suggested by ([1] p. 74).

Lemma 2.1: For any δ with $0 < \delta \leq \pi$ let S_δ be the sector

$$S_\delta : (|\operatorname{Arg} w| \leq \delta)$$

and let T_δ be the sector

$$T_\delta : (|\operatorname{Arg} w| < \delta + \pi/2)$$

where, if $\delta > \pi/2$, T_δ is to be regarded as a Riemann surface covering a region about the negative real axis twice.

Suppose $a(w)$ is holomorphic in S_δ and uniformly of exponential type zero there; that is

$$\lim_{w \to \infty} a(w)\, e^{-\epsilon |w|} = 0$$

uniformly in S_δ for any positive ϵ.

Then $\hat{a}(w)$ extends analytically to T_δ, and in the case $\delta > \pi/2$ is single-valued on each sheet of T_δ. The extension is given by the formula

$$\hat{a}(re^{-i\theta}) = \hat{a}_\theta(r) \tag{2.2}$$

where

$$\hat{a}_\theta(r) = e^{-i\theta} \int_0^\infty e^{-rx} a(xe^{i\theta})\,dx \ , \ \operatorname{Re} r > 0 \ , \ |\theta| < \delta \ . \tag{2.2'}$$

Furthermore, if $a(0) = 0$, we have

$$\hat{a}(w) = 0(|w|^{-2}) \tag{2.3}$$

as $|w| \to \infty$, uniformly in any sector T_η with $\eta < \delta$.

Proof: Let $s > 0$, and substitute $u = sx$ in the integral which defines $\hat{a}(s)$ to get

$$\hat{a}(s) = \frac{1}{s} \int_0^\infty e^{-u} a(u/s)\,du \ . \tag{2.4}$$

Under our hypothesis on $a(w)$ the integral (2.4) is an analytic function of s for s in S_δ. Now in (2.4) set $s = re^{i\theta}$ with $r > 0$ and $|\theta| \leq \delta$, then put $rx = u$ in the result, to get (2.2) for positive real r.

But the right-hand side of (2.2') converges for $\operatorname{Re} r > 0$, and is an analytic function of r. Thus (2.2) provides an analytic continuation of $\hat{a}|w|$ to T_δ.

Now we verify (2.3). By Cauchy's integral, $a''(w)$ is also uniformly of exponential type zero in S_η for any $\eta < \delta$. If we integrate by parts twice in the integral which defines $\hat{a}_\theta(r)$ for $\operatorname{Re} r > 0$, and recall that $a(0) = 0$, we have

$$\hat{a}_\theta(r) = \frac{e^{-i\theta}}{r} \left\{ \frac{-a'(0)}{r} + \frac{1}{r} \int_0^\infty e^{-i\theta} a''(xe^{i\theta}) e^{-rx}\,dx \right\} = 0(|r|^{-2}) \ .$$

We see that, under the hypotheses of Lemma 2.1, the right-hand side of (2.1) is an analytic function of w in the region obtained by removing from the w-plane the sets

$$CT_\delta + 2\pi m i \ , \quad m = 0, \pm 1, \pm 2, \ldots ,$$

CT_δ the complement of T_δ. We see by considering the mapping $w = -\log z$ that $f(z)$ is extended into one of the following regions:

(a) if $\delta = \frac{\pi}{2}$, $f(z)$ is extended to the z-plane cut along the ray $z \geq 1$,

(b) if $0 < \delta < \pi/2$, $f(z)$ is extended to the interior of the bounded closed curve having the following equation in polar coordinates.

$$\rho = e^{-A_\delta |\phi|}, \quad |\phi| \leq \pi$$

where

$$A_\delta = \frac{1}{\pi}\cot\delta .$$

(c) If $\delta > \pi/2$, $f(z)$ is extended to a possibly multiple-valued function in the plane with the point $z = 1$ deleted.

3. NOTATION FOR THE N-DIMENSIONAL CASE

It is helpful to use the arrow notation to distinguish between vectors and scalars. Thus

$$\vec{z} = (z_1, z_2, \ldots, z_N).$$

We call z_j the jth coordinate of $\vec{z}$. We write

$$\|\vec{z}\| = \sqrt{(|z_1|^2 + \cdots + |z_N|^2)}$$

$\mathbb{C}^N$ and E^N are, respectively, N-dimensional complex and real space. $d\vec{x}$ is the element of volume in E^N. $\vec{0}$ and $\vec{1}$ are respectively the vectors with all coordinates zero and all coordinates 1. We define

$$\vec{z}^{\vec{j}} = (z_1^{j_1}, z_2^{j_2}, \cdots, z_N^{j_N}) .$$

If $\vec{\theta}$ is a real vector with coordinates θ_j,

$$e^{\vec{\theta}} = (e^{\theta_1}, e^{\theta_2}, \cdots, e^{\theta_N})$$

with obvious analogous definitions for $\cos\vec{\theta}$ and $\sin\vec{\theta}$.

The following sets will be used:

Z^N, the subset of E^N in which all coordinates are integers, and Z_+^N the subset of Z^N in which all coordinates are non-negative. R^N, L^N, I^N, and H^N are, respectively, the Cartesian products of N copies of the positive real axis, the real axis, the interval $0 \leq x \leq 1$, and the right half-plane.

P is the unit polydisc, $|z_j| < 1$, $\underline{j = 1 - N}$.

Let $\vec{\delta}$ be a vector with coordinates satisfying $0 < \delta_j \leq \pi$.. Then

$$S_{\vec{\delta}} = S_{\delta_1} \times S_{\delta_2} \times \cdots \times S_{\delta_N}$$

$$T_{\vec{\delta}} = T_{\delta_1} \times T_{\delta_2} \cdots \times T_{\delta_N}$$

As in Section 2, $T_{\vec{\delta}}$ may cover some portions of $\mathbb{C}^N$ more than once.

Let S^N be the set

$$S^N = \{x / \|x\| = 1\}$$

in E^N, S^N_+ the subset of S^N for which all coordinates are positive.

Finally, let $d\,\Omega_N(\vec{u})$ denote the element of surface area on S^N. It is possible to introduce polar coordinates in E^N (cf. [10], p.149) such that if $\lambda(\vec{x})$ is any function from E^N to $\mathbb{C}^1$,

$$\int_{E^N} \lambda(\vec{x})\, d\vec{x} = \int_{S^N} \int_0^\infty \lambda(r\,\vec{u})\, r^{N-1} dr\, d\,\Omega_N(\vec{u}),$$

where $r = \|\vec{x}\|$.

In our notation, if $z_n = r_n e^{i\theta_n}$, we have

$$\vec{z} = \vec{r} \cdot e^{i\vec{\theta}} = e^{\log \vec{r} + i\vec{\theta}}$$

$$\vec{z}^{\vec{j}} = e^{\vec{j} \cdot (\log \vec{r} + i\vec{\theta})}$$

the dot meaning the usual inner product.

4. ANALYTIC CONTINUATION BY THE POISSON TRANSFORMATION IN THE N-DIMENSIONAL CASE

There is an analogue of the Poisson transformation for N-dimensions (cf. [3], p.166). It can be derived in a straightforward manner by generalizing the development in ([2], p.76) to several variables. The Poisson summation formula for N-dimensions is

$$\sum_{\vec{j}\in Z^N} \varphi(\vec{j}) = \sum_{\vec{k}\in Z^N} \int_{L^N} \varphi(\vec{t})\, e^{-2\pi i \vec{k}\cdot\vec{t}}\, d\vec{t}.$$

Let $f(\vec{z})$ be defined in the unit polydisc P by the series

$$f(\vec{z}) = \sum_{\vec{j}\in Z_+^N} a(\vec{j})\, \vec{z}^{\vec{j}}.$$

Setting

$$\varphi(\vec{r},\vec{\theta},\vec{j}) = \begin{cases} a(\vec{j})\, e^{\vec{j}\cdot(\log \vec{r} + i\vec{\theta})}, & \vec{j}\in Z_+^N \\ 0, & \vec{j}\notin Z_+^N \end{cases}$$

the Poisson transformation gives

$$f(\vec{z}) = \sum_{\vec{k}\in Z^N} \hat{a}(\vec{w} + 2\pi i\vec{k}) \tag{4.1}$$

where

$$\vec{w} = \log 1/\vec{z} = (-\log z_1, -\log z_2, \cdots, -\log z_N)$$

and $\hat{a}$ is the N-dimensional Laplace transform;

$$\hat{a}(\vec{y}) = \int_{R^N} a(\vec{t})\, e^{-\vec{t}\cdot\vec{y}}\, d\vec{t}. \tag{4.2}$$

(4.1) holds for z in P by absolute convergence of the series and integrals involved.

As in Section 2, we extend f by showing that the right-hand side of (4.1) is a uniformly convergent series of holomorphic functions. Let us make the following:

Definition: $a(\vec{z})$ is uniformly of exponential type zero in $S_{\vec{\delta}}$ if, for any $\epsilon > 0$, $a(\vec{z})\, e^{-\epsilon \|\vec{z}\|}$ tends uniformly to zero as $\|\vec{z}\|$ tends to infinity, $\vec{z}$ in $S_{\vec{\delta}}$.

Lemma 4.1: If $a(\vec{z})$ is holomorphic and uniformly of exponential type zero in $S_{\vec{\delta}}$, $\hat{a}(\vec{w})$ extends holomorphically to $T_{\vec{\delta}}$.

Proof: Let $\vec{y}$ be a vector with all coordinates positive real, (that is $\vec{y}\in R^N$) substitute $u_n = y_n t_n$, $(n = 1,2,\ldots,N)$ in (4.2). The Jacobian of this transformation is $y_1 y_2 \ldots y_N$ and we have

$$\hat{a}(\vec{y}) = \frac{1}{y_1 y_2 \cdots y_N} \int_{R^N} e^{-\vec{u}} a\left(\frac{u_1}{y_1}, \ldots, \frac{u_N}{y_N}\right) d\vec{u} \tag{4.3}$$

We see from this that $\hat{a}(\vec{y})$ is analytic in each y_n for $|\mathrm{Arg}\ s_n| < \delta_n$, hence by Hartog's Theorem ([7], p.27) $\hat{a}(\vec{y})$ is a holomorphic function of $\vec{y}$ in $S_{\vec{\delta}}$.

Now in (4.3) we put $\vec{y} = \vec{r} \cdot e^{i\vec{\theta}}$, where $\vec{r} \in R^N$ and make the change of variables $x_n = u_n/r_n$, $n = 1,2 \ldots, N$, to get

$$\hat{a}(\vec{r} \cdot e^{i\vec{\theta}}) = (e^{-i\vec{\theta}} \cdot \vec{1}) \int_{R^N} e^{-\vec{r} \cdot \vec{x}} a(\vec{x} \cdot e^{-i\vec{\theta}}) d\vec{x} \tag{4.4}$$

(This, note, is the analogue for N-dimensions of (2.2)). Now in (4.4) we can let $\vec{r}$ vary over R^N. The resulting function of $\vec{r}$ is analytic in each coordinate of $\vec{r}$ and thus, again by Hartog's Theorem, holomorphic in $\vec{r}$. This proves the lemma.

We see that, under the above hypotheses on $a(\vec{z})$, the series (4.1) consists of terms holomorphic in $T_{\vec{\delta}}$. It is a little more difficult to prove the convergence of the series.

Lemma 4.2: Suppose that, for $\|k\|$ large,

$$y_{\vec{k}}(\vec{w}) = 0(\|k\|^{-(N+1)})$$

uniformly as $\vec{w}$ ranges throughout a set D. Then $\sum_{\vec{k} \in Z} y_{\vec{k}}(\vec{w})$ converges uniformly in D.

Proof: We need only prove convergence of the series

$$\sum_{\vec{k} \in Z_+^N \ \vec{k} \neq \vec{0}} \|\vec{k}\|^{-(N+1)} .$$

We use induction on N, the case $N = 1$ giving an obvious basis for induction. By our inductive hypothesis, we can reduce the problem to proving convergence of the sum over all $\vec{k}$ such that all coordinates are greater than 1, let Σ' denote this sum.

Now by induction on the number of variables it is easy to establish the following N - dimensional version of the integral test;

$$\Sigma' \ A(\vec{k}) < \int_{R_1^N} F(\vec{x}) d\vec{x} ,$$

where R_1^N is the Cartesian product of N copies of the ray $x > 1$, and $F(\vec{x})$ is a continuous function which is decreasing in each coordinate, and such that $F(\vec{k}) = A(\vec{k})$. For the series we are considering this gives us

$$\Sigma' \|k\|^{-(N+1)} < \int_{R_1^N} (x_1^2 + \cdots + x_N^2)^{-\frac{N+1}{2}} d\vec{x} = \int_{S_+^N} \int_1^\infty \frac{1}{r^2} \|u\|^{-\frac{N+1}{2}} dr \, \alpha \, \Omega_N(\vec{u}) ,$$

which is obviously finite.

<u>Lemma 4.3</u>: Let the multi-sector $T_{\vec{\eta}}$ be given, such that $\eta_j < \delta_j$ for all j. Suppose $a(\vec{z})$ is holomorphic and uniformly of exponential type zero in $S_{\vec{\delta}}$, and that $a(\vec{0}) = 0$.

Then the series (4.1) converges uniformly on the compact subsets of $T_{\vec{\eta}}$.

<u>Proof</u>: Let $\vec{w} \in T_{\vec{\eta}}$. We may write $\vec{w}$ in the form $\vec{w} = \vec{r} \cdot e^{i\vec{\theta}}$, where each coordinate r_j of $\vec{r}$ satisfies $|\text{Arg } r_j| \leq \frac{\pi}{2} - \mu_j$, and also $|\theta_j| \leq \delta_j - \mu_j$, with $\vec{\mu}$ defined by

$$2\mu_j = \frac{\pi}{2} + \delta_j - \eta_j , \quad j = 1,2,\ldots,N .$$

Now, introducing polar coordinates,

$$\hat{a}(w) = \int_{S_+^N} d\Omega_N(\vec{u}) \int_0^\infty \rho^{N-1} e^{-(\vec{r}\cdot\vec{u})\rho} \, b(\rho\vec{u}) \, d\rho , \tag{4.5}$$

where

$$b(\rho\vec{u}) = a(\rho\vec{u} \cdot e^{-i\theta}) , \tag{4.5'}$$

$$\vec{w} = \vec{r} \cdot e^{i\vec{\theta}} .$$

Now hold $\vec{u}$ fixed in (4.5) and consider the inner integral, which is simply the one-dimensional Laplace transform of the function $\rho^{N-1} b(\rho\vec{u})$, evaluated at $\vec{r}\cdot\vec{u}$. We can integrate by parts $N+1$ times and, remembering that $b(\vec{0}) = 0$, we have for the inner integral the expression

$$(\vec{r}\cdot\vec{u})^{-(N+1)} \int_0^\infty e^{-(\vec{r},\, \vec{u})\rho} \frac{\partial}{\partial\rho} [b(\rho\vec{u})] d\rho = 0(|\vec{r}\cdot\vec{u}|^{-(N+1)}) \tag{4.6}$$

the last step justified if we can prove boundedness of the integral in (4.6). But this follows easily from the observation that $\frac{\partial}{\partial\rho}[b(\rho\vec{u})]$ is just a directional derivative of $a(\vec{z})$, with $\vec{z}$ in $S_{\vec{\mu}}$, and thus is a linear combination of the partial derivatives of $a(\vec{z})$ with respect to the coordinates of $\vec{z}$. This Cauchy Formula for the derivatives ([7] p.27) shows that each function $\frac{\partial}{\partial z_j} a(\vec{z})$ is uniformly of exponential type zero in $S_{\vec{\mu}}$.

Using (4.6) in (4.5), we have

$$|\hat{a}(\vec{r}\cdot e^{i\vec{\theta}})| = 0(1)\int_{S_+^N} |\vec{r}\cdot\vec{u}|^{-(N+1)}\, d\Omega_N(u) = 0(\|\vec{r}\|^{-(N+1)})\int_{S_+^N} |e^{i\vec{\phi}\cdot\vec{u}}|^{-(N+1)}\, d\Omega_N(\vec{u}),$$

where $\vec{\phi}$ is defined by $\vec{r} = \|\vec{r}\|\, e^{i\vec{\phi}}$. Referring to the first paragraph of this proof, let

$$h = \min_{1\leq j\leq N} \cos\left(\frac{\pi}{2} - \mu_j\right).$$

Then

$$|e^{i\vec{\phi}\cdot\vec{u}}| = \left|\sum_{j=1}^{N} u_j \cos\phi_j\right| = \sum_{j=1}^{N} u_j \cos\phi_j \geq h\sum_{j=1}^{N} u_j \geq h\sum_{j=1}^{N} u_j^2 = h$$

We have then, for $\vec{w} = \vec{r}\cdot e^{i\vec{\theta}}$ in $S_{\vec{\eta}}$

$$\hat{u}(\vec{r}\cdot e^{i\vec{\theta}}) = 0(\|\vec{r}\|^{-(N+1)}),$$

a uniform estimate for fixed $\vec{\eta}$. We complete the proof of Lemma 4.3 by appealing to Lemma 4.2.

5. AN APPLICATION TO HARMONIC FUNCTIONS

In this section we indicate how methods developed by Bergman, Gilbert and others can be used to study harmonic functions in three variables, given strong analyticity assumptions on the coefficients. We refer to ([4] Chapter 2) for terminology. Instead of using the operator $\underline{B}_3$, we use $\underline{G}_3$ which although slightly more complicated has the advantage of giving a simpler relation between the coefficients of the harmonic function V and its associated holomorphic function f. (For a more detailed treatment of $\underline{G}_3$ cf. [5].)

Let

$$V(r,\theta,\varphi) = \sum_{n=0}^{\infty} r^n\{A_{no}P_n(\cos\ \theta) + \sum_{m=1}^{n} P_n^m(\cos\ \theta)(A_{nm}\ \cos m\varphi + B_{nm}\sin m\varphi)\}$$

$$= \mathrm{Re} \sum_{n=0}^{\infty} r^n \sum_{m=0}^{n} a_{nm}\ P_n^m\ (\cos\ \theta)\, e^{im\varphi}, \quad r < 1$$

where

$$a_{no} = A_{no}$$

$$a_{nm} = (A_{nm} - i\ B_{nm})/2\ .$$

With this we associate the function

$$f(s,\zeta) = \Sigma\, b_{nm}\, s^n\, \zeta^m\ ,$$

where

$$b_{nm} = a_{nm}\frac{(2m)!}{(2^m m!} = \frac{a_{nm}}{\sqrt{(\pi)}}\ 2^m\ \Gamma\ (m + \frac{1}{2})\ . \tag{5.1}$$

Then the formal relation is

$$V(r,\theta,\varphi) = V(x_1,x_2,x_3) = \mathrm{Re}\ \frac{1}{2\pi i}\int_L \frac{f(s,\sigma)\,ds}{(x_1^2+x_2^2+(x_3-s)^2)^{1/2}} \tag{5.2}$$

where

$$\sigma = \sigma(x_1,x_2,x_3;s) = \frac{-s\,(x_1+i\ x_2)}{x_1^2+x_2^2+(x_3-s)^2}$$

(5.2) actually represents V in a sufficiently small neighbourhood of the origin if we take L the unit circumference, and extends V if we can deform L to the unit circumference without passing through a singularity of the integrand. ([5]).

We are going to study the operator (5.2) under the assumption that the $b_{nm} = b(n,m)$ defined in (5.1) extend holomorphically. A small difficulty arises since Gilbert ([5]) takes

$$f(s,\sigma) = \sum_{n=0}^{\infty}\ \sum_{m=0}^{n} b_{nm}\, s^n\, \sigma^m\ ,$$

while we wish to take the summation on m extending from zero to infinity. However, inspection of Gilbert's derivation shows that, just as in the case of the operator $\underline{B}_3$, the two different holomorphic functions generate the same harmonic function. To see this, we refer to the steps leading from (3.1) to (3.4) of [5], and since by the calculus of residues

$$\int_{|\zeta|=1} \zeta^m \sum_{k=0}^{n} \frac{2^k k!}{(2k)!} P^k (\cos \theta) \left(\frac{e^{i\varphi}}{\zeta}\right)^k \frac{d\zeta}{\zeta}$$

vanishes for $m > n$, we find that the integral operator of (5.2) maps the biholomorphic function

$$\sum_{n=0}^{\infty} \sum_{m=n+1}^{\infty} b_{nm} s^n \sigma^m$$

to the trivial harmonic function.

Let us define the <u>trace of</u> (x_1, x_2, x_3), denoted $\mathrm{Tr}(x_1, x_2, x_3)$, to be the path in the s-plane given by

$$\sigma(x_1, x_2, x_3; s) \geq 1 .$$

We assume $f(\zeta, s)$ is holomorphic unless $\zeta \geq 1$ or $s \geq 1$, and consider the geometric problem of using (5.2) to extend the definition of V outside the unit sphere. For convenience, we set $x_1 + i x_2 = z$, and thus

$$\sigma = \frac{-sz}{r^2 - 2x_3 s + s^2} \tag{5.3}$$

r the spherical coordinate, $r^2 = x_1^2 + x_2^2 + x_3^2$. We see that we can use (5.2) to define a harmonic function locally at (X_1, X_2, X_3) if L does not meet $\mathrm{Tr}(X_1, X_2, X_3)$. Notice that in avoiding $\mathrm{Tr}(X_1, X_2, X_3)$, we also avoid the singularity of the factor $(|z|^2 + (x_3 - s)^2)^{-1/2}$ appearing in (5.2), for the trace passes through the poles of σ. The multiple-valued nature of the integrand causes no trouble, for if L does not contain a branch point, we can always regard the branch cut as having been made along a path which does not meet L.

To extend the function V from the unit sphere to (X_1,X_2,X_3), we lead out to (X_1,X_2,X_3) along a ray from the origin, $(x_1,x_2,x_3) = (tX_1,tX_2,tX_3)$ with $0 \leq t \leq 1$. For small t, $Tr(x_1,x_2,x_3)$ does not meet the unit circle, by validity of Gilbert's formula ([5], [4] p. 74) for small r. The trace consists of one or two arcs passing through the poles $x_3 \pm i\ |z|$ of σ, which are of modulus r and thus lie inside the unit circle for small t. If as t increases to 1 we can continuously vary L within the slit plane to avoid $Tr(x_1,x_2,x_3)$, we have extended V to (X_1,X_2,X_3). Two things can happen to prevent this extension; the path L may be forced onto $s \geq 1$ by an arc of the trace, or L may be trapped between two coalescing arcs of the trace.

From (5.3) it is clear that σ and s cannot be real together unless z is real. It is also clear from (5.3) that the trace varies throughout a bounded set. Thus for $X_2 \neq 0$ we can always avoid $Tr(tX_1,tX_2,tX_3)$ by choosing L to be a path consisting of a circular arc $|s| = R$ for $\in \leq \arg s \leq 2\pi - \in$ with R sufficiently large depending on t, together with two line segments joining the end points of this arc to the point $s = 1$. Thus we need only consider the curve $Tr(x_1,0,x_3)$, and we take $z = x_1$ in (5.3). We can assume $x_1 \neq 0$ since in this case σ reduces to the constant zero.

We now study the mapping $\sigma(x_1,0,x_3;s)$ in detail. It is helpful to use a decomposition as discussed in Nehari's book ([8] pp. 266, 267). $\sigma(s)$ as noted above has two simple poles, at $s = x_3 \pm i\ x_1$. $\sigma(s)$ maps the s-sphere onto a two-sheeted covering of the sphere. The branch points lie over $s = \pm r$, and the branched values are

$$\sigma(r) = \frac{-x_1}{2(r-x_3)}, \quad \sigma(-r) = \frac{x_1}{2(r+x_3)}$$

$\sigma(r)$ and $\sigma(-r)$ are both finite since $r > |x_3|$. The function $(\sigma - \sigma(r)/(\sigma-\sigma(-r))$ is the square of a bilinear function having its zero at $s = r$ and its pole at $s = -r$. Thus we have the following decomposition of the mapping $\sigma(s)$;

$$\sigma_1 = iA\,\frac{s-r}{s+r} \tag{5.4}$$

$$\sigma_2 = \sigma_1^2$$

$$\sigma = \frac{x_1}{2(r^2 - x_3^2} \quad \frac{(r - x_3)\,\sigma_2 + (r + x_3)}{\sigma_2 - 1}$$

and the corresponding chain of inverse mappings is

$$\sigma_2 = \frac{\sigma + \dfrac{x_1}{2(r - x_3)}}{\sigma - \dfrac{x_1}{2(r + x_3)}} \tag{5.4'}$$

$$\sigma_1 = \sqrt{(\sigma_2)}$$

$$s = r\,\frac{\sigma_1 + iA}{iA - \sigma_1}$$

In the second member of (5.4'), σ_1 must be regarded as double-valued. We still have to determine A from the condition that σ has poles at $x_3 \pm i\,x_1$. The poles of σ correspond to $\sigma_1 = \pm 1$. The function $\frac{s-r}{s+r}$ sends conjugate points to conjugate points and by a short computation (recalling that $r^2 = x_1^2 + x_3^2$) the values at $x_3 \pm i\,x_1$ are pure imaginary, thus A is real. The precise value of A is immaterial in the sequel. We find that σ_2 increases or decreases with real σ according as, respectively, x_1 is negative or positive.

Now first consider the case $x_1 > 0$. Since σ_2 decreases with real σ, and $\sigma_2(\infty) = 1$, the ray $\sigma \geq 1$ maps to a segment $1 \leq \sigma_2 \leq \alpha$, $\alpha < \infty$. The image in the σ_1-plane is the union of the segments $1 \leq \sigma_1 \leq \sqrt{(\alpha)}$ and $-\sqrt{(\alpha)} \leq \sigma_1 \leq -1$. It is readily verified that the real σ_1-axis maps to a circle, touching the real axis at $s = r$ (when $\sigma_1 = 0$)and $s = -1$ (when $\sigma_1 = \infty$). Thus neither of the above two segments in the σ-plane meets the real axis. We infer that when $x_1 > 0$, $\mathrm{Tr}\,(x_1, 0, x_3)$ does not meet the ray $s > 1$.

Now let $x_1 < 0$. The ray $\sigma \geq 1$ maps to a segment $\alpha \leq \sigma_2 \leq 1$, where α may be $-\infty$. The image in the σ_1-plane consists of two real segments if $\alpha > 0$, and if $\alpha \leq 0$ the image contains the interval $-1 \leq \sigma_1 \leq 1$, the image of which meets $s \geq 1$ at the point $s = 1$. Thus the condition that $\mathrm{Tr}\,(x_1, 0, x_3)$ touch $s \geq 1$ is simply $\alpha \leq 0$. Computing α explicitly, this is

$$\frac{2(r-x_3)+x_1}{2(r+x_3)-x_1}\;\frac{r+x_3}{r-x_3}\leq 0 .$$

Since $x_1 < 0$ the only factor which can vanish is the numerator of the first fraction.

$$2(r-x_3)+x_1 \leq 0, \quad x_1 < 0$$

yields

$$-3x_1 \leq 4x_3 .$$

We summarize the above discussion by stating our result.

Theorem: Let $V(x_1,x_2,x_3)$ be harmonic in $r < 1$, with expansion

$$V(r,\varphi,\theta) = \operatorname{Re} \sum_{n=0}^{\infty} \sum_{m=0}^{n} a_{nm} r^n P_n^m(\cos\theta) e^{im\varphi} .$$

Suppose that the function

$$b(n,m) = a_{nm} 2^m \Gamma(m+1/2)$$

can be extended to a function holomorphic and uniformly of exponential type zero in the Cartesian product of two right half-planes.

Then V can be extended to a function single-valued and harmonic in the domain obtained by excluding the set (x_1,x_2,x_3) characterized by the inequalities

$$x_2 = 0, \; x_1 < 0, \; -3x_1 \leq 4x_3, \; x_1^2 + x_3^2 \geq 1.$$

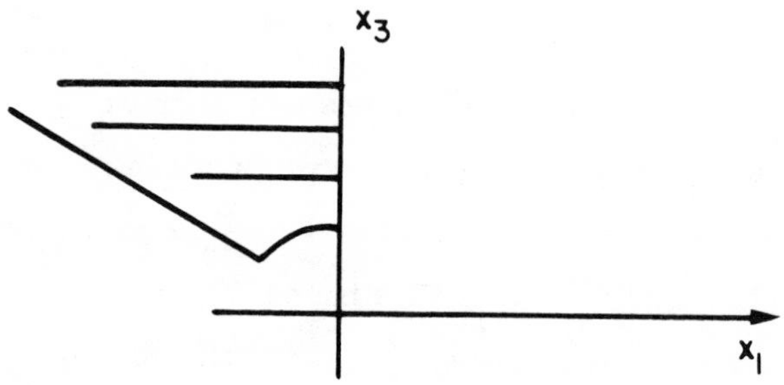

Fig. 5.1

The projection of the region in the $x_1 - x_3$ plane is pictured (Fig. 5.1), the shaded portion being the part to which V is perhaps not extendable.

Finally, we mention that if the two half-planes mentioned in the above theorem can be replaced by half-planes symmetric with respect to the positive real axis and of opening greater than π, $f(\sigma, s)$ has points $(\sigma, 1)$ and $(1, s)$ as the only singularities though it may no longer be single-valued off this set. Then V can be extended to be locally harmonic, (though perhaps multiple-valued) as long as we avoid (x_1, x_2, x_3) for which

$$\sigma(x_1, x_2, x_3; 1) = 1.$$

By (5.3) this yields

$$(x_1 + \frac{1}{2})^2 + (x_3 - 1)^2 = \frac{1}{4}.$$

From this circle, we exclude the open arc for which $x_1^2 + x_3^2 < 1$ to get the possible singularities of V.

References

1 R.P. Boas, Jr. Entire Function. Academic Press (1954).

2 Courant and Hilbert, Methods of Mathematical Physics, Vol. I, Interscience (1953).

3 W.F. Donoghue, Jr. Distributions and Fourier Transforms. Academic Press (1969).

4 R.P. Gilbert, Function Theoretic Methods in Partial Differential Equations. Academic Press (1969).

5 R.P. Gilbert, Operators which generate harmonic functions in three-variables. Scripta-Math. 27 (1964) 141-152.

6 E. Hille, Analytic Function Theory, Vol. II. Ginn (1962).

7 L. Hörmander, An Introduction to Complex Analysis in Several Variables.

D. Van Nostrand (1966).

8 Z. Nehari, Conformal Mapping. McGraw-Hill (1952).

9 W. Rudin, Real and Complex Analysis. McGraw-Hill (1966).

T.L. McCoy
Department of Mathematics
Michigan State University
East Lansing
Michigan
USA

L G MIKHAILOV

13 On some integral equations with homogeneous kernels

The following notations are used throughout: E_n is an n-dimensional Euclidian space of points

$$x = (x_1, x_2, \ldots, x_n),\ |x|^2 = \sum_1^n x_i^2,\ |x-y|^2 = \sum_1^n (x_i - y_i)^2,\ t \cdot x = (tx_1, \ldots, tx_n),$$

where t is a scalar, D is an arbitrary domain containing the origin. In the present paper we give a review of some recent results of the author and his co-workers, concerning the equation

$$f(x) = \int_D \Theta(x,y) \cdot f(y)\, dy + g(x), \quad x \in D\ (f = \Theta f + g). \tag{1}$$

Here the kernel satisfies the conditions of homogeneity ($\mathscr{H}$), summability ($\mathscr{S}$) and invariance ($\mathscr{I}$). The function $\Theta(x,y)$ is called homogeneous of the order $-n$, if

$$\Theta(t \cdot x, t \cdot y) = t^{-n} \cdot \Theta(x,y) \qquad (\mathscr{H})$$

for an arbitrary number $t > 0$. A homogeneous function, given in the neighbourhood of the origin, may be extended over the whole space by rays. The points $x = 0$ and $x = \infty$ are singular in the sense that the order of singularity equals the dimension of the space, and in general the kernel is non-integrable. It has been shown also, that the operator is not completely continuous, if the domain D contains the point $x = 0$ or $x = \infty$. The integrals in (1), and later integrals, are understood in the usual Riemann or Lebesgue sense; furthermore, we shall assume the existence of the real number β such that

$$\int\int_{E_n} |\Theta(j,u)| \cdot |u|^{-\beta} \cdot du < +\infty, \quad j = (1, 0, \ldots, 0). \qquad (\mathscr{S})$$

The number β determines some classes of functions in which the equation (1) is considered. Taking usual notations of Banach spaces M, C, L^p (series ϵ), by M_β, C_β, L^p_β (series ϵ_β) we are indicating isometric Banach spaces of functions $f(x) = |x|^{-\beta} \cdot F(x)$, $\|f\|_\beta = \|F\|$. It is possible to introduce subspaces where the points $x = 0$ or $x = \infty$ are isolated [1]. Substituting the functions we have

$$F(x) = \int_D T(x,y) \cdot F(y)\,dy + C(x)\,, \tag{2}$$

where $T(x,y) = |x|^\beta \cdot |y|^{-\beta} \cdot \Theta(x,y)$ satisfies the conditions $(\mathscr{H})$ and $(\mathscr{S})$ for $\beta = 0$ and $F, C \in \epsilon$. In the multidimensional case $n > 1$ our method requires in addition, the condition of the kernel invariance by arbitrary rotations of space E_n,

$$\Theta[\gamma(x), \gamma(y)] = \Theta(x,y)\,. \tag{$\mathscr{I}$}$$

This condition defines variety of the functions from scalar product (x,y) and from $|x|$, $|y|$, $|x-y|$ in particular, [2].

1. ONE-DIMENSIONAL EQUATIONS

The cases of interior or boundary points $x = 0$ (or $x = \infty$) are essentially distinguished as it has been shown from properties of the operators [3]. For multidimensional spaces $(n > 1)$ the equations with a singular point on the boundary have not yet been studied. If $n = 1$ the equations with end or interior singular points have been solved but the results are different.

<u>1.1 The equations with end singular point</u>. Without loss of generality the above equations can be written as

$$f(x) = \int_0^1 \Theta(x,y) \cdot f(y)\,dy + g(x)\,, \quad 0 < x < 1\,, \tag{2}$$

where $\Theta(x,y)$ is homogeneous of order -1 and $\int_0^\infty |\Theta(j,u)| \cdot u^{-\beta} \cdot du < +\infty$. Introducing new variables and indicating $x = e^{-s}$, $y = e^{-\tau}$, $f(e^{-s}) \equiv \Psi(s)$, $g(e^{-s}) \equiv d(s)$, we can transform (2) into

$$\Psi(s) = \int_0^\infty h(s-\tau) \cdot \Psi(\tau)\,d\tau + d(s)\,, \quad 0 < s < \infty\,, \tag{3}$$

where $h(x) = \exp[(1-\beta)x] \cdot \Theta(1, e^x)$. Thus, the equation with homogeneous kernel and end singular point is reduced to the equation with difference kernel in a semi-infinite interval, an equation of Wiener-Hopf type. There is no such interrelation in the multidimensional case. Many papers are devoted to this case and in this direction we obtained some results, some of which are given below [8].

Theorem 1: For the normal resolvability it is necessary and sufficient that

$$\Lambda - \mathscr{H}(x) \neq 0, \quad -\infty < x < \infty, \mathscr{H}(x) = \int_0^\infty \Theta(1,u) \cdot u^{ix-\beta} \cdot du. \tag{4}$$

Let $\mathscr{H} = \frac{-1}{2\pi} \{\arg[1-\mathscr{H}(x)]\}_{-\infty}^{\infty}$. If $\mathscr{H} \geq 0$, then the homogeneous equation has $\mathscr{H}$ linear independent solutions from ϵ_β and the non-homogeneous equations is solvable for an arbitrary $g(x)$. If $\mathscr{H} < 0$, then the homogeneous equation has no non-trivial solution, and for solvability of a non-homogeneous case $|\mathscr{H}|$ conditions of orthogonality $g(x)$ to the solutions of the transposed homogeneous equation from $\epsilon_{1-\beta}$ is necessary and sufficient.

The integral equations with variable upper limit,

$$f(x) = \int_0^x \Theta(x,y) \cdot f(y)\,dy + g(x), \tag{5}$$

which can be written in the form,

$$f(x) = \int_0^1 \Theta(1,t) \cdot f(x \cdot t) \cdot dt + g(x),$$

have a very old story. As mentioned by Holmgren [4], in 1823 Abel developing his method of conversion of integral equation, (known today under his name) was led to an equation (5), of the first kind, and announced that he had found the solution but it was too long to be described. In 1897 Volterra investigated it when $\Theta(1,t)$ is a polynomial. In 1913-1914 Brown, and later in 1922 Holmgren, obtained the solution by contour integration. Using the modern theory of linear operators in Banach spaces we can formulate theorems about solvability with a precise statement about the classes of functions and conditions.

Theorem 2: If we have the conditions of normal solvability (4) the equation (5) is always solvable and the homogeneous equation can have non-trivial solutions of the form

$$f(x) = x^{-\sigma} \cdot \Sigma\, C_{K_j} \cdot (\mathcal{L}n\, x)^j \cdot x^{iz_K}.$$

Let $K(x,y)$ be homogeneous of arbitrary order $\alpha \neq 0$. Then the equation of the first kind

$$\int_0^x K(x,y) \cdot f(y)\, dy = g(x), \quad 0 < x < 1$$

is reduced to (5). Numerous examples are contained in [4].

<u>1.2 The equation with interior singular point.</u> The equation considered is

$$f(x) = \int_a^b \Theta(x,y) \cdot f(y)\, dy + g(x), \qquad a < x < b \tag{6}$$

where $a < 0$, $b > 0$, the kernel satisfies the condition $(\mathcal{H})$ and a more general condition of the type $(\mathcal{S})$,

$$\int_{-\infty}^{+\infty} |\Theta(\pm 1,u)| \cdot |u|^{-\beta} \cdot du < +\infty.$$

Given (6) in intervals $(a,0)$, $(0,b)$ and setting $x = b\xi$ or $x = a\xi$, and $y = b\eta$, or $y = a\eta$, we arrive at a system of two equations with singular end point $x = 0$, on the segment $[0,1]$.

By the same method that was applied to the equation with singular end point, we shall arrive at a system of equations with difference kernels that have been studied in [7]. Though there exists a general theorem for such a system with a condition of normal solvability and index formula, in contrast to the one-dimensional case, the number of solutions of the homogeneous system (ℓ) and the conditions of solvability for the non-homogeneous system (P) are unknown. However, when the kernel satisfies, in addition, conditions of symmetry or evenness as, for example, $\Theta(-x,-y) = \Theta(x,y)$ we have obtained formulas for (ℓ) and P, [4].

2. THE EQUATIONS ON THE PLANE

We can write the conditions $(\mathcal{H})$ and $(\mathcal{J})$ as one complex condition by

$$\Theta(t \cdot z, t \cdot \zeta) = |t|^{-2} \cdot \Theta(z,\zeta) \tag{$\mathcal{H}\mathcal{J}$}$$

where t is an arbitrary complex number.

Let us consider the equation

$$f(z) = \int\limits_{|\zeta| \leq R}\int \Theta(z,\zeta) \cdot f(\zeta) \cdot d\mathscr{S}_\zeta + g(z), \quad |z| \leq R \tag{7}$$

and induce new variable $\zeta = \sigma \cdot Z$ and transform it by using polar coordinates $Z = re^{i\varphi}$, $\zeta = \rho e^{i\theta}$, $\sigma = \tau \cdot e^{i\alpha}$. Our method is analogous with the known partial differential equations method of separation of variables. We will construct the solution $f(z) \equiv f(r,\varphi)$ from its Fourier coefficients $f_K(r) = \frac{1}{2\pi} \cdot \int_0^{2\pi} f(r,\varphi) \cdot e^{-iK\varphi} \cdot d\varphi$. Multiplying (7) by $e^{-iK\varphi}$ and integrating we obtain

$$f_K(r) = \int_0^R T_K(r,\rho) \cdot f_K(\rho)\, d\rho + g_K(\tau), \quad K = 0, \pm 1, \pm 2, \ldots \tag{8}$$

where $T_K(r,\rho) = \frac{1}{r} \cdot \Omega_K(\frac{\rho}{r})$, $\Omega_K(\tau) = -i\tau \int\limits_{|t|=1} \Theta(1, \tau \cdot t) \cdot (\tau t)^{-\beta} \cdot t^{K-1} \cdot dt$.

The determined value has Fourier transformation of kernel

$$\mathscr{H}_K(x) = \int\limits_{E_2}\int \Theta(1,\sigma) \cdot |\sigma|^{ix-\beta} \cdot e^{iK\alpha} \cdot ds_\sigma .$$

From the Riemann-Lebesgue theorem $\lim\limits_{|K| \to \infty} |\Omega_K(\tau)| = 0$ and hence $\lim\limits_{|K| \to \infty} \|T_K\| = 0$, where T_K are operators from (8). There exists a number N such that for $|K| > N$ one has $\|T_K\| < 1$ and corresponding equations (8) have a unique solution with an arbitrary free term. Thus, two-dimensional equation (7) is reduced to a finite $(|K| \leq N)$ set of separate, independent, one-dimensional equations (8). Let us indicate

$$\mathscr{Y}_K(x) = 1 - \mathscr{H}_K(x), \quad \mathscr{H}_K = -\frac{1}{2\pi} \cdot \{\arg \mathscr{Y}_K(x)\}_{-\infty}^{\infty},$$

$$\mathscr{Y}(x) = \prod_{|K| \leq N} \mathscr{Y}_K(x), \quad \mathscr{H}_+ = \prod_{\mathscr{H}_K > 0} \mathscr{H}_K, \quad \mathscr{H}_- = -\sum_{\mathscr{H}_K < 0} \mathscr{H}_K$$

Theorem 3: Let the kernel of equation (7) satisfy $(\mathscr{H}\mathscr{I})$, $(\mathscr{S})$, the condition of normal solvability $\mathscr{Y}(x) \neq 0$, $-\infty < x < \infty$. Then the homogeneous equation has $\mathscr{H}_+$ linear independent solutions from ϵ_β, and $\mathscr{H}_-$ conditions of orthogonality

$g(z)$, to the solutions of the transposed, homogeneous equation from $\epsilon_{2-\beta}$. This hypothesis is necessary and sufficient.

The equation [9] is an example of this theory.

3. MULTI-DIMENSIONAL CASE $(n \geq 3)$

Let $D \equiv S_n$ be a sphere $|y| \leq 1$ and $\Theta(x,y)$ satisfy the conditions $(\mathscr{H})$, (S), $(\mathscr{I})$. Let us transform the spherical coordinates $x = (r, S_x)$, $y = (\rho, S_y)$, where S_x, S_y are the points on the hypersphere S_{n-1}. It has been proved that for an arbitrary function $\Theta(S_x, S_y)$, invariant under the rotation group and summable:

$$\int_{S_{n-1}} |\Theta(j,u)| \cdot du < +\infty, \quad \text{where } j \text{ is the vector } (1,0,0,\ldots,0),$$

there exists a full orthogonal system of continuous functions $\{t_k^{\ell}(S_x)\}$, $1 \leq \ell < \infty$, $1 \leq k \leq d_\ell$, that we have

$$\int_{S_{n-1}} \Theta(S_x, S_y) \cdot t_k^{\ell}(S_x) \cdot dS_x = t_k^{\ell}(S_y) \cdot \int_{S_{n-1}} \Theta(j, S_x \cdot t_1^{\ell}(S_x) \cdot dS_x$$

Writing (1) in spherical coordinates, multiplying by $t_k^{\ell}(S_x)$ and integrating, yields a system of the type (8). In the three-dimensional case $x = (r, \varphi, \Psi)$, $y = (\rho, \alpha, \gamma)$, $t_k^{\ell}(S_x) = P_k^{\ell}(\cos\Psi) \cdot e^{-i\ell\varphi} \cdot \sin\varphi$, where $P_k^{\ell}(\cos\Psi)$ are the associated Legendre functions. From the Hilbert [2] theory of invariance of it follows that $\Theta(x,y) = r^{-1} \cdot \rho^{-2} \cdot h(\frac{\rho}{r}, \cos\xi)$, where

$$\cos\xi = \cos\Psi \cdot \cos\gamma + \sin\Psi \cdot \sin\gamma \cdot \cos(\varphi - \alpha).$$

One-dimensional equations will be

$$f_k(r) = \int_0^1 \frac{1}{r} \cdot h_k\left(\frac{\rho}{r}\right) \cdot f_k^{\ell}(\rho) \cdot d\rho + g_k(\tau), \quad k = 0, \pm 1, \ldots, |\ell| \leq k$$

where $h_k(u) = \int_0^{\pi} h(u, \cos\Psi) \cdot P_k(\cos\Psi) \sin\Psi \cdot d\Psi$.

The stated methods are useful not only for calculating ℓ and P but for constructing solutions in useful forms (by integrals and series). There are many examples in [1], [3], [4], [9], and subsequent publications.

References

1 L.G. Mikhailov, A New Class of Singular Integral Equations and its Application to Differential Equations with Singular Coefficients. Academic-Verlag, Berlin (1970) or Wolters-Noordhoff Publishing, Groningen (1970).

2 H.Weyl, A classical group, their invariance and presentation. Gostechizdat, M. (1947).

3 L.G. Mikhailov, Dokl.Akad.Nauk SSSR 176 No. 2 (1967).

4 L.G. Mikhailov, Integral equations with a kernel homogeneous of degree -1. Donish, Dushanbe (1966).

5 E.C. Titchmarsh, Introduction in the theory of Fourier integrals. Gostechizdat, M. (1947).

6 M.G. Krein, Uspehi Math. Nauk XIII, 5 (83)(1958).

7 I.C. Gohberg and M.G. Krein, Uspehi Math.Nauk XII, 2 (80)(1958).

8 L.G. Mikhailov, Dokl.Akad.Nauk SSSR 190 No. 2 (1970).

9 L.G. Mikhailov, Dokl.Akad.Nauk SSSR 190 No. 3 (1970).

L.G. Mikhailov
Lenin Avenue 33
Academie of Sciences Tadjik SSSR
Dushanbe
USSR

T A CHANTURIA

14 On singular solutions of strongly nonlinear systems of ordinary differential equations

For the system of differential equations

$$\frac{dx_i}{dt} = f_i(t,x_1,\ldots,x_n) \quad (i = 1,\ldots,n)$$

we consider the question of existence of the solution $x_1(t),\ldots,x_n(t)$, defined on some interval $[t_0,t^*)$, $0 \le t_0 < t^* < +\infty$, and satisfying the condition $\lim_{t\to t^*} \sum_{i=1}^{n} |x_i(t)| = +\infty$.

Consider the system of differential equations

$$\frac{dx_i}{dt} = f_i(t,x_1,\ldots,x_n) \quad (i = 1,\ldots,n) \tag{1}$$

where the functions $f_i(t,x_1,\ldots,x_n)$ $(i = 1,\ldots,n)$ are defined on the domain $0 \le t < +\infty$, $-\infty < x_1,\ldots,x_n < +\infty$ and satisfy the local conditions of Caratheodory.

The solution $x_1(t),\ldots,x_n(t)$ of system (1) will be called singular if it is defined in the finite interval $[t_0,t^*)$ and

$$\lim_{t\to t^*} \sum_{i=1}^{n} |x_i(t)| = +\infty. \tag{2}$$

The point t^* is called a moving singular point (see [1] chap. XIII).

The solution $x_1(t),\ldots,x_n(t)$ of system (1) will be said to be non-continuable to the right either when it is defined in some infinite interval or when it is singular.

The present paper deals with the question of existence of singular solutions to system (1).

From the works on similar problems which are known to us we should mention first of all [1], in which the behaviour of solutions to systems of two differential equations near a moving singular point is studied, and also [2] and [3], where sufficient conditions are established for the existence of singular solutions of second order differential equations.

Theorem 1: For $t \geq 0$, $(-1)^{\nu_k} x_k \geq x_0$ $(k = 1,\dots,n)$ let the following inequalities

$$(-1)^{\nu_i} f_i\ (t, |x_1|,\dots,|x_n|) \geq a_i(t)\,|x_{i+1}|^{\lambda_i} \quad (i = 1,\dots,n) \tag{3}$$

be fulfilled, where $x_0 \geq 0$, $x_{n+1} = x_1$, $\nu_i = 0$ or 1, $\lambda_i > 0$ $(i = 1,\dots,n)$, $\prod_{i=1}^{n} \lambda_i > 1$ and $a_i(t)$ $(i = 1,\dots,n)$ are continuous and positive functions in the interval $[0, +\infty)$. Then system (1) has a singular solution $x_1(t),\dots,x_n(t)$ such that $(-1)^{\nu_i} x_i(t) > 0$ when $t \in [t_0, t^*)$ $(i = 1,\dots,n)$.

We will have to use some lemmas in order to prove this theorem.

Lemma 1: Let the conditions of Theorem 1 be satisfied and let $x_1(t),\dots,x_n(t)$ be a solution of system (1) and $y_1(t),\dots,y_n(t)$ - the solution of the system

$$\frac{dy_i}{dt} = a_i(t)\, y_{i+1}^{\lambda_i} \quad (i = 1,\dots,n; y_{n+1} = y_1),$$

defined on some interval $[t_0, t_1)$. Then if

$$(-1)^{\nu_i} x_i(t_0) > y_i(t_0) \geq 0, \quad (-1)^{\nu_i} x_i(t_0) > x_0 \quad (i = 1,\dots,n),$$

we have

$$(-1)^{\nu_i} x_i(t) > y_i(t) \quad \text{when} \quad t \in [t_0, t_1) \quad (i = 1,\dots,n).$$

This lemma is a consequence of a theorem on differential inequalities (see, e.g. [4] Chap. II).

Lemma 2: Let $\mathscr{S} > 0$, $\gamma > 0$, $\lambda_i > 0$ $(i = 1,\dots,n)$ and let $y_1(t),\dots,y_n(t)$ be the solution of the problem

$$\frac{dy_i}{dt} = \mathscr{S} y_{i+1}^{\lambda_i} \quad (i = 1,\dots,n, y_{n+1} = y_1), \tag{4}$$

$$y_1(t_0) = \gamma, \quad y_2(t_0) = \dots = y_n(t_0) = 0, \tag{5}$$

defined in some interval $[t_0, t_2)$. Then for $t \in [t_0, t_2)$ the following inequality

$$y_n(t) \le c_0 [y_1(t)]^{\alpha_1 - \frac{\alpha_{n-1}}{\beta_{n-1}}} [y_2(t)]^{\frac{1}{\beta_{n-1}}} \tag{6}$$

is fulfilled, where

$$\alpha_i = \prod_{k=n-i+1}^{n} \lambda_k \quad (i = 1,\dots,n), \quad \beta_i = 1 + \lambda_{n-i+1}\beta_{i-1} \quad (i = 2,\dots,n), \quad \beta_1 = 1,$$

$$c_0 = \prod_{k=1}^{n-2} c_k^{\frac{1}{\beta_k}}, c_i = \left(\frac{\beta_{i+1}}{\beta_i} c_{i-1}^{\lambda_{n-i+1}}\right)^{\frac{\beta_i}{\beta_{i+1}}} \quad (i = 2,\dots,n-2), \quad c_1 = \beta_2^{\frac{1}{\beta_2}}. \tag{7}$$

<u>Proof</u>: First we shall prove that

$$y_{n-i+1}(t) \le c_i [y_1(t)]^{\frac{\alpha_i}{\beta_{i+1}}} [y_{n-i}(t)]^{\frac{\beta_i}{\beta_{i+1}}} \text{ when } t \in [t_0, t_2) \quad (i = 1,\dots,n-2). \tag{8}$$

Multiplying the n-th equation of system (4) by $[y_n(t)]^{\lambda_{n-1}}$ and integrating over the interval (t_0, t) we obtain

$$y_n(t) \le (\lambda_{n-1} + 1)^{\frac{1}{\lambda_{n-1}+1}} [y_1(t)]^{\frac{\lambda_n}{\lambda_{n-1}+1}} [y_{n-1}(t)]^{\frac{1}{\lambda_{n-1}+1}},$$

i.e. the inequality (8) is valid for $i = 1$. Suppose, it is valid for some $i \in \{1,\dots,n-3\}$. If we raise both parts of inequality (8) to the power λ_{n-i}, then multiply them by $[y_{n-i}(t)]^{\frac{\beta_{i+2}}{\beta_{i+1}} - 1}$ and integrate from t_0 to t, since $\frac{\lambda_{n-i}\beta_i + \beta_{i+2}}{\beta_{i+1}} - 1 = \lambda_{n-i-1}$, according to (4) and (5) we shall have

$$y_{n-i}(t) \le \left(\frac{\beta_{i+2}}{\beta_{i+1}} c_i^{\lambda_{n-i}}\right)^{\frac{\beta_{i+1}}{\beta_{i+2}}} [y_1(t)]^{\frac{\alpha_{i+1}}{\beta_{i+2}}} [y_{n-i-1}(t)]^{\frac{\beta_{i+1}}{\beta_{i+2}}}.$$

Thus (8) is proved.

From (8) we easily obtain

$$y_n(t) \le \prod_{i=1}^{n-2} c_i^{\frac{1}{\beta_i}} [y_1(t)]^{\sum_{i=1}^{n-2} \frac{\alpha_i}{\beta_i \beta_{i+1}}} [y_2(t)]^{\frac{1}{\beta_{n-1}}}.$$

But in virtue of the equality $\frac{\alpha_i}{\beta_i \beta_{i+1}} = \frac{\alpha_i}{\beta_i} - \frac{\alpha_{i+1}}{\beta_{i+1}}$, this means that inequality (6) is satisfied. The lemma is proved.

Lemma 3: If $\mathscr{S} > 0$, $\lambda_i > 0$ $(i = 1,\dots,n)$, $\prod_{i=1}^{n} \lambda_i > 1$, $t_1 \in (t_0, +\infty)$ and $y_1(t),\dots,y_n(t)$ is the non-continuable to the right solution of problem (4), (5), where

$$\gamma = c\left(\frac{2c\beta_n}{\mathscr{S}(\alpha_n - 1)(t_1 - t_0)}\right)^{\frac{\beta_n}{\alpha_{n-1}}} + \left(\frac{1}{c_n^*}\left(\frac{2}{t_1 - t_0}\right)^{\beta_n}\right)^{\frac{1}{\alpha_{n-1}}}, \tag{9}$$

$$c = ((\lambda_n + 1)c_0)^{\frac{\beta_{n-1}}{\alpha_{n-1} + \beta_{n-1}}}, \; c_{i+1}^* = \frac{\mathscr{S}}{\beta_{i+1}}(c_i^*)^{\lambda_{n-i}} \; (i = 1,\dots,n-1), \; c_i^* = \mathscr{S},$$

and c_0, α_i, β_i are defined by means of equalities (7), then there exists a number $\tilde{t} \in (t_0, t_1)$, such that

$$\lim_{t \to \tilde{t}} y_1(t) = +\infty \quad (i = 1,\dots,n). \tag{10}$$

Proof: Assume that $y_1(t),\dots,y_n(t)$ is defined in the interval $[t_0, t_1)$. Then from (4) it readily follows that

$$y_{n-i+1}(t) \geq c_i^* \gamma^{\alpha_i}(t - t_0)^{\beta_i} \quad \text{when } t \in [t_0, t_1) \; (i = 1,\dots,n-1)$$

and

$$y_1(t) = \gamma + \mathscr{S}\int_{t_0}^{t} [y_2(\tau)]^{\lambda_1}\, d\tau \geq$$

$$\geq \gamma + \mathscr{S}(c_{n-1}^*)^{\lambda_1}\gamma^{\alpha_n}\int_{t_0}^{\frac{t_0+t_1}{2}} (\tau - t_0)^{\lambda_1\beta_{n-1}}\, d\tau + \mathscr{S}\int_{\frac{t_0+t_1}{2}}^{t} [y_2(\tau)]^{\lambda_1} d\tau =$$

$$= \gamma + c_n^*\left(\frac{t_1 - t_0}{2}\right)^{\beta_n}\gamma^{\alpha_n} + \mathscr{S}\int_{\frac{t_0+t_1}{2}}^{t} [y_2(\tau)]^{\lambda_1} d\tau \quad \text{when } \frac{t_0+t_1}{2} \leq t < t_1.$$

Hence, by (9), we have

$$y_1(t) \geq 2\gamma + \mathscr{S}\int_{\frac{t_0+t_1}{2}}^{t} [y_2(\tau)]^{\lambda_1} d\tau \quad \text{for } \frac{t_0+t_1}{2} \leq t < t_1. \tag{11}$$

On the other hand, due to (6)

$$[y_1(t)]^{\lambda_n+1} = \gamma^{\lambda_n+1} + (\lambda_n+1)\int_{t_0}^{t}[y_2(\tau)]^{\lambda_1}\,dy_n(\tau) \le \gamma^{\lambda_n+1} + (\lambda_n+1)[y_2(t)]^{\lambda_1}y_n(t) \le$$

$$\le [\gamma^{1+\frac{\alpha_{n-1}}{\beta_{n-1}}} + c^{1+\frac{\alpha_{n-1}}{\beta_{n-1}}}[y_2(t)]^{\lambda_1+\frac{1}{\beta_{n-1}}}]\,[y_1(t)]^{\lambda_n-\frac{\alpha_{n-1}}{\beta_{n-1}}} \quad \text{when} \quad t_0 \le t < t_1 .$$

Therefore

$$y_1(t) \le \gamma + c\,[y_2(t)]^{\frac{\beta_n}{\beta_{n-1}+\alpha_{n-1}}} \quad \text{for} \quad t_0 \le t < t_1 .$$

Due to this and (11) we obtain

$$[y_2(t)]^{\frac{\beta_n}{\beta_{n-1}+\alpha_{n-1}}} \ge \frac{\gamma}{c} + \frac{\mathscr{S}}{c}\int_{\frac{t_0+t_1}{2}}^{t}[y_2(\tau)]^{\lambda_1}\,d\tau \quad \text{for} \quad \frac{t_0+t_1}{2} \le t < t_1 ,$$

i.e.

$$[y_2(t)]^{\lambda_1}\Big[\frac{\gamma}{c} + \frac{\mathscr{S}}{c}\int_{\frac{t_0+t_1}{2}}^{t}[y_2(\tau)]^{\lambda_1}\,d\tau\Big]^{-1-\frac{\alpha_{n-1}}{\beta_n}} \ge 1 \quad \text{for} \quad \frac{t_0+t_1}{2} \le t < t_1 .$$

By integrating this inequality from $\frac{t_0+t_1}{2}$ to t we have

$$\left(\frac{\gamma}{c}\right)^{\frac{1-\alpha_n}{\beta_n}} - \left(\frac{\gamma}{c} + \frac{\mathscr{S}}{c}\int_{\frac{t_0+t_1}{2}}^{t}[y_2(\tau)]^{\lambda_1}\,d\tau\right)^{\frac{1-\alpha_n}{\beta_n}} \ge \frac{\mathscr{S}(\alpha_n-1)}{c\;\beta_n}\left(t-\frac{t_0+t_1}{2}\right)$$

$$\text{when} \quad \frac{t_0+t_1}{2} \le t < t_1 .$$

Consequently,

$$\left(\frac{\gamma}{c}\right)^{\frac{1-\alpha_n}{\beta_n}} \ge \frac{\mathscr{S}(\alpha_n-1)(t_1-t_0)}{2c\;\beta_n} ,$$

which contradicts (9). The contradiction thus obtained proves the lemma.

Proof of Theorem 1: Let us take arbitrary numbers t_0 and t_1, $0 \le t_0 < t_1 < +\infty$ and assume that $\mathscr{S} = \min\{a_i(t) : t_0 \le t \le t_1,\ i = 1,\dots,n\}$. Then due to Lemmas 1 and 3, it is evident that for the non-continuable to the right solution $x_1(t),\dots,x_n(t)$ of system (1), satisfying the condition

$$(-1)^{\nu_1}x_1(t_0) > \gamma + x_0 ,\quad (-1)^{\nu_i}x_i(t_0) > x_0 \quad (i = 2,\dots,n) ,$$

where γ is defined by means of equality (9), for some $t^* \in (t_0, t_1)$ the equality (2) is fulfilled. The theorem is proved.

From Theorem 1 we easily obtain

Theorem 1': For some $m \in \{1,\ldots,n\}$, $t \geq 0$, $(-1)^{\nu_i} x_i \geq x_0$ $(i = m,\ldots,n)$, $|x_i| < +\infty$ $(i = 1,\ldots,m-1)$, let the inequalities $(-1)^{\nu_i} f_i(t, x_1,\ldots,x_{m-1}, |x_m|,\ldots,|x_n|) \geq a(t)|x_{i+1}|^{\lambda_1}$ $(i = m,\ldots,n)$ be fulfilled where $x_0 \geq 0, x_{n+1} = x_m$, $\nu_i = 0$ or 1, $\lambda_i > 0$ $(i = m,\ldots,n)$, $\prod_{i=m}^{n} \lambda_i > 1$ and $a(t)$ is a continuous and positive function in the interval $[0, +\infty)$. The system (1) has a singular solution $x_1(t),\ldots,x_n(t)$ such that $\lim_{t\to t^*} \sum_{i=m}^{n} |x_i(t)| = +\infty$

Theorem 2: For $t \geq 0$, $(-1)^{\nu_k} x_k \geq x_0 \geq 0$ $(k = 1,\ldots,n)$ let the inequalities (3) be fulfilled, where $\nu_i = 0$ or 1, $\lambda_i > 0$ $(i = 1,\ldots,n)$, $\prod_{i=1}^{n} \lambda_i > 1$ and $a_1(t) = \ldots = a_n(t) = a(t)$ is a continuous positive function in the interval $[0, +\infty)$, satisfying the condition

$$\int_0^{+\infty} a(t)\,dt = +\infty .$$

Then for any $t_0 \geq 0$, the arbitrary non-continuable to the right solution $x_1(t),\ldots,x_n(t)$ of system (1), satisfying the condition

$$(-1)^{\nu_i} x_i(t_0) > x_0 \quad (i = 1,\ldots,n)$$

is singular.

Proof: On account of Lemma 1, to prove the theorem it suffices to prove that any non-continuable to the right solution $y_1(t),\ldots,y_n(t)$ of the system

$$\frac{dy_i}{dt} = a(t)\, y_{i+1}^{\lambda_i} \quad (i = 1,\ldots,n;\ y_{n+1} = y_1), \tag{12}$$

satisfying the condition $y_i(t_0) > 0$ $(i = 1,\ldots,n)$, for some $\tilde{t}_2 > t_0$ satisfies also condition (10).

Introduce a new independent variable

$$\tau = \int_{t_0}^{t} a(s)\,ds$$

and assume that $y_i(t) = z_i(\tau)$ $(i = 1, \ldots, n)$. Then system (12) will have the form

$$\frac{dz_i}{dt} = z_{i+1}^{\lambda_i} \quad (i = 1, \ldots, n), \tag{13}$$

where $z_{n+1} = z_1$ Therefore the validity of this theorem follows from the following lemma.

Lemma 4: Any non-continuable to the right solution $z_1(\tau), \ldots, z_n(\tau)$ of system (13) satisfying the condition

$$z_i(\tau_0) > 0 \quad (i = 1, \ldots, n; \tau_0 \geq 0), \tag{14}$$

for some $\tau_1 > \tau_0$ satisfies also the condition

$$\lim_{\tau \to \tau_1} z_i(\tau) = +\infty \quad (i = 1, \ldots, n).$$

Proof: Suppose the contrary: system (13) has the solution $z_1(\tau), \ldots, z_n(\tau)$ defined in the interval $[\tau_0, +\infty)$ and satisfying condition (14). It is clear that

$$\lim_{\tau \to +\infty} z_i(\tau) = +\infty \quad (i = 1, \ldots, n).$$

Besides,

$$[z_2(\tau)]^{\lambda_1} z_n(\tau) \geq \int_{\tau_0}^{\tau} [z_2(s)]^{\lambda_1} dz_n(s) = \int_{\tau_0}^{\tau} [z_1(s)]^{\lambda_n} dz_1(s) \geq \frac{1}{2(\lambda_n+1)} [z_1(\tau)]^{\lambda_n+1}$$

when $\tau \geq \tau_2$,

where $\tau_2 > \tau_0$ is sufficiently large. In a similar way for sufficiently large τ_3 we have

$$[z_2(\tau)^{\lambda_1(\lambda_{n-1}+1)} z_{n-1}(\tau) \geq \int_{\tau_0}^{\tau} [z_2(s)]^{\lambda_1(\lambda_{n-1}+1)} dz_{n-1}(s) =$$

$$= \int_{\tau_0}^{\tau} [z_2(s)]^{\lambda_1(\lambda_{n-1}+1)} [z_n(s)]^{\lambda_{n-1}} ds \geq \frac{1}{[2(\lambda_n+1)]^{\lambda_{n-1}}} \int_{\tau_0}^{\tau} [z_1(s)]^{\lambda_{n-1}(\lambda_n+1)} \times$$

$$\times \, dz_1(s) \geq \frac{1}{2[2(\lambda_n+1)]^{\lambda_{n-1}}(\lambda_n\lambda_{n-1}+\lambda_{n-1}+1)} [z_1(\tau)]^{\lambda_n\lambda_{n-1}+\lambda_{n-1}+1} \quad \text{when } \tau \geq \tau_3.$$

By repeating this process n-1 times, we shall prove the existence of numbers $\tau_{n+1} > \tau_0$ and $\eta > 0$, such that

$$[z_2(\tau)]^{\lambda_1\beta_n} \geq \eta\,[z_1(\tau)]^{\alpha_n+\beta_n-1} \quad \text{for} \quad \tau \geq \tau_{n+1}$$

Therefore

$$[z_1(\tau)]^{-\frac{\alpha_n+\beta_n-1}{\beta_n}} \frac{dz_1(\tau)}{d\tau} \geq \eta \quad \text{when} \quad \tau \geq \tau_{n+1} \,.$$

Since $\alpha_n = \prod_{i=1}^{n} \lambda_i > 1$, by integrating this inequality from τ_{n+1} to τ we have

$$[z_1(\tau_{n+1})]^{-\frac{\alpha_n-1}{\beta_n}} \geq \frac{\alpha_n-1}{\beta_n}\,\eta(\tau-\tau_{n+1}) \quad \text{for} \quad \tau \geq \tau_{n+1}$$

The contradiction thus obtained proves the lemma.

Theorem 3: For $t \geq a$, $(-1)^{\nu_k} x_k \geq x_0$ $(k = 1,\ldots,n)$ let the following inequalities

$$(-1)^{\nu_i} f_i(t,x_1,\ldots,x_n) \geq \mathscr{S} t^{\mu_i} |x_{i+1}|^{\lambda_i} \quad (i = 1,\ldots,n)$$

be fulfilled, where x_0, a and $\mathscr{S}$ are some positive numbers, $\nu_i = 0$ or 1, $\lambda_i > 0$ $(i = 1,\ldots,n)$, $\prod_{i=1}^{n} \lambda_i > 1$, $x_{n+1} = x_1$,

$$1 + \mu_\ell + \sum_{j=1}^{n-1} (1+\mu_{j+\ell}) \prod_{k=\ell}^{j+\ell-1} \lambda_k > 0 \quad \text{for some} \quad \ell \in \{1,\ldots,n\} \tag{15}$$

(here $\mu_{n+i} = \mu_i$, $\lambda_{n+i} = \lambda_i$ $(i = 1,\ldots,n-1)$). Then for any $t_0 \in [a,+\infty)$, the arbitrary non-continuable to the right solution $x_1(t),\ldots,x_n(t)$ of system (1) satisfying the condition

$$(-1)^{\nu_i} x_i(t_0) > x_0 \quad (i = 1,\ldots,n)$$

is singular.

Proof: On account of Lemma 1 to prove the theorem it is sufficient to prove that any non-continuable to the right solution $y_1(t),\ldots,y_n(t)$ of the system

$$\frac{dy_i}{dt} = \mathscr{S} t^{\mu_i} y_{i+1}^{\lambda_i} \quad (i = 1,\ldots,n;\, y_{n+1} = y_1), \tag{16}$$

satisfying the condition $y_i(t_0) > 0$ $(i = 1,\dots,n)$, for some $\tilde{t} > t_0$, satisfies also the condition (10).

Assume the contrary: the system (16) has the solution $y_1(t),\dots,y_n(t)$ defined in the interval $[t_0, +\infty)$, for which $y_i(t_0) > 0$ $(i = 1,\dots,n)$. From (15) and (16) there follows the existence of a number $t_r > t_0$, such that

$$y_i(t) \geq t^r \quad \text{when} \quad t \geq t_r \quad (i = 1,\dots,n).$$

Therefore we may say that r is chosen so large that for $t \geq t_r$ we have

$$\frac{dy_i(t)}{dt} \geq [y_{i+1}(t)]^{\sigma_i} \quad (i = 1,\dots,n),$$

where $0 < \sigma_i < \lambda_i$ $(i = 1,\dots,n)$ and $\prod_{i=1}^{n} \sigma_i > 1$. From Lemmas 1 and 4 it easily follows that for some $\tilde{t} > t_r$ $\lim_{t \to \tilde{t}} y_i(t) = +\infty$ $(i = 1,\dots,n)$. The contradiction thus obtained proves the theorem.

From Theorems 2 and 3, theorems analogous to Theorem 1' can be obtained.

References

1 N.P. Erugin, The book for reading in differential equations. Nauka i Technika, Minsk (1972).

2 I.T. Kiguradze, The asymptotic properties of solutions of a nonlinear differential equation of Emden-Fowler type. Izv.Akad.Nauk SSSR, Ser.mat. 29 N5 (1965) 965-986.

3 A.V. Kostin, The existence and behaviour of solutions that have asymptotes in the case of second order nonlinear equations. Differencial'nye Uravnenija, 6 N12 (1970)2182-2192.

4 W. Walter, Differential and Integral Inequalities. Springer-Verlag Berlin, Heidelberg, New York (1970).

T.A. Chanturia
Institute of Applied Mathematics
Tbilisi University
Tbilisi U S S R

S CHRISTIANSEN

15 On Kupradze's functional equations for plane harmonic problems

ABSTRACT

When a plane, interior Dirichlet problem for the Laplace's equation is solved by means of Kupradze's functional equation, we have found that non-unique results may be obtained when the exterior mapping radius of the auxiliary curve is equal to one. However, we have demonstrated how the non-uniqueness can be eliminated by adding a supplementary integral condition. When the non-unique case has been removed the location of the auxiliary curve must be chosen carefully in order to obtain accurate results. The investigations are illustrated by several numerical examples.

1. INTRODUCTION

During the last decade Kupradze and his co-workers have developed the method of canonical functional equations for solving boundary value problems for different equations of mathematical physics, especially the equations from the theory of elasticity: Kupradze [23]-[26]; Kupradze and Aleksidze [27]; Kupradze et al. [26]. The method can also be applied to Laplace's equation, and much interest has been devoted to boundary value problems for this equation: [23] Chap. VIII, Section 14-17; [24] Chap. X, Section 12-27; [25] Section 3; [26] Section 2; [27] pp. 84-95, 105-121; [28] Section 12.10; Aleksidze [1], and Szefer [38].

It is the purpose of this paper to investigate the method of canonical functional equations when applied to an interior Dirichlet problem for the Laplace equation. We shall investigate whether this method of solving the problem always gives a unique solution to the problem, which we know has a unique solution.

The rest of the paper is organized as follows: The relevant functional equations (Section 2) are considered with respect to uniqueness (Section 3). Using a normalized conformal mapping a criterion for uniqueness has been found together with a supplementary condition which is to be imposed on an intermediate quantity in order to ensure that the solution is always unique (Section 4). An elliptical boundary is considered specifically (Section 5) and the equations are in this case replaced by a system of linear algebraic equations (Section 6); a circular boundary is treated separately (Section 7). Several numerical examples (Section 8) illustrate that the non-uniqueness is encountered in practical cases, but that it can be eliminated using the supplementary condition. Some problems about the choice of the auxiliary curve are considered (Section 9) and a discussion (Section 10) concludes the main part of the paper.

2. THE METHOD OF FUNCTIONAL EQUATIONS

We consider an interior, plane Dirichlet problem for Laplace's equation for a simply-connected domain D with a sufficiently smooth boundary curve Γ, along which the boundary value f of an harmonic function u is prescribed:

$$\Delta u(\bar{r}) = 0, \quad \bar{r} \in D \tag{2.1a}$$

$$u(\bar{r}) = f(\bar{r}), \quad \bar{r} \in \Gamma. \tag{2.1b}$$

The solution $u = u(\bar{r})$ of this problem exists and is unique, and the normal derivative of u along the exterior normal to Γ exists and is unique. We denote it by

$$\frac{\partial u}{\partial \nu}(\bar{r}) \equiv \psi(\bar{r}), \quad \bar{r} \in \Gamma. \tag{2.2}$$

Kupradze's method for solving harmonic problems starts from Green's third identity (Courant and Hilbert, [12] pp. 256-257), which for an harmonic function, $u = u(\bar{r})$ depending upon two space variables, $\bar{r} = (x, y)$, states that

$$\int_\Gamma \{G(\bar{r}_1, \bar{r}) \frac{\partial u}{\partial \nu}(\bar{r}) - \frac{\partial G}{\partial \nu}(\bar{r}_1, \bar{r})\, u(\bar{r})\}\, ds =$$

$$\begin{cases} u(\bar{r}_1), & \bar{r}_1 \text{ inside } \Gamma \quad (2.3a) \\ 0, & \bar{r}_1 \text{ outside } \Gamma \quad (2.3b) \end{cases},$$

where the elementary solution of the equation $\Delta u = 0$ in two dimensions is

$$G(\bar{r}_1, \bar{r}) = \frac{1}{2\pi} \ln \frac{1}{\rho} \tag{2.4a}$$

$$\rho = |\bar{r}_1 - \bar{r}| \, . \tag{2.4b}$$

Here $\frac{\partial}{\partial \nu}$ denotes differentiation with respect to $\bar{r}$ in the direction of the outward normal to the curve Γ in the point of integration $\bar{r}$ on Γ When $u(\bar{r}) = f(\bar{r})$ is prescribed, equation (2.3b), where $\bar{r}_1$ is placed outside Γ, is an integral relation ('functional equation') which can be used to determine $\frac{\partial u}{\partial \nu}(\bar{r}) = \psi(\bar{r})$, (2.2) When $\frac{\partial u}{\partial \nu}(\bar{r})$ in this way has been determined, $u(\bar{r})$ and $\frac{\partial u}{\partial \nu}(\bar{r})$, $\bar{r} \in \Gamma$, are inserted in (2.3a), and by means of quadrature the value of u in a point inside Γ can be determined.

3. UNIQUENESS OF THE FUNCTION ψ

A crucial step in the application of the Kupradze-method is the determination of the function $\psi(\bar{r})$ from the relation (2.3b). In the derivation of the identity (2.3) one assumes that the boundary values of both u and $\frac{\partial u}{\partial \nu}$ are known beforehand everywhere on Γ. This pair of functions is inserted in (2.3) under the sign of integration and one obtains the results as shown, especially the result zero when $\bar{r}_1$ is outside Γ, cf. (2.3b). When Green's third identity as here is used as a basis for derivation of functional equations, one should at this stage pose the following 'opposite' question: Given the value of the integral (2.3b) equal to zero, and let the function $u(\bar{r}) = f(\bar{r})$, $\bar{r} \in \Gamma$, be known, is it then possible to determine the integrand $\psi(\bar{r})$, $\bar{r} \in \Gamma$, uniquely? In order to make the calculations easier Kupradze confines the point $\bar{r}_1$ to lie on an auxiliary curve Γ_1 which completely surrounds the curve Γ without touching it. Kupradze states, that this curve Γ_1 may be chosen arbitrarily: [24] p.264, [27] p.115, p.118. Here we shall investigate whether $\psi(\bar{r})$ always can be determined uniquely when the auxiliary curve Γ_1 is chosen arbitrarily.

To this end we consider the relation (2.3b) in the following way: The second integral is a double layer potential $W(\bar{r}_1)$ with a known source density $f(\bar{r}) = u(\bar{r})$

$$W(\bar{r}_1) = \int_{\Gamma} \frac{\partial G}{\partial \nu}(\bar{r}_1, \bar{r}) \, f(\bar{r}) \, ds \, , \tag{3.1}$$

such that $W(\bar{r}_1)$ can be computed everywhere outside Γ. The first integral is a logarithmic single layer potential $V(\overline{r}_1)$ with the unknown source density $\psi(\overline{r})$ ($= \frac{\partial u}{\partial \nu}(\overline{r})$?)

$$V(\overline{r}_1) = \int_\Gamma G(\overline{r}_1, \overline{r}) \, \psi(\overline{r}) \, ds \tag{3.2}$$

such that $V(\overline{r}_1)$ could be computed everywhere outside Γ, if $\psi(\overline{r})$ were known.

Now the problem is not whether $W(\overline{r}_1)$ and $V(\overline{r}_1)$ coincide everywhere outside Γ. But the problem has here been reduced to the requirement that $V(\overline{r}_1)$ along some arbitrary curve Γ_1 attains the value of $W(\overline{r})$, $\overline{r}_1 \in \Gamma_1$. The problem is now whether this method gives a unique determination of the source density $\psi(\overline{r})$. Or, with other words, whether a non-trivial single layer source density $\tilde{\psi}(\overline{r})$ ($\not\equiv 0$) exist, such that the corresponding single layer potential $\tilde{V}(\overline{r}_1)$

$$\tilde{V}(\overline{r}_1) = \int_\Gamma G(\overline{r}_1, \overline{r}) \, \tilde{\psi}(\overline{r}) \, ds \tag{3.3}$$

attains the value zero along some curve $\tilde{\Gamma}_1$, i.e.: $\tilde{V}(\overline{r}_1) \equiv 0$, $\overline{r}_1 \in \tilde{\Gamma}_1$. If such a curve $\tilde{\Gamma}_1$ exists, and the two potentials are compared along such a curve, the value of $\psi(\overline{r})$, determined in this way, need not to be correct, because a multiple of the spurious source density $\tilde{\psi}(\overline{r})$ can always be added to $\psi(\overline{r})$. If a multiple of $\tilde{\psi}(\overline{r})$ is added to $\psi(\overline{r})$ in the equation (2.3a) the result of the quadrature need not to be the correct value of $u(\overline{r}_1)$, $\overline{r}_1$ inside Γ.

When the boundary curve Γ is a circle C the calculations of Sections 2-3 can be carried out explicitly; see Appendix A.

4. SINGLE LAYER POTENTIAL AND CONFORMAL MAPPING

In order to find whether a curve $\tilde{\Gamma}_1$ exists along which a single layer potential takes on the value zero, we consider the features of a single layer potential v with density ϕ

$$v(\overline{r}_1) = \int_\Gamma \ln|\overline{r}_1 - \overline{r}| \, \phi(\overline{r}) \, ds . \tag{4.1}$$

The function $v(\overline{r}_1)$ is harmonic outside Γ, except possibly at $|\overline{r}_1| = \infty$, and for $|\overline{r}_1| \to \infty$, v behaves like

$$\lim_{|\bar{r}_1| \to \infty} v(\bar{r}_1) = \ln |\bar{r}_1| \int_{\Gamma} \phi(\bar{r})\, ds + 0(|\bar{r}_1|^{-1}) \tag{4.2}$$

(Muschelischwili [33] p.237, Note 1) which shows that an exterior single layer potential never contains a constant term. Only when the total source density is different from zero, i.e. when

$$\int_{\Gamma} \phi(\bar{r})\, ds \neq 0 , \tag{4.3}$$

a logarithmic component does appear.

The problem about harmonic functions generated by different single layer potentials can better be investigated when the domain outside the curve Γ in the complex z-plane is mapped onto the domain outside a circle C with radius d in the complex w-plane by means of the so-called normalized conformal mapping function, which for $|z| \to \infty$ has the form (Pólya and Szegö [37] p.17)

$$w = w(z) = z + k_0 + \frac{k_1}{z} + \frac{k_2}{z^2} + \dots . \tag{4.4}$$

The radius d is called the 'exterior (mapping)radius', which is equal to the 'transfinite diameter' (Pólya and Szegö [35] Section 1; [36]; Hille [19] Sections 16,17). A family of concentric circles C_1' with radii greater than d in the w-plane corresponds to a family of closed, non-intersecting curves Γ_1' in the z-plane enclosing the curve Γ.

An harmonic function in the w-plane which outside a circle with radius d behaves like v (4.2) can be expressed as a linear combination of terms of the following form:

$$\ln r \tag{4.5a}$$

$$r^{-n} \cos n\theta , \; r^{-n} \sin n\theta ; \; n = 1,2,3,\dots , \tag{4.5b}$$

where $(r,\theta) = (|w|, \mathrm{Arg} w)$ are polar coordinates.

The logarithmic term (4.5a) takes on constant values along circles in the w-plane, and no combination of decaying terms (4.5b) takes on a constant value along some simple, closed curve in the w-plane. The only possibility to obtain a constant solution along a curve is to have solely a logarithm, and a logarithm is generated by an exterior single layer potential if and only if the total source density is different from zero, cf. (4.3).

In the w-plane a logarithmic term $\ln|w|$ takes on constant values along the circles C_1', and on the corresponding curves Γ_1' in the z-plane, the function

$$\ln|w(z)|, \tag{4.6}$$

where $w(z)$ is the mapping function (4.4), takes on constant values. The function (4.6) corresponds with the logarithmic term (4.5a). The potential $\tilde{V}$ is characterized thereby that it takes on a constant value (namely zero) along a curve $\tilde{\Gamma}_1$ surrounding Γ. Therefore, if the spurious exterior single layer potential $\tilde{V}(\bar{r}_1)$, (3.3), is present, it can only be made up of a multiple of $\ln|w(z)|$, where $\bar{r}_1 \sim z$, cf. (4.6). Such a logarithmic component is generated by the single layer $\tilde{V}(\bar{r}_1)$, (3.3), only when the total source density of $\tilde{V}(\bar{r}_1)$ is different from zero, cf. (4.3). This characterizes the potential $\tilde{V}(\bar{r}_1)$. Due to the simple structure of the mapping function $w(z)$, (4.4) we have

$$\lim_{|z|\to\infty} \ln|w(z)| = \ln|z| + 0(|z|^{-1}),$$

and noticing the structure of a single layer potential (4.2), when the total source density is different from zero, we infer that the spurious single layer potential $\tilde{V}(\bar{r}_1)$, if it exists, can be written, cf. (3.3)

$$\tilde{V}(\bar{r}_1) = \frac{-1}{2\pi} \ln|w(z)| \int_\Gamma \tilde{\psi}(\bar{r})\, ds, \quad \text{for } \bar{r}_1 \sim z \text{ outside } \Gamma. \tag{4.7}$$

Using this simple formula, it is now - in principle - a simple matter to find whether a curve $\tilde{\Gamma}_1$ exists where $\tilde{V}(\bar{r}_1)$ is equal to zero, when $\bar{r}_1 \in \tilde{\Gamma}_1$:

The curve Γ is mapped onto C using (4.4), and when the radius d is smaller than one, there exists a circle $\tilde{C}_1$ in the w-plane with radius 1, which corresponds to a curve $\tilde{\Gamma}_1$ in the z-plane. If $d \geq 1$ no curve $\tilde{\Gamma}_1$ exists. On the circle C the value of the harmonic function $\ln|w|$ is equal to $\ln d$, and on $\tilde{C}_1$ it is $\ln 1 = 0$. This means that on the curve $\tilde{\Gamma}_1$, corresponding to $\tilde{C}_1$, the value of the potential $\tilde{V}(\bar{r}_1)$ is equal to zero, according to the relation (4.7).

If the comparison curve Γ_1 is chosen different from the critical auxiliary curve $\tilde{\Gamma}_1$ (if it exists), the determination of ψ can be carried out uniquely from a theoretical point of view (but from a practical-computational viewpoint there may still be difficulties when Γ_1 is chosen in the neighbourhood of $\tilde{\Gamma}_1$). For a given curve, Γ, it is possible to determine the exterior mapping radius, d, numerically, but it is a complicated process to carry out in practice (Gaier [15] IV, 4.2); then furthermore to find the critical curve $\tilde{\Gamma}_1$ (if $d < 1$) is also troublesome.

However, the problems with a critical auxiliary curve, and how to choose the auxiliary curve sufficiently far away from the critical, can easily be circumvented:

The unwanted single layer potential $\tilde{V}(\bar{r}_1)$, (3.3) is characterized by the fact that its source density has a total magnitude different from zero:

$$\int_{\Gamma} \tilde{\psi}(r)\, ds \neq 0$$

cf. (4.3), while the correct single layer potential $V(\bar{r}_1)$, (3.2) is generated by a source density ψ, which <u>must</u> be equal to the normal derivative of a function u, cf. (2.2), which is harmonic inside the curve Γ, cf. (2.1a), such that Gauss' theorem gives

$$\int_{\Gamma} \psi(\bar{r})\, ds = \int_{\Gamma} \frac{\partial u}{\partial \nu}(\bar{r})\, ds = \iint_{D} \Delta u(\bar{r})\, da = 0 . \qquad (4.8)$$

Thus the correct source density $\psi(\bar{r})$ generates an exterior single layer potential which behaves like

$$\lim_{|\bar{r}_1| \to \infty} V(\bar{r}_1) = 0(|\bar{r}_1|^{-1}), \qquad (4.9)$$

cf. (4.2). We notice the difference between (4.7) and (4.9). This difference we use to derive a procedure by which we exclude the unwanted spurious solution while the wanted solution is retained: When Kupradze's functional equation is solved, we simply require that the intermediate solution ψ must satisfy the <u>supplementary</u>

condition

$$\int_{\Gamma} \psi(\bar{r})\, ds = 0 \,. \tag{4.10}$$

Notice, however, since (4.10) always has to be satisfied - cf. (4.8) - and since it is automatically satisfied by the solution to the functional equation if the auxiliary curve does not coincide with the critical one, we can always impose (4.10).

When an interior Neumann boundary value problem is formulated, the prescribed boundary value $\frac{\partial u}{\partial \nu}$ has always to satisfy a condition like (4.10). Such a Neumann problem can be solved in the usual manner starting from a single layer potential and arriving at a Fredholm integral equation. In this case the prescribed known function always has to satisfy a solvability condition which has the form of an orthogonality condition like (4.10). For the Dirichlet problem here considered we arrive at a condition (4.10) which the unknown function must satisfy in some special case when a critical auxiliary curve is used.

5. AN ELLIPTICAL BOUNDARY CURVE

Most of the published numerical examples for Laplace's equation are concerned with a Dirichlet problem for an elliptical boundary curve ([23] Chap. VIII, Section 17; [24] Chap. X, Section 16; [27] pp. 112-121; [28] pp. 515-517). Therefore we here introduce an elliptical boundary curve:

$$\Gamma: \quad x = a\cos\theta \,, \quad y = b\sin\theta \,, \quad 0 \le \theta \le 2\pi \tag{5.1a}$$

and an elliptical auxiliary curve Γ_1, which is concentric with Γ, but not necessarily confocal with it:

$$\Gamma_1: \quad x_1 = a_1\cos\theta_1 \,, \quad y_1 = b_1\sin\theta_1 \,, \quad 0 \le \theta_1 \le 2\pi \,. \tag{5.1b}$$

Elliptical coordinates (μ, θ) are introduced

$$\left.\begin{aligned} x &= c\cosh\mu\cos\theta \\ y &= c\sinh\mu\sin\theta \end{aligned}\right\} \quad \left\{\begin{aligned} &0 < \mu \\ &0 \le \theta \le 2\pi \,. \end{aligned}\right. \tag{5.2}$$

An exterior solution of Laplace's equation in elliptical coordinates (which can be found using Moon and Spencer [32] pp. 17-19) contains a term independent of θ, namely $(\mu + D)A_0$, where D and A_0 are constants. Besides the constant term DA_0 also the term μA_0 contains a constant component: if we let $r = (x^2 + y^2)^{\frac{1}{2}}$ tend to infinity in (5.2), we see that μ behaves like

$$\mu = \ln r - \ln \frac{c}{2} + \cdots .$$

Consequently an exterior solution generated by a single layer potential, i.e. a solution without a constant term, has the form:

$$(\mu + \ln \frac{c}{2})A_0 + \sum_{n=1}^{\infty} e^{-n\mu}(A_n \cos n\theta + B_n \sin n\theta) . \qquad (5.3)$$

The sum $\mu + \ln \frac{c}{2}$, in (5.3), takes on constant values along confocal ellipses, i.e. along ellipses with semiaxes (a', b') where

$$a' = c \cosh\mu , \qquad b' = c \sinh\mu$$

and

$$\mu = \frac{1}{2} \ln \frac{a' + b'}{a' - b'} ,$$

and the value is

$$\mu + \ln \frac{c}{2} = \ln \frac{a' + b'}{2} .$$

This shows us that the unwanted exterior potential, cf. (3.3), is

$$\tilde{V}(\bar{r}_1) = A_0 \ln \frac{a' + b'}{2}$$

while the sought exterior potential, generated by a source distribution with an integral equal to zero, cf. (3.2), is

$$V(\bar{r}_1) = \sum_{n=1}^{\infty} e^{-n\mu}(A_n \cos n\theta + B_n \sin n\theta) .$$

From the well-known expression for a conformal mapping of the exterior of an ellipse with semiaxes (a', b') on to the exterior of a unit circle (Lawrentjew and Schabat [30] pp.186-187), the normalized mapping is found to be

$$w(z) = \frac{z + \sqrt{(z^2 - c^2)}}{a' + b'} \cdot \frac{a' + b'}{2} , \tag{5.4}$$

with $c^2 = a'^2 - b'^2$. For $|z| \to \infty$ this mapping function has the expansion, cf. (4.4)

$$w(z) = z - \frac{c^2/4}{z} + \dots .$$

The function $w(z)$ maps the ellipse (a', b') on to a circle with radius d' equal to

$$d' = \frac{a' + b'}{2}$$

in accordance with the expression for the transfinite diameter or the exterior mapping radius of an ellipse ([35] pp.8-9; [36] pp.247 f.). By means of the mapping (5.4) confocal ellipses in the z-plane are connected with concentric circles in the w-plane. When for the boundary ellipse Γ the sum of the semiaxes, $a+b$, is smaller than 2, there exists an auxiliary ellipse $\tilde{\Gamma}_1$, confocal with Γ, with semiaxes $\tilde{a}_1$ and $\tilde{b}_1$ where

$$\tilde{a}_1 + \tilde{b}_1 = 2 ; \tag{5.5}$$

thereby the critical auxiliary ellipse is characterized.

Finally we write the (functional) equation (2.3b) and the supplementary condition (4.10), using explicitly the elliptical geometry (5.1), and we obtain (cf. [23] p.252; [24] p.261; [27] p.115)

$$\frac{1}{2\pi} \int_0^{2\pi} \ln\rho^2 \cdot \Psi(\theta)\, d\theta = \frac{1}{2\pi} \int_0^{2\pi} \frac{2}{\rho^2} (ab - a\sin\theta \cdot y_1 - b\cos\theta \cdot x_1)\, F(\theta)\, d\theta , \tag{5.6a}$$

where

$$\rho^2 = (x_1 - a\cos\theta)^2 + (y_1 - b\sin\theta)^2 . \tag{5.6b}$$

It is to be noticed that the known function is $F(\theta) \equiv f(\bar{r})$, while the unknown function is $\Psi(\theta) = \frac{\partial u}{\partial \mu}$, different from the original unknown function

$$\psi(\bar{r}) = \frac{\partial u}{\partial \nu} = \frac{\partial u}{\partial \mu}\frac{\partial \mu}{\partial \nu} .$$

Similarly the supplementary condition (4.10) is transformed to

$$\int_0^{2\pi} \Psi(\theta)\, d\theta = 0 . \tag{5.7}$$

In (5.6) and (5.7) use has been made of the relation $\frac{\partial \mu}{\partial \nu}\frac{\partial s}{\partial \theta} = 1$, where the arc-length is $s = s(\theta)$.

6. NUMERICAL SOLUTION

The functional equation (5.6) may be solved by means of two different methods ([27] p.83):

(i) Some complete orthonormalized system of functions is constructed and the coefficients of the expansion of the unknown function in a Fourier series with respect to this system are found.

(ii) The functional equation is replaced by a system of algebraic equations, using some mechanical quadrature formulae, and the solution of this system gives approximate values of the unknown function at separate points of Γ. Kupradze has used a 16 point Gaussian quadrature ([23] Chap. VIII, Section 17; [24] Chap.X, Section 16; [27] pp. 112-121; [28] pp. 515-517).

Here we use the second method, the quadrature-method. Because the integrands are continuous and periodic functions, it is - in general - possible to obtain a high accuracy by means of the rectangular rule (the trapezoidal rule, Hämmerlin [18]). If the integrand has singularities of higher order, a Gaussian quadrature formula may give more accurate results (Davis [13]), but due to its simplicity we use the rectangular rule. Therefore on each of the two curves N points are chosen:

$$\left.\begin{aligned} &\text{On } \Gamma: \ \theta_j := (j-1)h, \quad j = 1,2,\ldots,N \\ &\text{On } \Gamma_1: \ \theta_{1j} := (i-1)h, \quad j = 1,2,\ldots,N \end{aligned}\right\} \quad h = \frac{2\pi}{N} . \tag{6.1}$$

The value of $F(\theta)$ is given as a Fourier series

$$F(\theta) = a_0 + \sum_{n=1}^{\infty} (a_n \cos n\theta + b_n \sin n\theta) , \tag{6.2}$$

and when the coefficients are prescribed, the value of the known quantities $F_j := F(\theta_j)$ can be determined.

In letting the point $\overline{r}_1 = (x_1, y_1)$ consecutively take on the values

$$x_{1i} = a_1 \cos\theta_{1i} , \quad y_{1i} = b_1 \sin\theta_{1i}$$

we arrive at the system of linear algebraic equations

$$\sum_{j=1}^{N} A_{ij} \Psi_j = k_i ; \qquad i = 1,2,\ldots,N , \tag{6.3}$$

where

$$A_{ij} = \ln\rho_{ij}^2 \tag{6.4a}$$

$$k_i = \sum_{j=1}^{N} \frac{2}{\rho_{ij}^2} (ab - a\sin\theta_j \cdot y_{1i} - b\cos\theta_j \cdot x_{1j}) F_j \tag{6.4b}$$

$$\rho_{ij}^2 = (x_{1j} - a\cos\theta_j)^2 + (y_{1j} - b\sin\theta_j)^2 \tag{6.4c}$$

The construction of the matrix $\{A_{ij}\}$ is very simple, when (a,b) and (a_1,b_1) are given, and at the same time the right-hand side k_i can be evaluated.

The supplementary condition (5.7) is transformed to the sum

$$h \sum_{j=1}^{N} \Psi_j = 0$$

or

$$\sum_{j=1}^{N} T \Psi_j = 0 , \tag{6.5}$$

where T is a constant to be chosen suitably; see below.

The complete system, i.e. (6.3) and (6.5), which is $N+1$ equations with N

unknowns, is solved by a standard least squares routine [20], which is based on [16].

The magnitude of T determines how strongly the condition (6.5) is taken into account: A small value of $|T|$ means that the condition has only a moderate effect, and a large value of $|T|$ means that the condition is forced strongly into play, which may give less accurate results because the complete system, (6.3) and (6.5), probably is incompatible due to quadrature errors. Therefore a compromise with respect to the magnitude of T is to be found. In Section 7 it is explained why the value of T

$$T = \sqrt{(N)} \cdot \frac{1}{N^2} \sum_{\substack{j=1\\i=1}}^{N} |A_{ij}| \tag{6.6}$$

is chosen.

The exact solution $\Psi(\theta)$ can be found as a Fourier-series

$$\Psi(\theta) = A_0 + \sum_{n=1}^{\infty} (A_n \cos n\theta + B_n \sin n\theta) \tag{6.7}$$

where the coefficients can be expressed in terms of the coefficients of $F(\theta)$, (6.2). The connection is derived in Appendix B. From $\Psi(\theta)$ the exact values $\Psi_j := \Psi(\theta_j)$ can be found and used for a comparison with the approximate values of Ψ_j determined by the least squares method from the linear algebraic equations.

7. A CIRCULAR BOUNDARY CURVE

Let the boundary curve Γ and the auxiliary curve Γ_1 be two concentric circles with the radii a and a_1 respectively $(a < a_1)$. On each curve N points are chosen according to (6.1). The boundary condition is assumed to be $F(\theta) \equiv 1$, which gives $\Psi(\theta) \equiv 0$. Due to the simplicity of the geometry the matrix has a rotationally symmetric structure, such that the sum of the elements in a row is the same for all rows, cf. (6.4a)

$$R_N := \sum_{j=1}^{N} A_{ij} \; ; \quad i = 1,2,\ldots,N, \tag{7.1a}$$

or

$$R_N = \sum_{j=1}^{N} \ln (a_1^2 + a^2 - 2a_1 a\cos (2\pi j/N)) \ , \tag{7.1b}$$

which may be summed in closed form ([22], Formula 1077)

$$R_N = 2 \ \ln (a_1^N - a^N) \tag{7.1c}$$

or when the ratio $\alpha = a_1/a \ (>1)$ is introduced we have

$$R_N = 2N \ \ln \ a_1 + 2 \ \ln (1 - \alpha^{-N}) \ . \tag{7.1d}$$

Due furthermore to the constant boundary condition all the elements on the right-hand side are equal: $k_1 = k_2 = \cdots = k_N$; the common value is denoted by k_N. The element is

$$k_N = 2a \sum_{j=1}^{N} \frac{a - a_1 \cos(2\pi j/N)}{a_1^2 + a^2 - 2a_1 a\cos(2\pi j/N)} \tag{7.2a}$$

which may be summed in closed form ([22], Formulas 489 and 490 †)

$$k_N = - \frac{2Na^N}{a_1^N - a^N} \ , \tag{7.2b}$$

or when the ratio $\alpha = a_1/a \ (>1)$ is introduced we have

$$k_N = - \frac{2N}{\alpha^N - 1} \ . \tag{7.2c}$$

Due to the symmetry of the system of equations (6.3) all the unknowns are equal: $\Psi_1 = \Psi_2 = \cdots = \Psi_N$; the common value is denoted by Ψ_N. The solution Ψ_N is

$$\Psi_N = \frac{k_N}{R_N} \ . \tag{7.3}$$

It is impossible to determine the solution Ψ_N uniquely when $R_N = 0$. For a fixed value of $\alpha > 1$ this case appears when $a_1 = a_1^*$, where - according to (7.1d)

---

† Professor Erik Hansen is acknowledged for pointing these formulas out.

$$a_1^* = \frac{\alpha}{(\alpha^N - 1)^{\frac{1}{N}}} \tag{7.4}$$

It is seen that a_1^* is greater than one, but - for N sufficiently large - a_1^* is very near one. This is in accordance with the fact that a circle with radius a_1 has an exterior mapping radius d equal to a_1 ([35] p.8; [36] pp. 247 f), and the critical auxiliary curve has exterior mapping radius equal to one. This result is illustrated further by elementary calculations in Appendix A.

If we solve the equations (6.3) together with the auxiliary condition (6.5) by means of the method of least squares, and we assume that all the unknowns are equal to Ψ_N, the sum of the squares of the residuals is

$$S = N(R_N \Psi_N - k_N)^2 + (NT\Psi_N)^2 .$$

For a fixed value of T the sum S is minimum for

$$\Psi_N = \frac{k_N R_N}{R_N^2 + NT^2} . \tag{7.5}$$

For a fixed value of T this approximate value of Ψ_N, which should have been equal to zero, attains its largest absolute value for

$$R_N = \pm \sqrt{(N)}\ |T| , \tag{7.6}$$

which gives an absolute maximal value of Ψ_N equal to

$$|\Psi_N|_{max} = \frac{k_N}{2R_N} , \tag{7.7}$$

which is half the value attained if the supplementary condition is disregarded, cf. (7.3).

When $T \neq 0$ the solution Ψ_N in (7.5) can be determined uniquely when $R_N = 0$, contrarily to the solution Ψ_N in (7.3), but a small value of $T\ (\neq 0)$ gives a large value of $|\Psi_N|_{max}$, cf. (7.6) and (7.7). Therefore a reasonably large value of $|T|$ should be chosen. However, numerical experiments show that a too large value of $|T|$ gives an alternating solution of the form $(1, -1, 1, -1, \cdots, -1)$. Such a solution gives a sum of the squares, which is greater than the least sum of the squares for the solution $(1,1,\dots,1)$. Therefore numerical instabilities seem to bring the alternating solution into play.

Consequently it seems wise to choose a value of T which is somewhat 'intermediate'; this requirement seems only to make sense when the value of T is related to the value of the matrix elements A_{ij}. The relation (7.6) leads to the choice

$$T = \sqrt{(N)} \cdot \frac{R_N}{N} = \sqrt{(N)} \cdot \frac{1}{N} \sum_{j=1}^{N} A_{ij} ,$$

but this expression does not neutralize the exceptional case $R_N = 0$. This is obtained by the formula

$$T = \sqrt{(N)} \cdot \frac{1}{N} \sum_{j=1}^{N} |A_{ij}|$$

which can be generalized to a matrix without symmetry, whereby we are lead to apply the value of T given in (6.6).

8. NUMERICAL EXAMPLES

The problems about the non-unique solution for certain auxiliary curves, and the method for solving this problem, is here illustrated by solving some different problems both with and without the supplementary condition (6.5).

8.1. Concentric circles: The boundary curve is a circle with radius a, and the auxiliary curve a circle with radius a_1. The ratio between a and a_1 is kept fixed: $a_1 = \alpha a$, $\alpha > 1$. The boundary condition is $F(\theta) \equiv 1$, which gives the exact solution $\Psi(\theta) \equiv 0$. The factor k_N is common for the two methods of solutions, (7.3) and 7.5), and it is here neglected. For a fixed value of α, the solutions with and without the supplementary condition are determined as a function of α_1. These 'solutions' can be found using elementary calculations (without having to form and solve linear algebraic equations).

Without $(-)$, cf. (7.3):

$$\Psi_{\bar{N}}/k_N = R_N^{-1} , \tag{8.1a}$$

With (+), cf. (7.5):

$$\Psi_N^+/k_N = \frac{R_N}{R_N^2 + NT^2} \quad , \tag{8.1b}$$

where

$$R_N = \sum_{j=1}^{N} E_j(\alpha, a_1)$$

$$T = \sqrt{(N)} \cdot N^{-1} \sum_{j=1}^{N} |E_j(\alpha, a_1)|$$

and

$$E_j(\alpha, a_1) = \ln(a_1^2 + (a_1/\alpha)^2 - 2a_1(a_1/\alpha)\cos(2\pi j/N)) \; .$$

In Fig. 1 the two results (8.1a) and (8.1b) are shown as a function of a_1, for $0.5 \leq a_1 \leq 1.5$, and for $N = 8$ and $\alpha = 10/9$. Here it is clearly seen that without the supplementary condition (6.5) the result (8.1a) becomes infinite for $a_1 = a_1^*$, which is equal to 1.07 determined from (7.4), while with the condition (6.5) the result (8.1b) is exactly equal to zero for $a_1 = a_1^*$.

8.2. Confocal ellipses: The boundary curve is an ellipse with semiaxes a and b, where $b = \epsilon a$ with $0 < \epsilon \leq 1$, and the auxiliary curve is a confocal ellipse with semiaxes a_1 and b_1, where $b_1 = \epsilon_1 a_1$, with $0 < \epsilon_1 \leq 1$ The ratio between the two curves is kept fixed, e.g. let $a_1 = \alpha a$, $\alpha > 1$. Then using $a^2 - b^2 = a_1^2 - b_1^2$ one has

$$\epsilon_1 = \sqrt{(1 - \frac{1 - \epsilon^2}{a^2})} \; ,$$

and the critical exterior major semiaxis $\tilde{a}_1$, for which $\tilde{a}_1 + \tilde{b}_1 = 2$, cf. (5.5) is

$$\tilde{a}_1 = \frac{2}{1 + \sqrt{(1 - \frac{1-\epsilon^2}{\alpha^2})}} \; . \tag{8.2}$$

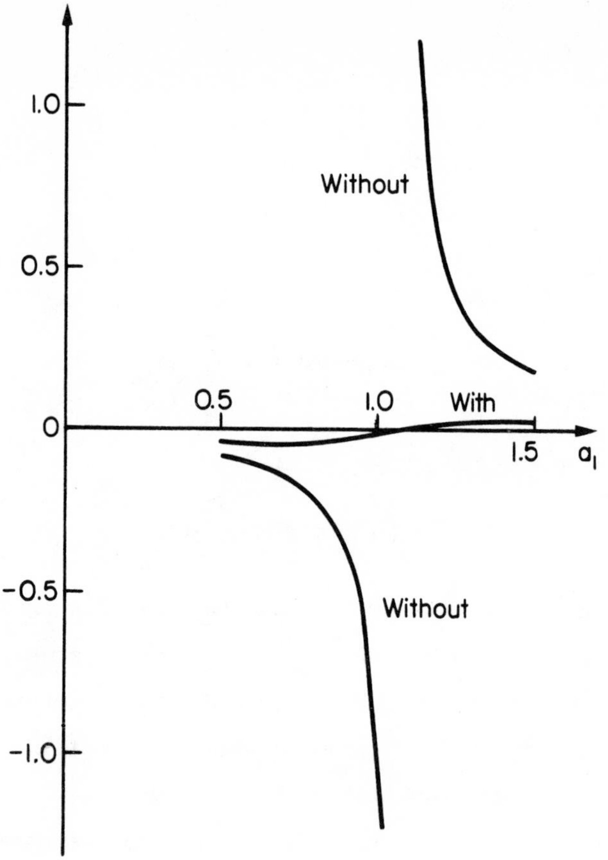

Fig. 1. Concentric circles with radii a and a_1, cf. Section 8.1. The result (8.1a) without the supplementary condition (6.5), and the result (8.1b) with the condition as a function of a_1 with $N = 8$, and $a_1/a = 10/9$.

The boundary condition is $F(\theta) \equiv 1$, which gives $\Psi(\theta) \equiv 0$ such that the approximate results are equal to the error. When $\epsilon = 0.5$ and $\alpha = 10/9$ the equations have been solved without and with the condition (6.5) for $N = 4,8,16$ and 32. It is found that the N approximate results obtained are equal except for rounding errors. When the approximate results without the condition are plotted as a function of a_1, the common feature is that they diverge for a_1 near the value $\tilde{a}_1 = 1.23$, which is found from (8.2). See the case $N = 16$ and $0.1 \leq a_1 \leq 2.5$ shown in Fig. 2.

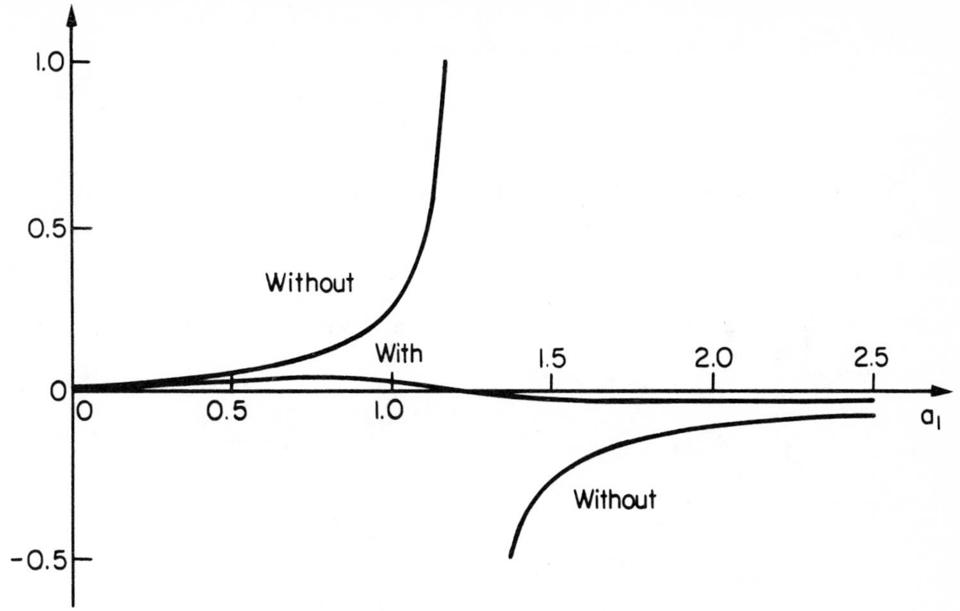

Fig. 2. Confocal ellipses with semiaxes (a,b) and (a_1, b_1), cf. Section 8.2. The result without and with the supplementary condition (6.5) when $b/a = 0.5$, and $a_1/a = 10/9$, as a function of a_1 with $N = 16$.

The effect of the critical magnitude is somewhat smeared out in the neighbourhood of the exact critical magnitude. The absolute value of the error goes to infinity in a zone around $\tilde{a}_1$, and this zone becomes narrower when N increases, and (outside the zone) the error decreases with N. For $N \to \infty$ the peak would be infinitely thin corresponding to the exact critical magnitude (if not other effects would come into play, cf. Section 9). The approximate solutions with the condition do not - as a function of a_1 - exhibit a critical peak; on the contrary the solution is equal to zero when the omission of the supplementary condition gives an infinite result. The zero-solution is obtained for slightly different values of a_1 due to the quadrature error:

N:	4	8	16	32
a_1:	1.45	1.28	1.23	1.23 .

8.3. Concentric ellipses: If the boundary curve is an ellipse we have shown (Section 5) that a critical auxiliary curve is an ellipse confocal with the boundary ellipse and satisfying the condition (5.5). This has been demonstrated in Section 8.2. But there it was also found that application of an auxiliary curve somewhat different from the critical one may give inaccurate results. As a series of similar examples we now consider an elliptical boundary with concentric but not confocal, auxiliary curve. Thereby no critical curve exists, even though some auxiliary curves may be only somewhat different from the confocal critical one, whereby bad numerical results are to be expected.

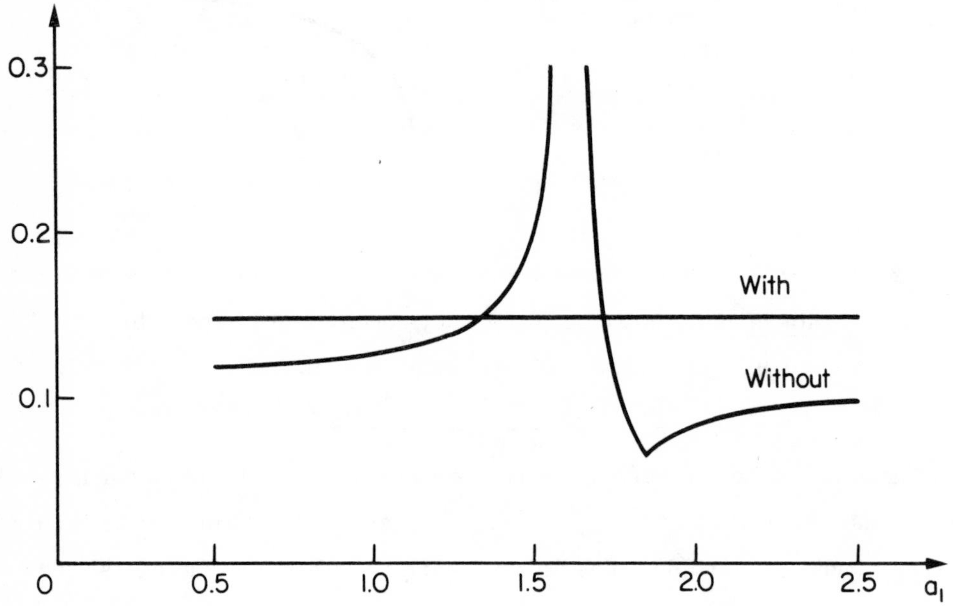

Fig. 3. Concentric ellipses, cf. Section 8.3.1. The absolute value of the absolute largest error without and with the supplementary condition (6.5) as a function of a_1, when $b/a = 0.5$, $b_1/a_1 = 0.5$, $a_1/a = 2$.

8.3.1. Example 1: Geometry: $\epsilon = 0.5$, $\epsilon_1 = 0.5$, $\alpha = 2$; cf. Section 8.2. When $a_1 = 2$ this gives a geometry which has been considered by Kupradze ([23] Chap. VIII, Section 17, [24] Chap.X,Section 16), Kupradze and Aleksidze ([27]

pp.112-121) and Kupradze et al. ([28] pp.515-517). We choose the boundary condition $F(\theta) \equiv 1$, which gives the exact result $\Psi(\theta) = 0$, such that the approximate results are equal to the error. The N approximate function values are not equal (as in the previous cases), and the absolute largest value is considered as representative. In Fig. 3 the absolute largest value is shown, both without and with the supplementary condition (6.5), as a function of a_1, where $0.5 \leq a_1 \leq 2.5$. When the supplementary condition is imposed the error is independent of a_1 (to within the accuracy of the figure); the problem is invariant to a scaling of all lengths. Contrary to this, when the problem is solved without the condition, the error depends strongly upon a_1: Somewhere in the interval $1.33 < a_1 < 1.72$ the error without condition becomes extremely large, and in this interval the error without the condition (6.5) is greater than the error with the condition.

8.3.2. Example 2: Geometry: $a = 1$, $b = 0.5$.

Boundary condition: $f = 0.5\,(x^2 + y^2)$, which gives the interior harmonic solution: $u = u(x,y) = 0.2 + 0.3\,(x^2 - y^2)$. This case has been considered by Kupradze ([23] Chap. VIII, Section 17, [24] Chap. X, Section 16), Kupradze and Aleksidze ([27] pp. 112-121) and Kupradze et al. ([28] pp.515-517).

Using elliptical coordinates, cf. (5.2), we find $\mu = \ln\sqrt{(3)}$ and $c = \frac{\sqrt{(3)}}{2}$, and the interior solution is

$$u = \frac{5}{16} + \frac{1.8}{16} \cosh 2\mu \cdot \cos 2\theta \; .$$

The prescribed boundary condition is

$$F(\theta) = \frac{5}{16} + \frac{3}{16} \cos 2\theta \qquad (8.3)$$

with the exact solution, cf. (6.7)

$$\Psi(\theta) = \frac{3}{10} \cos 2\theta \; , \qquad (8.4)$$

which also could be found from (8.3) using [7].

Kupradze has solved this problem using several different auxiliary curves ('auxiliary ellipses'), namely $(a_1,b_1) = (2,1), (5,3), (3,1.5), (4,2), (7,3.5)$. For the two first comparison curves there has been given intermediate results, i.e. values of a function ϕ (which seems to be equal to $\frac{1}{2} \cdot \frac{\partial u}{\partial \nu}$). This function ought to be independent of the auxiliary curve, because the function ϕ is related to the normal derivative of the potential problem evaluated at the boundary. This normal derivative is to be independent of the auxiliary curve. The intermediate results given ought to be the value of this function evaluated at some fixed points on the boundary.

These two sets of function values are given in several references ([23] Chap. VIII, Section 17, pp.254 and 256, [24] Chap.X, Section 16, pp. 263 and 264, [27] pp.117 and 118, [28] Section 12.10, pp.516 and 517). When two signs are changed ([27] p.117), the four pairs of tables are identical, but the results for the two different auxiliary curves are completely different.

Because the intermediate result, ϕ, is closely related to ψ, the integral of ψ, or the sum of results, ϕ_i, should not be too far from zero. The sum need not be exactly zero because the points are not equally distributed in θ while a Gaussian rule has been used, and because the factor $\frac{\partial \mu}{\partial \nu}$ is present in the function ϕ.

For the smallest auxiliary ellipse, $(a_1,b_1) = (2,1)$ with $a_1 + b_1 = 3$, the sum of 8 elements is 10.01465747, while for the next ellipse, $(a_1,b_1) = (5,3)$, with $a_1 + b_1 = 8$, the sum of 8 elements is -0.37517605. The absolute smallest sum is obtained for a comparison curve having its sum of the semiaxes far away from the critical value 2.

8.3.3. Example 3: We impose the boundary condition of Kupradze (8.3) on an ellipse with $\epsilon = 0.5$ ($= b/a$), where the auxiliary ellipse has $\epsilon_1 = 0.5$ ($= b_1/a_1$) (which holds for 4 out of 5 of the curves mentioned in Section 8.3.2). The relation between the two ellipses, characterized by $\alpha = a_1/a$, is kept fixed. The geometry of the problem can be scaled up and down solely by changing the length a_1. The equations have been solved without and with the condition (6.5) using $N = 16$, the approximate results have been compared with the corresponding exact results (8.4) and the relative errors have been found (neglecting, however, the results for $\theta = \pm \frac{\pi}{4}$ and $\pm \frac{3\pi}{4}$, because the exact solution there is zero); the greatest absolute relative error has been taken as representative for the N results found.

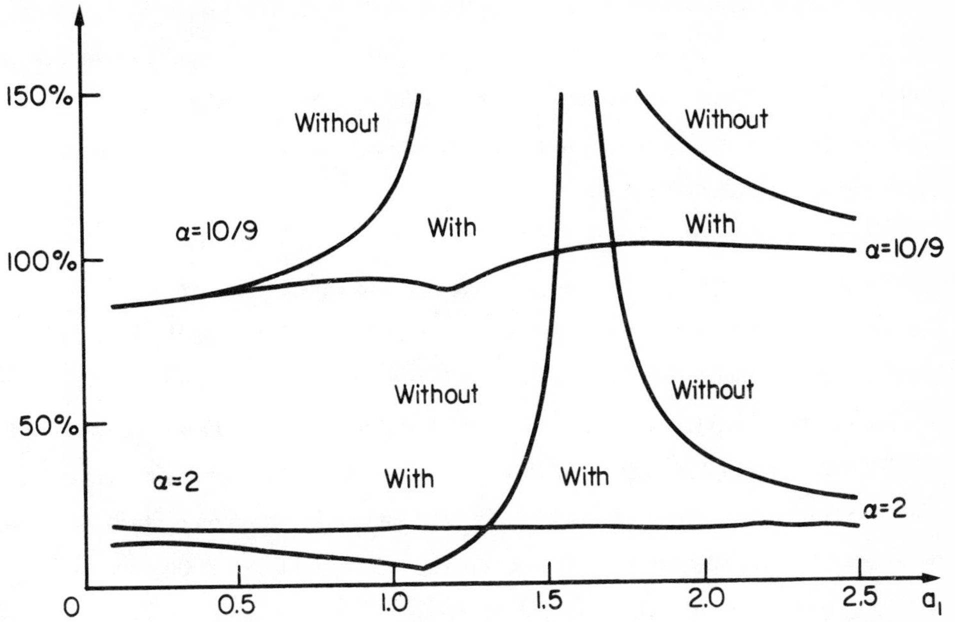

Fig. 4. Concentric ellipses, cf. Section 8.3.3. The absolute value of the absolute greatest relative error without and with the supplementary condition (6.5) as a function of a_1, when $b/a = 0.5$ and $b_1/a_1 = 0.5$. The parameter $\alpha = a_1/a$ is 10/9 or 2.

In Fig. 4 the absolute value of these errors are shown as a function of a_1, for $0.1 \leq a_1 \leq 2.5$, when $\alpha = 10/9$ and $\alpha = 2$.

When the problem is solved with the supplementary condition, the error is fairly independent of a_1: For $\alpha = 10/9$ the error varies between 86% and 105%, and for $\alpha = 2$ between 16%and 19%. Contrary to this, when the condition is left out, the error depends strongly upon a_1: For $\alpha = 10/9$ the error is greater than 150%, when $1.09 < a_1 < 1.79$, and for $\alpha = 2$ it is greater than 100% when $1.54 < a_1 < 1.73$.

Kupradze has considered the case $\alpha = 2$, $a_1 = 2$, cf. Example 2, where the error is 19% with condition, but 39% without the condition.

9. ON THE CHOICE OF THE AUXILIARY CURVE

Based upon results derived from the examples of Section 8.3.2, especially with the auxiliary ellipses $(a_1, b_1) = (2,1)$ and $(5,3)$, where the intermediate results were completely different, Kupradze states ([23] Chap. VIII, Section 17, pp. 254 and 256, [27] p. 118) that it may be advantageous to use an auxiliary curve not too near the boundary curve. But the first mentioned auxiliary ellipse is near the critical one, and the intermediate results are not reliable, as discussed in Section 8.3.2, such that it seems to be dangerous to base a conclusion upon these results.

A change of the auxiliary curve involves both a change of the absolute magnitude of the curve, and a change of its relative distance from the boundary curve. These two effects are not necessarily separated when the critical geometry has not been eliminated. Without the supplementary condition the results obtained are not invariant to a change of scale for the whole problem; it is first necessary to remedy this, and then it can be investigated how the results behave when the magnitude of the auxiliary curve is changed relative to the boundary curve.

This problem can be investigated in case of concentric circles (Section 7) when $F(\theta) \equiv 1$, in which case the exact solution is $\Psi(\theta) \equiv 0$. The approximate solution, when the supplementary condition is taken into account, is the product of k_N and a factor (8.1b) containing R_N and T. For a fixed value of $\alpha = a_1/a$ it is seen (for $\alpha = 10/9$) from Fig. 1, that this last factor is nearly completely independent of a_1; results for other values of α show the same feature.

The value of k_N depends only on α, and not on a_1, cf. (7.2c). The approximate solution for concentric circles is thus seen to be - in practice - independent of a_1 (when the supplementary condition is used; but strongly dependent upon a_1 when the condition is let out of consideration). (Also for more complicated geometries and boundary conditions it is seen that the solution is, in practice, independent of a_1; see Figs. 2, 3, and 4.)

The factor (8.1b) depends very weakly upon α, such that the dependence of the solution Ψ_N upon α mainly is described by the factor k_N. If k_N is zero, then the approximate solution is correct. The value of k_N depends upon α in such a way (7.2c) that $k_N \to 0$ very rapidly when α increases somewhat above 1.

These statements may lead to the supposition that it is wise to choose an auxiliary curve far away from the boundary curve, in accordance with the statements cited above.

But this conclusion is premature, because when α is large, the matrix-elements are nearly alike (they are nearly $\ln a_1^2$ all together), such that "...the determinacy of the matrix ... deteriorates ..." ([24] Chap.X, Section 16, p.267), or the matrix becomes badly conditioned.

In order to investigate the conditioning of the matrix $\{A_{ij}\}$ for concentric circles, without the supplementary condition we consider the eigenvalue problem

$$(\underline{\underline{A}} - \lambda \underline{\underline{I}})\underline{x} = 0 ,$$

for the matrix $\underline{\underline{A}} = \{A_{ij}\}$, (6.3a) with N even. Due to the symmetry, the following two eigenvectors exist:

$$\underline{x}^{(1)} = (1,1,1,\cdots 1)^T , \quad \text{with } \lambda = \lambda_1$$

and

$$\underline{x}^{(N)} = (1,-1,1,\cdots,-1)^T , \quad \text{with } \lambda = \lambda_N .$$

Here

$$\lambda_1 = \sum_{j=1}^{N} A_{ij} = R_N = 2 \ln (a_1^N - a^N) ,$$

cf. (7.1a) and (7.1c), and

$$\lambda_N = \sum_{j=1}^{N} -(-1)^j A_{1j} = 2 \sum_{j=1,3,\cdots} A_{1j} - \sum_{j=1}^{N} A_{1j}$$

$$= 2 R_{N/2} - R_N = 2 \ln \frac{a_1^{N/2} - a^{N/2}}{a_1^{N/2} + a^{N/2}}$$

or, if $\alpha = a_1/a$ is introduced,

$$\lambda_N = 2 \ln \frac{\alpha^{N/2} - 1}{\alpha^{N/2} + 1} = 2 \ln (1 - \frac{2}{\alpha^{N/2} + 1}) \tag{9.1}$$

or approximately for $\alpha^{N/2}$ large

$$\lambda_N \simeq - \frac{4}{\alpha^{N/2} + 1} . \tag{9.2}$$

When the absolute value of an eigenvalue is small, then the corresponding eigenvector may creep into the solution. When $\lambda_1 = 0$ which occurs when $a_1 = a_1^*$, cf. (7.4), then $\underline{x}^{(1)}$ comes in when the equations are solved without the condition (6.5), but the addition of this condition neutralizes the effect of $|\lambda_1| \simeq 0$. More serious is the presence of an alternating eigenvector $\underline{x}^{(N)}$ corresponding to λ_N, because $|\lambda_N|$ may become small. The approximation (9.2) shows that $\lambda_N \rightarrow 0$ when α and/or N increases. This shows that N should be kept as small as possible (the choice of N depends upon how complicated the prescribed boundary condition is), and α also must be kept as small as possible, but $\alpha > 1$ (the choice of α depends upon how accurate k_N, (7.2c) are to be evaluated).

These statements have been illustrated by two examples, the results of which are summarized in Table 1.

Kupradze's (functional) equation may be solved either by using a system of non-orthogonal functions or by applying some quadrature rule. What here has been shown by giving eigenvalues and eigenvectors explicitly, and demonstrating the practical effect of it on an example, may serve to supplement some investigations by Aleksidze [3] for the case where the equations are solved using a system of functions. The basic functions for the plane Dirichlet problem can be shown to be linearly independent and complete ([3] Th. 3), in the class of functions, γ, satisfying the condition $\int \gamma \, ds = 0$ ([27] p.87). We know that this condition ought to be satisfied, cf. (4.10), but in the works cited it has not been secured that this condition really is satisfied.

Even though the basic-functions - strictly speaking - are linearly independent, they form - from a practical point of view - a nearly dependent system, an unreliable system in the sense of Mihlin [31]. This has been illustrated numerically by Aleksidze ([2], [4]) for a set of basic-functions derived from a variant of Kupradze's equation, but similar to the basic functions which are used in the equation of Kupradze under consideration here. Using a certain degree of precision it was

TABLE 1

Largest absolute relative error, without and with the supplementary condition (6.5), and the boundary condition (8.3).

Number of points N	Circles: $a = 1$, $a_1 = 2$, Without	$b = 1$, $b_1 = 2$, With	Ellipses: $a = 1$, $a_1 = 2$, Without	$b = 0.5$, $b_1 = 1$, With
8	0.49×10^{-1}	0.46×10^{-1}	0.60	-0.29
16	0.32×10^{-3}	0.38×10^{-3}	0.38	0.16
24	0.53×10^{-2}	0.58×10^{-2}	*	*
32	0.40	0.23	*	*
40	*	*	*	*

* The system was insoluble using [20] with eps $= 10^{-5}$.

impossible to orthonormalize more than about 10 ([2]) to 20 ([4]) functions depending upon the method used. (It may be better to use non-orthogonal systems of functions [5], or approximately orthonormalized functions [6].)

The small number of linearly independent functions came out when the orthogonalization was performed using the Grammian determinant, and it was found that this determinant increases (i.e. the functions become 'more' linearly independent) when the auxiliary curve is near the boundary curve ([2] p. 844). It also turned out ([4] p. 921) that the number of functions to be used depends upon the smoothness of the boundary condition, and the distance of (what corresponds to) the auxiliary curve from the boundary curve depends upon the number of functions used.

These results hold for a variant of the Fourier-method, mentioned in the beginning of Section 6, while our results, presented above, are found for the quadrature method for solving Kupradze's equation.

The difficulties with establishing a system of linearly independent functions, or the deterioration of the system of linear algebraic equations, seems to be inherent in the (functional) equation of Kupradze. The crucial equation (2.3b) has some similarity with an integral equation of the first kind with bounded kernel, and a solution of such an equation is not stable ([7], [14], [34], [44]). However such ill-posed problems can be solved using the regularization method of Tikhonov ([39], [40], [21]), which also has been used specifically for solving integral equations of the first kind ([8], [43]). Bakushinskii [10] has shown how the regularization-method can be applied to the Kupradze's equation, when it is solved using systems of functions, to obtain a better stability of the calculations, i.e. to eliminate (to a certain degree) the weakness in the form of lack of linear independency.

The regularization-method may also be used to solve (nearly) singular systems of linear algebraic equation ([41], [42]). It seems to be an open question whether the method can be used to improve the system under consideration and eliminate the deficiency which appears when the number of points increases and/or when the auxiliary curve is moved away from the boundary curve.

10. DISCUSSION

The main purpose of the present work was to investigate Kupradze's functional equation when it is applied to the solution of an interior plane Dirichlet problem for the Laplace-equation. And here specifically to find whether the auxiliary curve, placed outside the boundary curve, may be chosen arbitrarily without losing the uniqueness of the solution determined by means of the equation.

We found that when the exterior mapping radius of the boundary curve is smaller than one, there exists a critical curve with exterior mapping radius equal to one, and with a shape related to the boundary curve through a conformal mapping. The solution loses its uniqueness when the critical curve is used as auxiliary curve. We found furthermore how this lack of uniqueness could be removed by imposing a simple integral condition on an intermediate unknown function used in the course of the computations.

Kupradze's equation was solved approximately by transforming it to a set of linear algebraic equations, and in the frames of this approach it turned out to be easy to incorporate the integral condition necessary to ensure uniqueness, and thereby obtain a system of equations which easily could be solved by a least square method.

The numerical results computed showed a deficiency where it is predicted by the analytical investigations and, due to the finite number of equations, also in some neighbourhood of the theoretical critical curve it was numerical difficult to obtain accurate results. All these difficulties were remedied when taking the supplementary integral condition into account.

The linear algebraic equations corresponding to an auxiliary curve near the critical one are (nearly) singular. This singularity is eliminated by taking the supplementary condition (equation) into account. Another way of elimination would be to use the regularization method of Tikhonov, which also can be used for linear algebraic equations ([41], [42]). Even though there exist methods for constructing regularization algorithms [9], application of the regularization method does not seem to be simpler than to apply the least square method used here.

APPENDIX A. Circular boundary curve

The investigations of Sections 2-3 can be carried out in an elementary manner, when the boundary curve Γ is a circle C with radius c:

$$C: \ x = c\cos\theta, \ y = c\sin\theta \ ; \ 0 \leq \theta \leq 2\pi \ .$$

Using polar coordinates the integration point $\bar{r}$ is characterized by (c,θ), while the parameterpoint $\bar{r}_1$ is characterized by (R,ϕ), where $0 \leq \phi \leq 2\pi$, $0 \leq R \leq \infty$. The definite integrals in (2.3) or, the double layer potential (3.1) and the single layer potential (3.2) can then by simple geometrical considerations be written $(\frac{\partial}{\partial\nu} = \frac{\partial}{\partial r})$:

$$W(R,\phi) = -\frac{c}{2\pi}\int_0^{2\pi} (c - R\cos(\theta-\phi))\rho^{-2}\cdot u(c,\theta)\,d\theta \tag{A.1a}$$

and

$$V(R,\phi) = -\frac{c}{2\pi}\int_0^{2\pi} \ln\rho\cdot\psi(c,\theta)\,d\theta \tag{A.1b}$$

where, cf. (2.4b)

$$\rho^2 = c^2 + R^2 - 2cR\cos(\theta - \phi) \; . \tag{A.1c}$$

Let the prescribed boundary condition be

$$u(c,\theta) = a_0 + \sum_{n=1}^{\infty} (a_n \cos n\theta + b_n \sin n\theta) \tag{A.2a}$$

and assume that the unknown quantity ψ can be written as

$$\psi(c,\theta) = c^{-1}[A_0 + \sum_{n=1}^{\infty} n(A_n \cos n\theta + B_n \sin n\theta)] \; . \tag{A.2b}$$

In (A.2) it would be possible to impose the further condition that u is a solution of Laplace's equation within a circle with radius c and especially that $\psi(c,\theta)$ should be equal to $\frac{\partial u}{\partial r}(c,\theta)$, and if so, then separation of the variables would give the following relations among the coefficients:

$$A_0 = 0 \tag{A.3a}$$

$$A_n = a_n, \quad n = 1,2,\cdots \tag{A.3b}$$

$$B_n = b_n, \quad n = 1,2,\cdots \; . \tag{A.3c}$$

However, greater insight will be obtained if we go further before using the facts which are expressed in (A.3). We insert the series (A.2a) and (A.2b) in (A.1a) and (A.1b) respectively, carry out some elementary calculations, and apply the integrals (C.1b), (C.2b) and (C.3b), (C.4b), respectively, and obtain for $R > c$:

$$W(R,\phi) = 0 + \tfrac{1}{2} \sum_{n=1}^{\infty} (\tfrac{c}{R})^n \{a_n \cos n\phi + b_n \sin n\phi\} \tag{A.4a}$$

and

$$V(R,\phi) = -A_0 \ln R + \tfrac{1}{2} \sum_{n=1}^{\infty} (\tfrac{c}{R})^n \{A_n \cos n\phi + B_n \sin n\phi\} \; . \tag{A.4b}$$

Now the problem is to equate the two potentials (A.4a) and (A.4b) along some curve Γ_1; here we choose as comparison curves circles $r = R$, where $R > c$.

To obtain a unique result one has to find whether there exists a non-trivial single layer source density $\tilde{\psi}(\bar{r})$ $(\neq 0)$ such that the corresponding single-layer potential $\tilde{V}(\bar{r}_1)$ attains the value zero along some curve $\tilde{\Gamma}_1$. As a source density $\tilde{\psi}(\bar{r})$ one can choose

$$\tilde{\psi}(c,\theta) = \tilde{A}_0/c \tag{A.5}$$

which gives a potential for $R_1 > c$:

$$\tilde{V}(R_1,\phi) = -\tilde{A}_0 \ln R_1 ,$$

which, provided that $c < 1$, is equal to zero along the curve $\tilde{\Gamma}_1 : R_1 = 1$. If the comparison of the two potentials (A.4a) and (A.4b) is carried out along a curve $\Gamma_1 :$ $R_1 \neq 1$ then the known relations (A.3) are found. But when the comparison is carried out along $\tilde{\Gamma}_1$, then the relation (A.3a) cannot be derived.

The value of the function u inside the curve Γ is found from (2.3a) to be

$$u(\bar{r}_1) = V(\bar{r}_1) - W(\bar{r}_1) \tag{A.6}$$

with the appropriate source densities inserted. In (A.6) we insert the expressions (A.1) for V and W, and use the series (A.2). We carry out some elementary calculations and apply the integrals (C.1a), (C.2a), (C.3a), and (C.4a), and obtain finally for $R < c$:

$$u(R,\phi) = a_0 - A_0 \ln c + \tfrac{1}{2} \sum_{n=1}^{\infty} \left(\frac{R}{c}\right)^n [(a_n + A_n)\cos n\phi + (b_n + B_n)\sin n\phi] , \quad R < c .$$

We insert here the relations (A.3b) and (A.3c), which could be found unambiguously, and obtain

$$u(R,\phi) = a_0 - A_0 \ln c + \sum_{n=1}^{\infty} \left(\frac{R}{c}\right)^n [a_n \cos n\phi + b_n \sin n\phi] . \tag{A.7}$$

The interior solution of the harmonic problem with the prescribed boundary condition (A.2a) is, except for the term $-A_0 \ln c$, equal to the result (A.7) found above. The spurious component of the source density, $\tilde{A}_0/c$, (A.5), which may come in when $c < 1$, gives a contribution to the internal solution equal to $-\tilde{A}_0 \ln c$. Thereby the internal solution is not uniquely determined if a part of the spurious solution is present in the intermediate solution ψ.

APPENDIX B. Harmonic functions in elliptic coordinates

For use in Section 6 connections relating to a harmonic function defined inside an ellipse with semiaxes a and b are here derived. Introducing elliptical coordinates (μ, θ), and applying separation of variables the possible components of the harmonic function can be found ([32], pp.17-19). Now taking only such combinations which give continuous and differentiable functions everywhere inside the ellipse, one obtains the most general solution to be

$$u(\mu,\theta) = C_0 + \sum_{n=1}^{\infty} c_n \cosh n\mu \cdot \cos n\theta + D_n \sinh n\mu \cdot \sin n\theta ,$$

from which $\frac{\partial u}{\partial \mu}$ can be found

$$\frac{\partial u}{\partial \mu}(\mu,\theta) = \sum_{n=1}^{\infty} n\{C_n \sinh n\mu \cdot \cos n\theta + D_n \cosh n\mu \sin n\theta\} .$$

At the boundary, which is defined by $\mu = \mu_0$, where $\mu_0 = \frac{1}{2}\ln\frac{a+b}{a-b}$, $u(\mu_0,\theta)$ is put equal to

$$F(\theta) = a_0 + \sum_{n=1}^{\infty} (a_n \cos n\theta + b_n \sin n\theta)$$

and similarly $\frac{\partial u}{\partial \mu}(\mu_0,\theta)$ is put equal to

$$\Psi(\theta) = A_0 + \sum_{n=1}^{\infty} (A_n \cos n\theta + B_n \sin n\theta) .$$

Now using

$$\operatorname{tgh} n\mu_0 = \frac{1 - \left(\frac{a-b}{a+b}\right)^n}{1 + \left(\frac{a-b}{a+b}\right)^n}$$

we obtain the connections between the coefficients (a_n, b_n) and (A_n, B_n) to be

$$A_0 = 0 \tag{B.1a}$$

$$A_n = n\, a_n \cdot \frac{1-\left(\frac{a-b}{a+b}\right)^n}{1-\left(\frac{a-b}{a+b}\right)^n}, \quad n = 1,2,\cdots \tag{B.1b}$$

$$B_n = n\, b_n \cdot \frac{1+\left(\frac{a-b}{a+b}\right)^n}{1-\left(\frac{a-b}{a+b}\right)^n}, \quad n = 1,2,\cdots . \tag{B.1c}$$

These relations are used in connecting (6.7) with (6.2).

APPENDIX C. Some definite integrals

The following integrals can be derived from [17]. The numbers of the formulas used are written in square brackets.

$$0 \le x \le 2\pi, \quad 0 < c, \quad 0 \le R; \quad \rho^2 = c^2 + R^2 - 2cR\cos x .$$

$$\frac{c}{2\pi}\int_0^{2\pi} (c - R\cos x)\,\rho^{-2}\,dx = \begin{Bmatrix} 1, & R < c \\ 0, & R > c \end{Bmatrix}; \quad [3.613-2] \tag{C.1^a_b}$$

$$\frac{c}{2\pi}\int_0^{2\pi} (c - R\cos x)\,\rho^{-2}\cos nx\,dx = \frac{1}{2}\begin{Bmatrix} +(R/c)^n, & R < c \\ -(c/R)^n, & R > c \end{Bmatrix}, \quad n=1,2,\cdots; \quad [3.613-2]^{\dagger} \tag{C.2^a_b}$$

$$\frac{1}{2\pi}\int_0^{2\pi} \ln\rho\,dx = \begin{Bmatrix} \ln c, & R < c \\ \ln R, & R > c \end{Bmatrix}; \quad [4.224-14] \tag{C.3^a_b}$$

$$\frac{1}{2\pi}\int_0^{2\pi} \ln\rho\cos nx\,dx = -\frac{1}{2n}\begin{Bmatrix} (R/c)^n, & R < c \\ (c/R)^n, & R > c \end{Bmatrix}, \quad n = 1,2,\cdots . \quad [4.397-6] \tag{C.4^a_b}$$

---

† an application of $\cos x \cos nx = \frac{1}{2}[\cos(n+1)x + \cos(n-1)x]$

Acknowledgements

The present work was begun at the Institute of Technical Mechanics, The Technical University of Aachen, The Federal Republik of Germany, and was finished at the Laboratory of Applied Mathematical Physics, The Technical University of Denmark. For many stimulating and helpful discussions Professor, Dr. Georg Rieder, The Technical University of Aachen, Professor, Dr. Erik Hansen, The Technical University of Denmark, and Professor, Dr. Adam Piskorek, The University of Warszawa, are thanked. For reading the manuscript and giving valuable comments on it Professor Erik Hansen and Henning Rasmussen, Ph.D., The Technical University of Denmark, are thanked.

The work at the Institute of Technical Mechanics was financially supported by two grants, which are gratefully acknowledged:

1. A research scholarship from 'Heinrich-Hertz-Stiftung' in Nordrhein-Westfalen, The Federal Republic of Germany.
2. A scholarship from 'NATO Science Fellowship Programme', Denmark.

The numerical calculations were carried out at the Northern Europe University Computing Center (NEUCC), The Technical University of Denmark, using an IBM 370/165.

A very short version of a part of this paper has been presented on April 3, 1974, at the Scientific Meeting of 'Gesellschaft für Angewandte Mathematik und Mechanik' (GAMM), which was held in Bochum (Germany); cf. Christiansen [11].

References

1 M.A. Aleksidze, On approximate solutions of a certain mixed boundary value problem in the theory of harmonic functions. Differential Equations 2 (1966) 515-518. (Differencial'nye Uravnenija 2 (1966) 988-994).

2 M.A. Aleksidze, The practical application of a new approximation method. Differential Equations 2 (1966) 843-845. (Differencial'nye Uravnenija 2 (1966) 1625-1629.

3 M.A. Aleksidze, Notes on an approximate method of solving boundary value

problems. Soviet Math. Dokl. 8 (1967) 297-299. (Dokl. Akad. Nauk SSSR 173 (1967) 9-11.)

4 M.A. Aleksidze, On the completeness of certain systems of functions. Differential Equations 3 (1967) 919-921. (Differencial'nye Uravnenija 3 (1967) 1766-1771.)

5 M.A. Aleksidze, Series in nonorthogonal systems of functions. USSR Computational Math. and Math. Physics 8, 5 (1968) 40-68. (Zh. vȳchisl. Mat. mat. Fiz. 8, 5 (1968) 965-987.)

6 M.A. Aleksidze, N.M. Arveladze and N.L. Lekisvili, Cislennaja realizacija odnogo novogo priblizennogo metoda resenija granicnych zadac. Izdatel'stvo "Mecniereba", Tbilisi (1969).

7 C.T.H. Baker, L.Fox, D.F. Mayers and K. Wright, Numerical solution of Fredholm integral equations of first kind. Comput J.7 (1964) 141-148.

8 A.B. Bakushinskii, A numerical method for solving Fredholm integral equations of the 1st kind. USSR Computational Math. and Math. Physics 5, 4 (1965) 226-233. (Zh. vȳchisl. Mat. mat. Fiz. 5, 4 (1965) 744-749.)

9 A.B. Bakushinskii, A general method of constructing regularizing algorithms for a linear ill-posed equation in Hilbert space. USSR Computational Math. and Math. Physics 7, 3 (1967) 279-287. (Zh. vȳchisl. Mat. mat. Fiz. 7, 3 (1967) 672-677.)

10 A.B. Bakushinskii, The Kupradze-Aleksidze method. Differential Equations 6 (1970) 989-991. (Differencial'nye Uravnenija 6 (1970) 1298-1301.)

11 S. Christiansen, On Kupradze's functional equations for plane harmonic problems. Z. Angew. Math. Mech. 55 (1975) T197 - T199.

12 R. Courant and D. Hilbert, Methods of Mathematical Physics, Vol. II (Partial Differential Equations, R. Courant), Interscience

Publishers, New York (1962).

13 P.J. Davis, On the Numerical Integration of Periodic Analytic Functions. pp. 45-60 in [29].

14 J.N. Franklin, Well-posed stochastic extensions of ill-posed linear problems. J. Math. Anal. Appl. 31 (1970) 682-716.

15 D. Gaier, Konstruktive Methoden der konformen Abbildung. Springer-Verlag, Berlin (1964).

16 G. Golub, Numerical method for solving linear least squares problems. Numer. Math. 7 (1965) 206-216.

17 I.S. Gradshteyn and I.M. Ryzhik, Table of Integrals, Series, and Products. 4th ed., Academic Press, New York (1965).

18 G. Hämmerlin, Zur numerischen Integration periodischer Funktionen. Z. Angew. Math. Mech. 39 (1959) 80-82.

19 E. Hille, Analytic Function Theory, Vol. II. Ginn and Company, Boston, (1962).

20 IBM Application Program, System/360 Scientific Subroutine Package, Version III. Programmer's Manual, Program No. 360 A-CM-03x, Fifth ed. August 1970, pp. 160-164, LLSQ: solution of linear least-squares problems.

21 V.A. Il'in, Tikhonov's work on methods of solving ill-posed problems. Russian Math. Surveys 22, 2 (1967) 142-149. (Uspehi Mat. Nauk.)

22 L.B.W. Jolley, Summation of Series. Dover Publications, New York, 2nd rev. ed. (1961).

23 V.D. Kupradze, Dynamical Problems in Elasticity (in Progress in Solid Mechanics, III, eds. J.N. Sneddon and R. Hill). North-Holland, Amsterdam (1963).

24 V.D. Kupradze, Potential Methods in the Theory of Elasticity. Israel Program for Scientific Translations, Jerusalem (1965). (Metody

potentsiala v teorii uprugosti, Gosudarstvennoe Izdatel'stvo Fiziko-Matematicheskoi Literatury, Moskva (1963).)

25 V.D. Kupradze, On the approximate solution of problems in mathematical physics. Russian Math. Surveys 22, 2 (1967) 58-108. (Uspehi Mat. Nauk 22, 2 (1967) 59-107.)

26 V.D. Kupradze, Über numerische Verfahren der Potential - und der Elastizitätstheorie. Z Angew. Math. Mech. 49 (1969) 1-9.

27 V.D. Kupradze and M.A. Aleksidze, The method of functional equations for the approximate solution of certain boundary value problems. USSR Computational Math. and Math. Physics 4, 4 (1964) 82-126. (Zh. vȳchisl. Mat. mat. Fiz. 4,4 (1964) 683-715.)

28 V.D Kupradze, T.G. Gegelia, M.O. Baselejsvili and T.V. Burculadze, Trechmernye Zadaci Matematiceskoj Teorii Uprugesti, Izdatel'stvo Tbilisskogo Universiteta, Tbilisi (1968).

29 R.E. Langer (ed.), On Numerical Approximation. The University of Wisconsin Press, Madison (1959).

30 M.A. Lawrentjew and B.W. Schabat, Methoden der komplexen Funktionentheorie. VEB Deutscher Verlag der Wissenschaften, Berlin (1967).

31 S.G. Mihlin, The stability of the Ritz method. Soviet Math. Dokl. 1 (1960), 1230-1233. (Dokl. Akad. Nauk SSSR 135 (1960) 16-19.)

32 P. Moon and D.E. Spencer, Field Theory Handbook. Springer-Verlag, Berlin (1961).

33 N.I. Muschelischwili, Singuläre Integralgleichungen. Akademie-Verlag, Berlin, (1965)

34 D.L. Phillips, A technique for the numerical solution of certain integral equations of the first kind. J. Assoc. Comput. Mach. 9 (1962) 84-97.

35 G. Pólya and G. Szegö, Über den transfiniten Durchmesser (Kapazitätskonstante)

von ebenen und räumlichen Punktmengen. J. Reine Angew. Math. 165 (1931) 4-49

36 G. Pólya and G. Szegö, Isoperimetric Inequalities in Mathematical Physics. Princeton University Press, Princeton (1951).

37 G. Pólya and G. Szegö, Aufgaben und Lehrsätze aus der Analysis, Zweiter Band. Springer-Verlag, Berlin (1971).

38 G. Szefer, On a certain method of the potential theory for unbounded regions. Arch. Mech. Stos. 19 (1967) 367-383.

39 A.N. Tihonov, Solution of incorrectly formulated problems and the regularization method. Soviet Math. Dokl. 4 (1963) 1035-1038. (Dokl. Akad. Nauk SSSR 151 (1963) 501-504.)

40 A.N. Tihonov, Regularization of incorrectly posed problems. Soviet Math. Dokl. 4 (1963) 1624-1627. (Dokl. Akad. Nauk SSSR 153 (1963) 49-52.)

41 A.N. Tihonov, Incorrect problems of linear algebra and a stable method for their solution. Soviet Math. Dokl. 6 (1965) 988-991. (Dokl. Akad. Nauk SSSR 163 (1965) 591-594.)

42 A.N. Tikhonov, The stability of algorithms for the solution of degenerate systems of linear algebraic equations. USSR Computational Math. and Math. Physics 5, 4 (1965) 181-188. (Zh. vȳchisl. Mat. mat. Fiz. 5, 4 (1965) 718-722.)

43 A.N. Tikhonov and V.B. Glasko, The approximate solution of Fredholm integral equations of the first kind. USSR Computational Math. and Math. Physics 4, 3 (1964) 236-247. (Zh. vȳchisl. Mat. mat. Fiz. 4, 3 (1964) 564-571.)

44 S. Twomey, On the numerical solution of Fredholm integral equations of the first kind by the inversion of the linear system produced by quadrature. J. Assoc. Comput. Mach. 10 (1963) 97-101.

Søren Christiansen

Laboratory of Applied Mathematical Physics

The Technical University of Denmark

Lyngby

Denmark

R HUSS, H KAGIWADA and R KALABA

16 Reduction of a family of nonlinear boundary value problems to a Cauchy system

1. INTRODUCTION

Some biological processes lead to nonlinear boundary value problems: determination of the chemical potential of solutions containing charged particles [1] and mathematical description of the standing osmotic gradient model of water transport in epithelia [2]. The numerical solution of such problems is frequently a formidable task, often requiring iterative techniques [3]. In this paper we demonstrate how such boundary-value problems can be converted into initial-value problems, which are more readily solved by numerical techniques [4], [5].

In previous papers in the series [6], [7], it was shown how certain linear two-point boundary value problems could be converted into Cauchy systems, and a Cauchy system for the Green's function was also found. In this paper we extend the previous results to nonlinear two-point boundary value problems. It is further shown that the solution of the Cauchy problem provides a solution of the original nonlinear boundary value problem. Finally, a numerical example is presented. Numerical results for the standing gradient model of water transport in epithelia will be presented subsequently.

2. A NONLINEAR TWO POINT BOUNDARY VALUE PROBLEM

In this paper we consider the general nonlinear boundary value problem

$$\ddot{u}(t) = -g(u,\lambda), \qquad 0 \le t \le 1, \qquad 0 \le \lambda \le \Lambda, \tag{1}$$

with boundary conditions

$$u(0) = 0 \tag{2}$$

and

$$u(1) = 0 \tag{3}$$

We shall first show how a Cauchy system can be obtained for the special case

$$g(u,\lambda) = \lambda f(u) , \tag{4}$$

and then we shall indicate how this leads directly to the solution of the general case.

3. CONVERSION OF THE BOUNDARY VALUE PROBLEM TO A CAUCHY SYSTEM

Consider the nonlinear two point boundary value problem

$$\ddot{u}(t) = -\lambda f(u) , \qquad 0 \leq t \leq 1 , \tag{5}$$

$$0 \leq \lambda \leq \Lambda ,$$

with boundary conditions

$$u(0) = 0 \tag{6}$$

and

$$u(1) = 0 . \tag{7}$$

Here we assume that a unique differentiable solution exists for $0 \leq \lambda \leq \Lambda$. Since it is clear that the function u depends on the value of λ, we shall write, where needed,

$$u = u(t,\lambda) . \tag{8}$$

First, differentiate both sides of equation (5) with respect to λ to obtain

$$(u_\lambda)'' = -\lambda f'[u(t,\lambda)]\, u_\lambda - f[u(t,\lambda)] . \tag{9}$$

We regard u_λ as a new function of t and λ, and write

$$u_\lambda = u_\lambda(t,\lambda) . \tag{10}$$

The boundary conditions for equation (9) are

$$u_\lambda(0,\lambda) = 0 \tag{11}$$

and

$$u_\lambda(1,\lambda) = 0 \,. \tag{12}$$

Next we introduce the second order equation for a function w

$$\ddot{W}(t,\lambda) = -\lambda f'(u)\, w - \varphi(t,\lambda)\,, \qquad 0 \le t \le 1\,, \quad 0 \le \lambda \le \Lambda\,, \tag{13}$$

with boundary conditions

$$w(0,\lambda) = 0 \tag{14}$$

and

$$w(0,\lambda) = 0\,, \tag{15}$$

where φ is an arbitrary function.

Now introduce the Green's function G, in terms of which the solution of equations (13)-(15) is

$$w(t,\lambda) = \int_0^1 G(t,y,\lambda)\,\varphi(y,\lambda)\,dy\,, \qquad 0 \le t \le 1\,, \quad 0 \le \lambda \le \Lambda\,. \tag{16}$$

In terms of this Green's function, the solution of equations (9)-(12) may be written

$$u_\lambda(t,\lambda) = \int_0^1 G(t,y,\lambda)\, f[u(y,\lambda)]\,dy\,. \tag{17}$$

This is one of the basic differential equations of the Cauchy system. The initial condition on the function u at $\lambda = 0$ is

$$u(t,0) = 0\,, \qquad 0 \le t \le 1\,. \tag{18}$$

We next shall find a Cauchy system for the Green's function, G. First, differentiate both sides of equation (13) with respect to λ to obtain

$$(w_\lambda)'' = -[f'(u) + \lambda f''(u)\, u_\lambda]\, w(t,\lambda) - \lambda f'(u)\, w_\lambda - \varphi_\lambda(t,\lambda), \quad 0 \le t \le 1 \tag{19}$$

$$0 \le \lambda \le \Lambda,$$

with boundary conditions

$$w_\lambda(0,\lambda) = 0 \tag{20}$$

and

$$w_\lambda(1,\lambda) = 0. \tag{21}$$

We now define the function $R(t,\lambda)$ to be

$$R(t,\lambda) = f'[u(t,\lambda)] + \lambda f''[u(t,\lambda)]\, u_\lambda(t,\lambda). \tag{22}$$

In terms of the Green's function, the function $w_\lambda(t,\lambda)$ can be written

$$w_\lambda(t,\lambda) = \int_0^1 G(t,y,\lambda)\, [R(y,\lambda)\, w(y,\lambda) + \varphi_\lambda(y,\lambda)]\, dy. \tag{23}$$

Differentiation of both sides of equation (16) yields a second equation for the function w_λ namely,

$$w_\lambda(t,\lambda) = \int_0^1 [G,(t,y,\lambda)\, \varphi_\lambda(y,\lambda) + G_\lambda(t,y,\lambda)\, \varphi(y,\lambda)]\, dy. \tag{24}$$

Equating the two expressions for w_λ yields the relation

$$\int_0^1 G_\lambda(t,y,\lambda)\, \varphi(y,\lambda)\, dy = \int_0^1 G(t,y,\lambda)\, R(y,\lambda)\, w(y,\lambda)\, dy. \tag{25}$$

If we now substitute the expression for $w(t,\lambda)$ from equation (16) into the right-hand side of equation (25) we find that

$$\int_0^1 G_\lambda(t,y,\lambda)\, \varphi(y,\lambda)\, dy = \int_0^1 G(t,y,\lambda)\, R(y,\lambda) \int_0^1 G(y,z,\lambda)\, \varphi(z,\lambda)\, dz dy. \tag{26}$$

Since the function φ is arbitrary, it follows that

$$G_\lambda(t,y,\lambda) = \int_0^1 G(t,z,\lambda)\, R(z,\lambda)\, G(z,y,\lambda)\, dz. \tag{27}$$

This is the second basic differential equation of the Cauchy system. The initial condition on the function G at $\lambda = 0$ is known to be

$$G(t,y,0) = \begin{cases} t(1-y), & y \geq t, \\ y(1-t), & y \leq t. \end{cases} \tag{28}$$

4. STATEMENT OF THE CAUCHY SYSTEM

The functions u and G are now defined to be solutions of the differential equations

$$u_\lambda(t,\lambda) = \int_0^1 G(t,z,\lambda)\, f[u(z,\lambda)]\, dz, \tag{29}$$

$$G_\lambda(t,y,\lambda) = \int_0^1 G(t,z,\lambda)\, R(z,\lambda)\, G(z,y,\lambda)\, dz, \quad \text{for } 0 \leq \lambda \leq \Lambda, \ 0 \leq t \leq 1, \tag{30}$$

where

$$R(t,\lambda) = f'[u(t,\lambda)] + \lambda f''[u(t,\lambda)] \int_0^1 G(t,z,\lambda)\, f[u(z,\lambda)]\, dz. \tag{31}$$

The initial conditions on u and G at $\lambda = 0$ are

$$u(t,0) = 0, \qquad 0 \leq t \leq 1, \tag{32}$$

and

$$G(t,y,0) = \begin{cases} y(1-t), & y \leq t, \\ t(1-y), & y \geq t. \end{cases} \tag{33}$$

We assume that this Cauchy system has a unique solution for $0 \leq \lambda \leq \Lambda$, that the function u has a continuous second derivative in t, that the function f is differentiable in u and possesses a continuous second derivative, and that the function G is continuous in t. As we shall see, there is a jump discontinuity in $\dot{G}$ at $t = y$ for $\lambda \geq 0$.

5. VERIFICATION OF THE CAUCHY SYSTEM

We first obtain some properties of the function G. Equation (27) is rewritten as

$$G_\lambda(t,y,\lambda) = \int_0^t G(t,z,\lambda)\, R(z,\lambda)\, G(z,y,\lambda)\, dz + \tag{34}$$

$$+ \int_t^1 G(t,z,\lambda)\ R(z,\lambda)\ G(z,y,\lambda)\ dz$$

Letting $\dot{G}$ and $\ddot{G}$ represent the first and second derivatives of G with respect to its first argument, we find that

$$\dot{G}_\lambda(t,y,\lambda) = \int_0^t \dot{G}(t,z,\lambda)\ R(z,\lambda)\ G(z,y,\lambda)\ dz \tag{35}$$

$$+ \int_t^1 \dot{G}(t,z,\lambda)\ R(z,\lambda)\ G(z,y,\lambda)\ dz$$

and

$$\ddot{G}_\lambda(t,y,\lambda) = \dot{G}(t,t-,\lambda)\ R(t,\lambda)\ G(t,y,\lambda) - \dot{G}(t,t+,\lambda)\ R(t,\lambda)\ G(t,y,\lambda) + \tag{36}$$

$$+ \int_0^1 \ddot{G}(t,z,\lambda)\ R(z,\lambda)\ G(z,y,\lambda)\ dz\ .$$

Here we have used the notation

$$\dot{G}(t,t-,\lambda) = \lim_{\epsilon \to 0^+} \dot{G}(t,t-\epsilon,\lambda) \tag{37}$$

and

$$\dot{G}(t,t+,\lambda) = \lim_{\epsilon \to 0^+} \dot{G}(t,t+\epsilon,\lambda)\ . \tag{38}$$

Now we introduce the function ψ to be

$$\psi(t,\lambda) = \dot{G}(t,t-,\lambda) - \dot{G}(t,t+,\lambda)\ , \qquad 0 \le t \le 1\ , \quad 0 \le \lambda \le \Lambda\ . \tag{39}$$

From equation (33) we see that

$$\psi(t,0) = -1\ , \qquad 0 \le t \le 1. \tag{40}$$

Differentiation of the function ψ with respect to λ shows that

$$\psi_\lambda(t,\lambda) = \dot{G}_\lambda(t,t-,\lambda) - \dot{G}_\lambda(t,t+,\lambda) = \int_0^1 \dot{G}(t,z,\lambda)\, R(z,\lambda)\, G(z,t-,\lambda)\, dz - \tag{41}$$

$$- \int_0^1 \dot{G}(t,z,\lambda)\, R(z,\lambda)\, G(z,t+,\lambda)\, dz = 0 \qquad 0 \le t \le 1,$$

$$0 \le \lambda \le \Lambda .$$

In the last equation we make use of the continuity of G in its second argument. From the last two equations we conclude that

$$\psi(t,\lambda) \equiv -1 , \qquad 0 \le t \le 1, \tag{42}$$

$$0 \le \lambda \le \Lambda ,$$

so that the Cauchy system for the function $\ddot{G}$ is

$$\ddot{G}_\lambda(t,y,\lambda) = -R(t,\lambda)\, G(t,y,\lambda) + \int_0^1 \ddot{G}(t,z,\lambda)\, R(z,\lambda)\, G(z,y,\lambda)\, dz , \tag{43}$$

$$0 \le \lambda \le \Lambda ,$$

with the initial condition

$$\ddot{G}(t,y,0) = 0 \tag{44}$$

Next we introduce the function v to be

$$v(t,y,\lambda) = -\lambda f'[u(t,\lambda)]\, G(t,y,\lambda) , \qquad 0 \le t \le 1, \tag{45}$$

$$0 \le y \le 1,$$

$$0 \le \lambda \le \Lambda .$$

Differentiate both sides of equation (45) with respect to λ to obtain

$$v_\lambda(t,y,\lambda) = -\{f'[u(t,\lambda)] + \lambda f''[u(t,\lambda)]u_\lambda(t,\lambda)\}\, G(t,y,\lambda) - \lambda f'[u(t,\lambda)]\, G_\lambda(t,y,\lambda). \tag{46}$$

Substituting the expression for G_λ from equation (30), utilizing the definition of R from equation (31), and simplifying, we find that

$$v_\lambda(t,y,\lambda) = -R(t,\lambda)\, G(t,y,\lambda) + \int_0^1 v(t,z,\lambda)\, R(z,\lambda)\, G(z,y,\lambda)\, dz . \tag{47}$$

Since the function v clearly satisfies the initial condition

$$v(t,y,0) = 0 , \qquad 0 \le t \le 1 , \quad 0 \le y \le 1 , \tag{48}$$

it follows that

$$\ddot{G}(t,y,\lambda) = -\lambda f'[u(t,\lambda)]\, G(t,y,\lambda) , \quad t \ne y . \tag{49}$$

Since

$$G(0,y,0) = 0 \tag{50}$$

and

$$G_\lambda(0,y,\lambda) = \int_0^1 G(0,z,\lambda)\, R(z,\lambda)\, G(z,y,\lambda)\, dz , \tag{51}$$

it follows that

$$G(0,y,\lambda) = 0 , \qquad 0 \le y \le 1 \tag{52}$$
$$0 \le \lambda \le \Lambda .$$

Similarly, we see that

$$G(1,y,\lambda) = 0 , \qquad 0 \le y \le 1 , \tag{53}$$
$$0 \le \lambda \le \Lambda .$$

Having established these properties of the function G, defined in Section 4, we now consider the function u. The validation will be complete when we show that the function obtained from the initial value problem of equations (31) and (32) satisfies the differential equation of equations (5)-(7). First, we see that

$$u(0,0) = 0 \tag{54}$$

and

$$u_\lambda(0,\lambda) = \int_0^1 G(0,z,\lambda)\, f[u(z,\lambda)]\, dz = 0 , \qquad 0 \le \lambda \le \Lambda \tag{55}$$

It follows that

$$u(0,\lambda) = 0\,, \qquad 0 \le \lambda \le \Lambda \tag{56}$$

In the same manner we find that

$$u(1,\lambda) = 0\,, \qquad 0 \le \lambda \le \Lambda\,. \tag{57}$$

By differentiation of both sides of equation (29) with respect to t we see that

$$\dot{u}_\lambda(t,\lambda) = \int_0^1 \dot{G}(t,z,\lambda)\, f[u(z,\lambda)]\, dz\,, \tag{58}$$

and

$$\ddot{u}_\lambda(t,\lambda) = \dot{G}(t,t-,\lambda)\, f[u(t,\lambda)] - \dot{G}(t,t+,\lambda)\, f[u(t,\lambda)] + \int_0^1 \ddot{G}(t,z,\lambda)\, f[u(z,\lambda)]\, dz. \tag{59}$$

Utilizing the function ψ evaluated in equations (39) and (42), and substituting ior $\ddot{G}$ the expression in equation (49), we find that the function $\ddot{u}$ satisfies the equation

$$\ddot{u}_\lambda(t,\lambda) = -f[u(t,\lambda)] - \int_0^1 \lambda f'[u(t,\lambda)]\, G(t,z,\lambda)\, R(z,\lambda)\, G(z,y,\lambda)\, dz\,, \qquad 0 \le \lambda \le \Lambda\,. \tag{60}$$

The function $\ddot{u}$ also satisfies the initial condition at $\lambda = 0$

$$\ddot{u}(t,0) = 0\,, \qquad 0 \le t \le 1\,. \tag{61}$$

Define the auxiliary function q by the equation

$$q(t,\lambda) = -\lambda f[u(t,\lambda)]\,, \qquad 0 \le t \le 1,\quad 0 \le \lambda \le \Lambda\,. \tag{62}$$

We first observe that

$$q(t,0) = 0\,, \qquad 0 \le t \le 1\,, \tag{63}$$

and then by differentiation we see that

$$q_\lambda(t,\lambda) = -f[u(t,\lambda)] - \lambda f'[u(t,\lambda)]\, u_\lambda(t,\lambda) \tag{64}$$

Substituting for u_λ from equation (29), we obtain

$$q_{\lambda}(t,\lambda) = -f[u(t,\lambda)] - \lambda f'[u(t,\lambda)] \int_0^1 G(t,z,\lambda)\, R(z,\lambda)\, G(z,y,\lambda)\, dz \,. \tag{65}$$

Since the function $q(t,\lambda)$ in equation (65) satisfies the same differential equation as the function $\ddot{u}(t,\lambda)$ in equation (60), and the initial conditions are identical it follows that

$$z(t,\lambda) = \ddot{u}(t,\lambda) \,, \tag{66}$$

or

$$\ddot{u}(t,\lambda) = -\lambda f(u) \,, \qquad 0 \leq t \leq 1 \,, \quad 0 \leq \lambda \leq \Lambda \,. \tag{67}$$

We have now shown that the function u, defined by the Cauchy system in equations (29) and (32), does satisfy the nonlinear two point boundary value problems of equations (5)-(7). Furthermore, we have verified that the function G, defined by the Cauchy system of equations (30) and (33), is the Green's function [8].

6. A MORE GENERAL PROBLEM

In the previous sections we have been concerned with a special case of the nonlinear problem of equations (1)-(3), namely

$$\ddot{u}(t,\mu) = -f[u(t,\mu),\mu] \,, \qquad 0 \leq t \leq 1 \,, \quad 0 \leq \mu \leq M \,, \tag{68}$$

with boundary conditions

$$u(0,\mu) = 0 \tag{69}$$

and

$$u(1,\mu) = 0 \,. \tag{70}$$

We may solve this problem in one of two ways. First, if we wish to know the solution for only one value of μ, namely $\mu = M$, we introduce the parameter λ and write

$$\ddot{u}(t,M,\lambda) = -\lambda f[u(t,M,\lambda),M], \qquad 0 \le t \le 1, \quad 0 \le \lambda \le 1, \tag{71}$$

with boundary conditions

$$u(0,M,\lambda) = 0 \tag{72}$$

and

$$u(1,M,\lambda) = 0. \tag{73}$$

We have thus reduced the general problem to the special case, whose Cauchy system has already been found, provided the solution exists for $0 \le \lambda \le 1$.

If it is the case that we wish to know the solution for certain values of μ, $0 \le \mu \le M$, we proceed as follows. First, differentiate equation (68) with respect to μ and obtain

$$(u_\mu)'' = -f_u[u(t,\mu),\mu]\,u_\mu(t,\mu) - f_\mu[u(t,\mu),\mu]. \tag{74}$$

The boundary conditions are

$$u_\mu(0,\mu) = 0 \tag{75}$$

and

$$u_\mu(1,\mu) = 0. \tag{76}$$

It is easily shown, using the method of the previous sections, that the Cauchy system for the function u and for the appropriate Green's function G is

$$u_\mu(t,\mu) = \int_0^1 G(t,y,\mu)\,f_\mu[u(y,\mu),\mu]\,dy \tag{77}$$

and

$$G_\mu(t,y,\mu) = \int_0^1 G(t,z,\mu)\,S(z,\mu)\,G(z,y,\mu)\,dz, \tag{78}$$

where

$$S(t,\mu) = f_{uu}[u(t,\mu),\mu]\,u_\mu(t,\mu) + f_{u\mu}[u(t,\mu),\mu]. \tag{79}$$

The initial conditions, $u(t,0)$ and $G(t,y,0)$, are obtained from a second Cauchy system. This system is obtained by setting $M = 0$ in equations (71)-(73). This system, when integrated numerically over the range $0 \leq \lambda \leq 1$, yields the initial conditions on the functions u and G in the system of equations (77)-(79).

7. COMPUTATIONAL PROCEDURE

The technique to be used for the numerical evaluation of the Green's function and the function u is the well-known method of lines [5]. Equations (29) and (30) are integrated over the range $0 \leq \lambda \leq \Lambda$ along lines of constant values of t. At each step in the numerical integration, the right-hand sides of these equations are evaluated by a quadrature formula using the values of the functions along the lines of constant t. It is clear that the type and order of the quadrature formula dictate the locations of the lines of constant t.

In the example to be discussed, Simpson's rule for various interval sizes is used to evaluate the definite integrals, and a fourth-order Adams-Moulton procedure is used to integrate the differential equations. The programs were written in FORTRAN IV, and all computations were carried out in single precision on an IBM 360/44 digital computer.

8. A NUMERICAL EXAMPLE

We consider the boundary value problem

$$\ddot{u} = e^{u}, \qquad 0 \leq t \leq 1, \tag{80}$$

with boundary conditions

$$u(0) = 0 \tag{81}$$

and

$$u(1) = 0. \tag{82}$$

Equation (80) is a special case of the Poisson-Boltzmann equation, which arises in the study of the chemical potential of solutions containing charged particles [1].

In terms of the notation of equation (5), we see that

$$f(u) = -e^u \tag{83}$$

and

$$\Lambda = 1 . \tag{84}$$

The function $R(t,\lambda)$ as defined by equation (31) is

$$R(t,\lambda) = -e^u - \lambda e^u u_\lambda (t,\lambda) . \tag{85}$$

The Cauchy system of equations (29)-(33) is then written as

$$u_\lambda (t,\lambda) = -\int_0^1 G(t,y,\lambda)\, e^{u(y,\lambda)}\, dy \tag{86}$$

$$G_\lambda (t,y,\lambda) = -\int_0^1 G(t,z,\lambda)\,[e^u + \lambda e^u u_\lambda (t,\lambda)]\, G(z,y,\lambda)\, dz \tag{87}$$

$$0 \le \lambda \le 1$$

$$0 \le t,\ y \le 1 ,$$

with boundary conditions

$$u(t,0) = 0 \tag{88}$$

and

$$G(t,y,0) = \begin{cases} y(1-t) , & y \le t , \\ t(1-y) , & y \ge t . \end{cases} \tag{89}$$

The analytic solution of this problem, for $\Lambda = 1$, is (from [9])

$$u(t) = -\ln 2 + 2\ln\{c \sec [c(t-0.5)/2]\} , \tag{90}$$

where c is the root of

$$\sqrt{(2)} = c \sec (\frac{c}{4}) \tag{91}$$

which lies between 0 and $\pi/2$, namely, $c = 1.3360557$, to eight figures.

Table 1 shows the exact values of $u(t)$ obtained from equation (90) along with the values obtained using the Cauchy system with 20-interval Simpson's rule integration. The differential equations were integrated using a fourth-order Adams-Moulton scheme, with a step size of 0.05. The execution time, on the IBM 360/44 computer, was approximately four minutes, including compilation.

TABLE 1

Comparison of exact and approximate values of u(t)

t	u(t) Analytical	u(t) Numerical
0.0	0.000000	0.000000
0.1	-0.041436	-0.041429
0.2	-0.073269	-0.073258
0.3	-0.095800	-0.095787
0.4	-0.109238	-0.109223
0.5	-0.113704	-0.113689
0.6	-0.109238	-0.109223
0.7	-0.095800	-0.095787
0.8	-0.073269	-0.073258
0.9	-0.041436	-0.041429
1.0	0.000000	0.000000

9. DISCUSSION

We have shown that a class of nonlinear boundary value problems can be transformed into a Cauchy system, and we have verified that this Cauchy system satisfies the original boundary value problem. Further, we have presented a specific numerical example which shows its computational efficacy.

Subsequent papers in this series will deal with specific biological problems, including inverse problems [9].

References

1 R. Plonsey, Bioelectrical Phenomena. McGraw-Hill, New York (1969).

2 J. Diamond and W.H. Bossert, J. Gen. Physiol. 50:2061-2083 (1967).

3 K. Kunz, Numerical Analysis. McGraw-Hill, New York (1957).

4 G. Bekey and W. Karplus, Hybrid Computation. John Wiley and Sons, New York (1968).

5 I.S. Beregin and N.P. Zhidkov, Computing Methods. Addison Wesley, Reading, Massachusetts (1965).

6 R. Huss and R. Kalaba, J. Opt. Thy. Applic. 6:415-423 (1970).

7 R. Huss and H. Kagiwada and R. Kalaba, J. Franklin Inst. (to appear).

8 R. Courant and D. Hilbert, Methods of Mathematical Physics. Interscience Publishers, New York (1953).

9 R. Bellman and R. Kalaba, Quasilinearization and Nonlinear Boundary Value Problems. American Elsevier, New York (1965).

R. Huss
H. Kagiwada
R. Kalaba
Department of Electrical Engineering
University of Southern California
Los Angeles
California 90007
USA

I T KIGURADZE and N R LEZHAVA

17 On a nonlinear boundary value problem

The purpose of this paper is to establish sufficient conditions for the existence and uniqueness of a solution of the differential equation

$$u'' = f(t,u,u')$$

satisfying the boundary conditions

$$u'(0) = \varphi_1[u(0)], \quad u'(1) = \varphi_2[u(1)].$$

As an application we consider the boundary value problem arising in chemical reactor theory [1], [2].

1. GENERAL CASE

Consider the boundary value problem

$$u'' = f(t,u,u') \tag{1.1}$$

$$u'(0) = \varphi_1[u(0)], \quad u'(1) = \varphi_2[u(1)] \tag{1.2}$$

where the function $f(t,x,y)$ is defined on the domain

$$\mathscr{D} = \{(t,x,y) : 0 < t < 1, \quad -\infty < x,y < +\infty\}$$

and satisfies Carathéodory's local conditions, i.e. is measurable in t on $(0,1)$ for any x and y, continuous in x,y on the domain $-\infty < x,y < +\infty$ for almost all $t \in (0,1)$ and

$$\sup\{|f(t,x,y)| : |x| + |y| \le \tau\} \in L(0,1) \quad \text{for any} \quad \tau \in (0,+\infty).$$

As to the functions $\varphi_1(x)$ and $\varphi_2(x)$ they are defined and continuous on the interval $(-\infty, +\infty)$.

Under a solution of the problem (1.1), (1.2) we mean a function $u(t)$, which, together with $u'(t)$, is absolutely continuous on the segment $[0,1]$ and almost everywhere on this segment satisfies the equation (1.1) and boundary conditions (1.2).

It is convenient to introduce the following definition:

Definition: Let $\sigma(t)$ be a function which, together with its derivative, is absolutely continuous on $[0,1]$. If

$$f(t,\sigma(t),\sigma'(t)) \geq \sigma''(t) \qquad \text{for} \quad 0 < t < 1$$

and

$$\varphi_1[\sigma(0)] \geq \sigma'(0), \quad \varphi_2[\sigma(1)] \leq \sigma'(1),$$

then we say that $\sigma(t)$ is the upper function of the problem (1.1), (1.2). And if

$$f(t,\sigma(t),\sigma'(t)) \leq \sigma''(t) \qquad \text{for} \quad 0 < t < 1$$

and

$$\varphi_1[\sigma(0)] \leq \sigma'(0), \quad \varphi_2[\sigma(1)] \geq \sigma'(1),$$

then we say that $\sigma(t)$ is the lower function of the problem (1.1), (1.2).

Theorem 1.1: Let $\sigma_1(t)$ be the lower function, and $\sigma_2(t)$ the upper function of the problem (1.1), (1.2); $\sigma_1(t) \leq \sigma_2(t)$ for $0 \leq t \leq 1$ and let for some $t_0 \in [0,1]$ on the domain $0 < t < 1$, $\sigma_1(t) \leq x \leq \sigma_2(t)$, $|y| < +\infty$ the inequality

$$f(t,x,y)\,\mathrm{sign}\,[(t_0-t)\,y \leq \omega(|y|) \sum_{k=1}^{m} h_k(t)\,(1+|y|)^{\frac{1}{q_k}}, \dagger \tag{1.3}$$

holds, where

$$1 \leq q \leq +\infty, \quad \frac{1}{q_k} + \frac{1}{p_k} = 1, \quad h_k(t) \in L^{p_k}(0,1) \quad (k = 1,\dots,m), \tag{1.4}$$

and the function $\omega(t)$ be positive and continuous on the interval $[0,+\infty)$ and

† It is supposed that if $q_k = +\infty$, then $\frac{1}{q_k} = 0$.

$$\int_0^{+\infty} \frac{dt}{\omega(t)} = +\infty . \tag{1.5}$$

Then the problem (1.1), (1.2) has the solution $u(t)$ such that

$$\sigma_1(t) \le u(t) \le \sigma_2(t) \qquad \text{for} \quad 0 \le t \le 1 . \tag{1.6}$$

Corollary 1: Let σ_1 and σ_2 be some constants, $\sigma_1 < \sigma_2$,

$$(-1)^i f(t,\sigma_i,0) \ge 0 \text{ for } 0 < t < 1, \ (-1)^i \varphi_1(\sigma_i) \ge 0, \ (-1)^i \varphi_2(\sigma_i) \le 0 \ (i = 1,2)$$

and let for some $t_0 \in [0,1]$ on the domain $0 < t < 1$, $\sigma_1 < x < \sigma_2$, $|y| < +\infty$ the inequality (1.3) hold, where q_k, p_k, $h_k(t)$ $(k = 1,\ldots,m)$ and $\omega(t)$ satisfy the conditions of Theorem 1.1. Then the problem (1.1), (1.2) has the solution $u(t)$ such that

$$\sigma_1 \le u(t) \le \sigma_2 \qquad \text{for} \quad 0 \le t \le 1 .$$

Corollary 2: Assume that

$$\liminf_{|x| \to +\infty} \varphi_1(x) \operatorname{sign} x > -\infty, \qquad \lim_{|x| \to +\infty} \varphi_2(x) \operatorname{sign} x = -\infty \tag{1.7}$$

and

$$f(t,x,y) \operatorname{sign} y \le h(t)\,\omega(|y|) \qquad \text{for} \quad (t,x,y) \in \mathscr{D}, \tag{1.8}$$

or

$$\lim_{|x| \to +\infty} \varphi_1(x) \operatorname{sign} x = +\infty, \qquad \limsup_{|x| \to +\infty} \varphi_2(x) \operatorname{sign} x < +\infty \tag{1.9}$$

and

$$f(t,x,y) \operatorname{sign} y \ge -h(t)\,\omega(|y|) \qquad \text{for} \quad (t,x,y) \in \mathscr{D}, \tag{1.10}$$

where $h(t) \in L(0,1)$, and $\omega(t)$ is the function which is positive and continuous on the interval $[0,+\infty)$ and satisfies the condition (1.5). Then the problem (1.1), (1.2) is solvable.

Theorem 1.2: Assume that

$$\varphi_1(x_2) \geq \varphi_1(x_1)\ , \quad \varphi_2(x_2) < \varphi_2(x_1) \qquad \text{for} \quad x_2 > x_1 \tag{1.11}$$

and for any $\tau \in (0,+\infty)$

$$f(t,x_2,y_2) - f(t,x_1,y_1) \geq \ell_\tau(t)(y_2 - y_1) \tag{1.12}$$

$$\text{for} \quad x_2 > x_1\ , \ y_2 < y_1\ , \ |x_i| + |y_i| < \tau \quad (i = 1,2)$$

or

$$\varphi_1(x_2) > \varphi_1(x_1)\ , \quad \varphi_2(x_2) \leq \varphi_2(x_1) \qquad \text{for} \quad x_2 > x_1 \tag{1.13}$$

and for any $\tau \in (0,+\infty)$

$$f(t,x_2,y_2) - f(t,x_1,y_1) \geq -\ell_\tau(t)(y_2 - y_1) \tag{1.14}$$

$$\text{for} \quad x_2 > x_1\ , \ y_2 > y_1\ , \ |x_i| + |y_i| \leq \tau \quad (i = 1,2)\ ,$$

where $\ell_\tau(t)$ is the non-negative function, summable on $(0,1)$. Then the problem (1.1), (1.2) has at most one solution.

In order to give the proof of these theorems, some lemmas will be needed.

Lemma 1.1: Assume that $t_0 \in [0,1]$ and the functions $h_k(t)$ $(k = 1,\ldots,m)$ and $\omega(t)$ satisfy the conditions of Theorem 1.1. Then for any $\tau \in (0,+\infty)$ we find a positive number $c = c(\tau)$, such that

$$|u'(t)| \leq c \quad \text{for} \quad 0 \leq t \leq 1 \tag{1.15}$$

for any function $u(t)$ which, together with $u'(t)$, is absolutely continuous on $[0,1]$ and satisfies the inequalities

$$|u(t)| \leq \tau \quad \text{for} \quad 0 \leq t \leq 1 \tag{1.16}$$

$$|u'(0)| \leq \tau\ , \quad |u'(1)| \leq \tau \tag{1.17}$$

and

$$u''(t)\,\mathrm{sign}\,[(t_0 - t)\,u'(t)] \le \omega(|u'(t)|) \sum_{k=1}^{m} h_k(t)(1 + |u'(t)|)^{\frac{1}{q_k}} \tag{1.18}$$

for $0 \le t \le 1$.

Proof: According to (1.4) and (1.5) for given $\tau > 0$ we can find a positive number $c = c(\tau)$, such that

$$\int_{\tau}^{c(\tau)} \frac{ds}{\omega(s)} > \sum_{k=1}^{m} (1+2\tau)^{\frac{1}{q_k}} \|h_k\|_{p_k}, \tag{1.19}$$

where $\|h_k\|_{p_k}$ is the norm of the function $h_k(t)$ on the space $L^{p_k}(0,1)$

Let us now suppose that we can find the function $u(t)$ which, together with $u'(t)$, is absolutely continuous on $[0,1]$, satisfies the conditions (1.16)-(1.18) and does not satisfy the inequality (1.15), i.e.

$$|u'(t^*)| = \max\{|u'(t)| : 0 \le t \le 1\} > c. \tag{1.20}$$

Assume that $t^* \in [0,t_0]$. Then by (1.17) we can find the point $t_* \in [0,t^*]$ such that

$$|u'(t_*)| = \tau \tag{1.21}$$

and

$$u'(t) \ne 0 \quad \text{for} \quad t_* \le t \le t^* . \tag{1.22}$$

Dividing both parts of the inequality (1.8) by $\omega(|u'(t)|)$ and integrating from t_* to t^*, by (1.20) and (1.21) we obtain

$$\int_{\tau}^{c} \frac{ds}{\omega(s)} \le \sum_{k=1}^{m} \int_{t_*}^{t^*} h_k(t)(1 + |u'(t)|)^{\frac{1}{q_k}} dt . \tag{1.23}$$

According to Hölder's inequality and conditions (1.16) and (1.22)

$$\int_{t_*}^{t^*} h_k(t)(1+|u'(t)|)^{\frac{1}{q_k}} dt \le (t^*-t_*+|u(t^*)-u(t_*)|)^{\frac{1}{q_k}} \|h_k\|_{p_k} \le (1+2\tau)^{\frac{1}{q_k}} \|h_k\|_{p_k}$$

$(k = 1, \dots, m)$

Consequently from (1.23) we obtain the inequality which contradicts the inequality (1.19). This contradiction proves that $t^* \notin [0, t_0]$.

In an entirely analogous way it can be easily shown that $t^* \notin [t_0, 1]$.

Hence the inequality (1.20) is not valid. Thus the lemma is proved.

Note: Assume that

$$1 < q < +\infty\,, \quad \frac{1}{p} + \frac{1}{q} = 1\,, \quad 0 < \epsilon < p\,, \quad 0 < \lambda < 1 \quad \text{and} \quad \frac{\lambda}{q} < \frac{\epsilon}{p(p-\epsilon)}\,.$$

Then

$$h(t) = 2\lambda^{-\frac{1}{q}}(1-\lambda)\,|2t-1|^{-\frac{1}{p}-\frac{\lambda}{q}} \in L^{p-\epsilon}(0,1)$$

and for any $\eta > 0$ the function

$$u_\eta(t) = 1 - \lambda \int_0^t [\eta + |2\tau - 1|^{\frac{1-\lambda}{q}}]^{-q}\, d\tau$$

satisfies the conditions

$$|u_\eta(t)| \le 1\,, \quad |u''_\eta(t)| = h(t)\,|u'_\eta(t)|^{1+\frac{1}{q}} \quad \text{for} \quad 0 \le t \le 1\,,$$

$$|u'_\eta(0)| \le \lambda\,, \quad \text{and} \quad |u'_\eta(1)| \le \lambda\,.$$

On the other hand

$$|u'_\eta(\tfrac{1}{2})| = \lambda\eta^{-q} \to +\infty \qquad \text{for} \quad \eta \to 0\,.$$

This example shows that if we replace the condition (1.4) by the condition

$$h_k(t) \in L^{p_k-\epsilon}(0,1) \quad (k = 1, \ldots, m)\,,$$

where $\epsilon > 0$ is an arbitrarily small number, then Lemma 1.1 will not be valid

Lemma 1.2: Assume that

$$|f(t,x,y)| \le h(t) \quad \text{for} \quad (t,x,y) \in \mathscr{D}, \; h(t) \in L(0,1) \tag{1.24}$$

and either the condition (1.7) or condition (1.9) holds. Then the problem (1.1), (1.2) is solvable.

Proof: For the sake of definiteness we suppose below that the condition (1.7) holds.

It is obvious that there exists the positive number γ_0 such that

$$\varphi_2(x) < \varphi_1(\gamma_0) - \rho_0 \quad \text{for} \quad x \geq \gamma_0 + \varphi_1(\gamma_0) - \rho_0 , \tag{1.25}$$

$$\varphi_2(x) > \varphi_1(-\gamma_0) + \rho_0 \quad \text{for} \quad x \leq -\gamma_0 + \varphi_1(-\gamma_0) + \rho_0 ,$$

where

$$\rho_0 = \int_0^1 h(t)\, dt .$$

We denote by H the set of all points of the form $(u(1), u'(1))$ where $u(t)$ is the solution of the equation (1.1) with the initial conditions

$$u(0) = \gamma , \quad u'(0) = \varphi_1(\gamma) \tag{1.26}$$

and γ runs through the interval $[-\gamma_0, \gamma_0]$. According to the theorem of Hukuhara [3] H is the connected and compact set.

Put

$$\varphi(x,y) = y - \varphi_2(x) .$$

By (1.24) for any solution $u(t)$ of problem (1.1), (1.2) we have

$$\varphi_1(\gamma) - \rho_0 \leq u'(1) \leq \varphi_1(\gamma) + \rho_0$$

and

$$\gamma + \varphi_1(\gamma) - \rho_0 \leq u(1) \leq \gamma + \varphi_1(\gamma) + \rho_0 .$$

Consequently from (1.25) it follows that

$$\varphi[u(1), u'(1)] > 0 \qquad \text{for} \quad \gamma = \gamma_0$$

and

$$\varphi[u(1), u'(1)] < 0 \qquad \text{for} \quad \gamma = -\gamma_0 .$$

Therefore on the connected and compact set H the continuous function $\varphi(x,y)$

takes the values of different signs. Hence we can find a point $(x_0, y_0) \in H$ such that $\varphi(x_0, y_0) = 0$, i.e.

$$y_0 = \varphi_2(x_0) .$$

Therefore, according to definition of H, it follows that there exists a number $\gamma \in [-\gamma_0, \gamma_0]$ and a solution $u(t)$ of problem (1.1), (1.26) such that

$$u'(1) = \varphi_2[u(1)] .$$

It is clear that $u(t)$ is the solution of problem (1.1), (1.2). Thus the lemma is proved.

Proof of Theorem 1.1: Assume that

$$\tau_0 = \max\{|\sigma_1(t)| + |\sigma_2(t)| : 0 \le t \le 1\}, \ \tau = \tau_0 + \max\{|\varphi_1(t)| + |\varphi_2(t)| : |t| \le \tau_0\} \tag{1.27}$$

and $c = c(\tau)$ is a number chosen according to Lemma 1.1. Put

$$\rho = c + \max\{|\sigma_1'(t)| + |\sigma_2'(t)| : 0 \le t \le 1\} ,$$

$$f^*(t,x,y) = \begin{cases} f(t,x,y) & \text{for } |y| \le \rho \\ 2\left(1 - \dfrac{|y|}{2\rho}\right) f(t,x,y) & \text{for } \rho < |y| \le 2\rho , \\ 0 & \text{for } |y| > 2\rho \end{cases} \tag{1.28}$$

$$f_k^*(t,x,y) = \begin{cases} f^*(t,\sigma_1(t),\sigma_1'(t)) & \text{for } x \le \sigma_1(t) - \dfrac{1}{k} \\ k(x-\sigma_1(t)+\dfrac{1}{k})f^*(t,\sigma_1(t),y) + k(\sigma_1(t)-x)f^*(t,\sigma_1(t),\sigma_1'(t)) & \\ & \text{for } \sigma_1(t) - \dfrac{1}{k} < x < \sigma_1(t) \\ f^*(t,x,y) & \text{for } \sigma_1(t) \le x \le \sigma_2(t) \\ k(\sigma_2(t)+\dfrac{1}{k}-x)f^*(t,\sigma_2(t),y) + k(x-\sigma_2(t))f^*(t,\sigma_2(t),\sigma_2'(t)) & \\ & \text{for } \sigma_2(t) < x < \sigma_2(t) + \dfrac{1}{k} \\ f^*(t,\sigma_2(t),\sigma_2'(t)) & \text{for } x \ge \sigma_2(t) + \dfrac{1}{k} \end{cases} \tag{1.29}$$

and

$$\varphi_1^*(x) = \begin{cases} \varphi_1[\sigma_1(0)] - \sigma_1(0) + x & \text{for } x < \sigma_1(0) \\ \varphi_1(x) & \text{for } \sigma_1(0) \le x \le \sigma_2(0), \\ \varphi_1[\sigma_2(0)] - \sigma_2(0) + x & \text{for } x > \sigma_2(0) \end{cases}$$

$$\varphi_2^*(x) = \begin{cases} \varphi_2[\sigma_1(1)] + \sigma_1(1) - x & \text{for } x < \sigma_1(1) \\ \varphi_2(x) & \text{for } \sigma_1(1) \le x \le \sigma_2(1) \\ \varphi_2[\sigma_2(1)] + \sigma_2(1) - x & \text{for } x > \sigma_2(1) \end{cases} \tag{1.30}$$

It is clear that

$$(-1)^{i-1} \lim_{|x| \mapsto +\infty} \varphi_i^*(x) \operatorname{sign} x = +\infty \quad (i = 1,2),$$

$$\begin{aligned} &(-1)^i [\varphi_1^*(x) - \sigma_i'(0)] > 0 \quad \text{for} \quad (-1)^i [x - \sigma_i(0)] > 0, \\ & \qquad\qquad\qquad\qquad\qquad\qquad\qquad\qquad (i = 1,2) \\ &(-1)^i [\varphi_2^*(x) - \sigma_i'(1)] < 0 \quad \text{for} \quad (-1)^i [x - \sigma_i(1)] > 0, \end{aligned} \tag{1.31}$$

$$(-1)^i [f_k^*(t,x,y) - \sigma_i''(t)] \ge 0 \quad \text{for} \quad (-1)^i [x - \sigma_i(t)] \ge \frac{1}{k} \quad (i = 1,2) \tag{1.32}$$

and

$$|f_k^*(t,x,y)| \le h(t) \quad \text{for} \quad (t,x,y) \in \mathscr{D}, \tag{1.33}$$

where

$$h(t) = \sup\{|f(t,x,y)| : |x| \le \tau, \quad |y| \le 2\rho\} \in L(0,1).$$

By Lemma 1.2 for any natural number k the problem

$$u'' = f_k^*(t,u,u')$$

$$u'(0) = \varphi_1^*[u(0)], \quad u'(1) = \varphi_2^*[u(1)]$$

has a solution $u_k(t)$. Let us prove that

$$\sigma_1(t) - \frac{1}{k} \le u_k(t) \le \sigma_2(t) + \frac{1}{k} \quad \text{for} \quad 0 \le t \le 1. \tag{1.34}$$

Assume the contrary. Then for some $i \in \{1,2\}$ and $t_0 \in (0,1)$ we have

$$v(t_0) > 0 \,, \tag{1.35}$$

where

$$v(t) = (-1)^i [u_k(t) - \sigma_i(t)] - \frac{1}{k} \,.$$

Let $(\alpha,\beta) \subset (0,1)$ be the maximal interval containing t_0 where $v(t) > 0$. As it follows from (1.31),

$$\text{if } \alpha > 0 \ (\beta < 1) \,, \text{ then } v(\alpha) = 0 \text{ and } v'(\alpha) \geq 0 \, [v(\beta) = 0 \text{ and } v'(\beta) \leq 0] \,, \tag{1.36}$$

and if $\alpha = 0 \ (\beta = 1)$, then $v'(0) > 0 \ [v'(1) < 0]$.
On the other hand, according to (1.32),

$$v''(t) \geq 0 \quad \text{for} \quad \alpha \leq t \leq \beta \,.$$

But this inequality contradicts conditions (1.35) and (1.36). The contradiction thus obtained proves the validity of the condition (1.34)

It follows from (1.33) and (1.34) that the sequences $\{u_k(t)\}_{k=1}^{+\infty}$ and $\{u_k'(t)\}_{k=1}^{+\infty}$ are uniformly bounded and equicontinuous on $[0,1]$. Hence, according to Arcela-Ascoli lemma, without loss of generality, it can be assumed that they converge uniformly. Due to (1.28), (1.29), (1.30) and (1.34) it can be easily concluded that $u(t) = \lim\limits_{k \to +\infty} u_k \, \omega$ is a solution of the equation

$$u'' = f^*(t,u,u')$$

satisfying the conditions (1.2) and (1.6).

By (1.2), (1.3), (1.6), (1.27) and (1.28) it is clear that $u(t)$ satisfies the inequalities (1.16)-(1.18). Therefore according to Lemma 1.1,

$$|u'(t)| \leq c \leq \rho \quad \text{for} \quad 0 \leq t \leq 1 \,.$$

By this condition it follows from (1.28), that $u(t)$ is a solution of equation (1.1).

Thus the theorem is proved.

For $\sigma_i(t) \equiv \sigma_i = \text{const}$ $(i = 1,2)$ the immediate consequence of the above theorem is Corollary 1.

Proof of Corollary 2 of Theorem 1.1: For the sake of definiteness it can be assumed that the conditions (1.7) and (1.8) are satisfied. Suppose that

$$\rho_{10} = \sup\{\varphi_1(x) + |\varphi_1(x)| : x \leq 0\},$$

$$\rho_{20} = \inf\{\varphi_1(x) - |\varphi_1(x)| : x \geq 0\},$$

and $\rho_i(t)$ $(i = 1,2)$ are the solutions of problems

$$\frac{d\rho}{dt} = (-1)^{i-1} h(t)\omega(|\rho|), \quad \rho(0) = \rho_{i0} \quad (i = 1,2).$$

According to (1.7) we can find numbers $\gamma_1 < 0$ and $\gamma_2 > 0$ such that the functions

$$\sigma_i(t) = \gamma_i + \int_0^t \rho_i(\tau)\,d\tau \quad (i = 1,2)$$

satisfy the conditions

$$\sigma_1(t) < \sigma_2(t) \quad \text{for} \quad 0 \leq t \leq 1$$

$$\varphi_2[\sigma_1(1)] > \rho_1(1) \quad \text{and} \quad \varphi_2[\sigma_2(1)] < \rho_2(1).$$

Using (1.8) it can be easily verified that $\sigma_1(t)$ is the lower function and $\sigma_2(t)$ is the upper function of problem (1.1), (1.2). Consequently all the conditions of Theorem 1.1 are satisfied. Thus the solvability of problem (1.1), (1.2) is proved.

Proof of Theorem 1.2: We consider only the case when the conditions (1.11) and (1.12) are satisfied because the case when the conditions (1.13) and (1.14) are satisfied can be considered in a similar way.

Assume that the problem (1.1), (1.2) has two solutions $u_1(t)$ and $u_2(t)$ and for some $t_0 \in (0,1)$ the inequality

$$u_2(t_0) > u_1(t_0)$$

holds. Let (α, β) be the maximal interval containing t_0 and such that on this interval the function $u(t) = u_2(t) - u_1(t)$ is positive. Then according to the first of inequalities (1.11) we have

$$u'(\alpha) \geq 0 .$$

Assume that

$$\min\{u'(t) : \alpha \leq t \leq \beta\} < 0 .$$

Then one can find the points $t_1 \in [\alpha, \beta)$ and $t_2 \in (t_1, \beta)$ such that

$$u'(t_1) = 0 , \quad u'(t) < 0 \quad \text{for} \quad \alpha < t \leq t_2 . \tag{1.37}$$

Due to this fact and condition (1.12) for sufficiently large $\tau > 0$ we shall have

$$\frac{d}{dt}\{u'(t)\exp[-\int_0^t \ell_\tau(\tau)\,d\tau]\} \geq 0 \quad \text{for} \quad \alpha \leq t \leq t_2 . \tag{1.38}$$

But the conditions (1.37) and (1.38) contradict each other. Consequently,

$$u'(t) \geq 0 \quad \text{for} \quad \alpha \leq t \leq \beta .$$

Since $u(t_0) > 0$, according to definition of β, it is clear from the last inequality that $\beta = 1$,

$$u(1) = u_2(1) - u_1(1) > 0 , \quad u'(1) \geq 0 .$$

On the other hand, the second inequality of (1.11) shows that

$$0 \leq u'(1) = \varphi_2[u_2(1)] - \varphi_2[u_1(1)] < 0 .$$

The contradiction thus obtained proves that problem (1.1), (1.2) cannot have two different solutions. Thus the theorem is proved.

2. ON A PROBLEM ARISING IN CHEMICAL REACTOR THEORY

In this section we consider the boundary value problem

$$u'' = f(t,u,u') \tag{2.1}$$

$$u'(0) = 0, \quad u'(1) = \varphi[u(1)], \tag{2.2}$$

arising in chemical reactor theory [1], [2].

Throughout what follows we shall assume that $0 < \sigma < +\infty$, the function $\varphi(x)$ is continuous on the interval $[0,\sigma]$, and the function $f(t,x,y)$ is defined on the domain

$$0 < t < 1, \quad 0 \le x \le \sigma, \quad -\infty < y \le 0, \tag{2.3}$$

measurable in t and for any $\tau \in (0,+\infty)$ we have

$$|f(t,x_1,y_1) - f(t,x_2,y_2)| \le \ell_\tau(t)(|x_1 - x_2| + |y_1 - y_2|)$$

$$\text{for} \quad 0 < t < 1, \ 0 \le x_i < \sigma, \ -\tau \ge y_i \le 0 \quad (i = 1,2),$$

where $\ell_\tau(t)$ is the function, summable on $(0,1)$.

<u>Theorem 2.1</u>: Assume that

$$\varphi(0) = 0, \quad \varphi(x_2) < \varphi(x_1) \quad \text{for} \quad 0 \le x_1 < x_2 \le \sigma, \tag{2.4}$$

and the function $f(t,x,y)$ is nondecreasing in x,

$$f(t,\sigma,0) = 0 \quad \text{for} \quad 0 \le t \le 1 \tag{2.5}$$

and for some $t_0 \in [0,1]$ on the domain (2.3) the inequality

$$f(t,x,y)\,\text{sign}(t - t_0) \le \sum_{k=1}^{m} h_k(t)(1 + |y|)^{1+\frac{1}{q_k}} \tag{2.6}$$

holds, where $q_k \ge 1$, $\frac{1}{q_k} + \frac{1}{p_k} = 1$ and $h_k(t) \in L^{p_k}(0,1)$ $(k = 1,\ldots,m)$. Then problem (2.1), (2.2) has one and only one solution $u(t)$ and

$$0 \le u(t) < \sigma, \quad u'(t) \le 0 \quad \text{for} \quad 0 \le t \le 1 . \tag{2.7}$$

Proof: Assume that

$$g_1(x) = \begin{cases} \sigma & \text{for } x > \sigma \\ x & \text{for } 0 \le x \le \sigma , \\ 0 & \text{for } x < 0 \end{cases} \qquad g_2(x) = \begin{cases} 0 & \text{for } x > 0 \\ x & \text{for } x \le 0 \end{cases}$$

and

$$\tilde{f}(t,x,y) = f(t, g_1(x), g_2(y)) . \tag{2.8}$$

Assume that the function $\varphi(x)$ is extended along all the interval $(-\infty, +\infty)$ and it retains continuity and strict monotonicity.

Consider the equation

$$u'' = \tilde{f}(t,u,u') . \tag{2.9}$$

It follows easily from (2.4)-(2.6) and (2.8) that $\sigma_1(t) \equiv 0$ is the lower function and $\sigma_2(t) \equiv \sigma$ is the upper function of problem (2.9), (2.2) and

$$\tilde{f}(t,x,y)\,\mathrm{sign}[(t_0 - t)\,y] \le \sum_{k=1}^{m+1} h_k(t)\,(1 + |y|)^{1+\frac{1}{q_k}}$$

$$\text{for} \quad 0 < t < 1, \quad 0 \le x \le \sigma, \quad |y| < +\infty ,$$

where $h_{m+1}(t) = |f(t,0,0)| \in L(0,1)$ and $q_{m+1} = +\infty$. Therefore, according to Corollary 1 of Theorem 1.1, using Theorem 1.2, the problem (2.9), (2.2) has one and only one solution $u(t)$ and

$$0 \le u(t) \le \sigma \quad \text{for} \quad 0 \le t \le 1 . \tag{2.10}$$

If we assume that $u(t_0) = \sigma$ for some $t_0 \in [0,1]$, then $u'(t_0) = 0$ and, since the Cauchy problem for the equation (2.9) is uniquely solvable, we shall have $u(t) \equiv \sigma$ for $0 \le t \le 1$. But this is impossible since $u(t) \equiv \sigma$ does not satisfy the boundary conditions (2.2). Thus

$$0 \le u(t) < \sigma \quad \text{for} \quad 0 \le t \le 1 .$$

It is easy to see that $u(t)$ is the solution of the linear equation

$$u'' = f_1(t)u' + f_2(t) ,$$

where $f_1(t)$ is some function, summable on $(0,1)$ and

$$f_2(t) = \tilde{f}(t,u(t),0) = f(t,u(t),0) \le 0 \quad \text{for} \quad 0 \le t \le 1 .$$

Therefore

$$u'(t) = \int_0^t \exp\left[\int_\tau^t f_1(s)\,ds\right] f_2(\tau)\,d\tau \le 0 \quad \text{for} \quad 0 \le t \le 1 .$$

Consequently, $u(t)$ satisfies the inequalities (2.7) and, since on the domain (2.3) the functions $\tilde{f}(t,x,y)$ and $f(t,x,y)$ coincide with each other, it is the solution of the equation (2.1) too. Thus the theorem is proved.

Theorem 2.2: Assume that

$$\varphi(0) = 0 , \quad x_1\varphi(x_2) < x_2\varphi(x_1) \quad \text{for} \quad 0 < x_1 < x_2 \le \sigma , \tag{2.11}$$

$$f(t,\sigma,0) = 0 , \quad f(t,x,0) < 0 \quad \text{for} \quad 0 < t < 1 , \quad 0 \le x < \sigma \tag{2.12}$$

and

$$x_2 f(t,x_1,y_1) \le f(t,x_2,y_2)x_1 \quad \text{for} \quad 0 < x_1 < x_2 \le \sigma , \quad x_1y_2 \le y_1x_2 \le 0 . \tag{2.13}$$

Then the problem (2.1), (2.2) has one and only one solution $u(t)$ and

$$0 < u(t) < \sigma \quad \text{for} \quad 0 \le t \le 1 , \quad u'(t) < 0 \quad \text{for} \quad 0 < t \le 1 . \tag{2.14}$$

Proof: Consider first of all the problem (2.9), (2.2), where $\tilde{f}(t,x,y)$ is the function defined by equality (2.8). From (2.11)-(2.13) it follows, that $\sigma_1(t) \equiv 0$ is the lower function and $\sigma_2(t) \equiv \sigma$ the upper function of problem (2.9), (2.2) and

$$\tilde{f}\,t,x,y)\,\mathrm{sign}\,y \le h(t) \quad \text{for} \quad 0 < t < 1 , \quad 0 \le x \le \sigma , \quad |y| < +\infty ,$$

where

$$h(t) = \sup\{ |f(t,x,0)| : 0 \leq x \leq \sigma\} \in L(0,1) .$$

Therefore according to Corollary 1 of Theorem 1.1, the problem (2.9), (2.2) has the solution $u(t)$ satisfying the inequality (2.10). Now applying here the reasoning, exactly analogous to that which we used to prove Theorem 2.1, it will be shown that $u(t)$ satisfies the inequalities (2.14) and, consequently, it is the solution of problem (2.1), (2.2).

Now let us suppose that the problem (2.1), (2.2) has two solutions $u_1(t)$ and $u_2(t)$. It can be easily verified, that

$$0 < u_i(t) < \sigma \quad \text{for} \quad 0 \leq t \leq 1 , \qquad u_i'(t) < 0 \quad \text{for} \quad 0 < t \leq 1 \qquad (i = 1,2) .$$

$u_1(1) \neq u_2(1)$ since the Cauchy problem for the equation (2.1) is uniquely solvable. Without loss of generality we can suppose that

$$u_2(1) > u_1(1) . \tag{2.15}$$

Then by (2.11),

$$u_2'(1)\, u_1(1) < u_2(1)\, u_1'(1) .$$

Since, furthermore, $u_i'(0) = 0 \ (i = 1,2)$, it is obvious that there exists a point $t_0 \in [0,1)$ such that

$$u_2'(t)\, u_1(t) < u_1'(t)\, u_2(t) \qquad \text{for} \qquad t_0 < t \leq 1 \tag{2.16}$$

and

$$u_2'(t_0)\, u_1(t_0) = u_1'(t_0)\, u_2(t_0) . \tag{2.17}$$

It follows from (2.13) and (2.16) that

$$u_2(t) > u_1(t) \qquad \text{for} \qquad t_0 \leq t \leq 1 . \tag{2.18}$$

According to (2.13) and (2.16)-(2.18)

$$0 > u_2'(1)u_1(1) - u_1'(1)u_2(1) = \int_{t_0}^{1} [f(t,u_2(t),u_2'(t))u_1(t) - f(t,u_1(t),u_1'(t))u_2(t)]dt \geq 0 .$$

The contradiction thus obtained proves the theorem.

Theorem 2.2': Assume that

$$\varphi(0) = 0, \quad x_1 \varphi(x_2) \le x_2 \varphi(x_1) < 0 \quad \text{for} \quad 0 < x_1 < x_2 \le \sigma, \tag{2.19}$$

the conditions (2.12) are satisfied and for any $\delta \in (0, \sigma/2)$ we can find $\epsilon > 0$ which depends only on δ and is such that

$$f(t, x_1, y_1) x_2 < f(t, x_2, y_2) x_1 \tag{2.20}$$

$$\text{for } 0 < t < 1, \ 2\delta \le x_1 + \delta \le x_2 \le \sigma, \ x_1 y_2 \le x_2 y_1 + \epsilon \le 0, \ y_i \ge -\frac{1}{\delta} \ (i = 1,2).$$

Then the problem (2.1), (2.2) has one and only one solution $u(t)$ satisfying the conditions (2.14).

Proof: Using the reasoning exactly analogous to that which we used above we shall show that the problem (2.1), (2.2) is solvable and each of its solutions satisfies the inequalities (2.14).

Now let us suppose that the problem (2.1), (2.2) has two solutions $u_1(t)$ and $u_2(t)$ satisfying the inequalities

$$u_i'(1) > -\frac{1}{\delta} \ (i = 1,2), \quad 2\delta < u_1(1) + \delta < u_2(1) \le \sigma, \tag{2.21}$$

where $2\delta \in (0, \sigma)$. We choose $\epsilon \in (0, |u_2(1) u_1'(1)|)$ in such a way that the condition (2.20) is satisfied. It is obvious that there exists a number $\alpha \in (0,1)$ such that

$$0 < u_1(t) - \delta < u_2(t) \le \sigma, \quad u_i'(t) > -\frac{1}{\delta} \quad (i = 1,2),$$

$$u_1(t) u_2'(t) < u_2(t) u_1'(t) + \epsilon < 0 \quad \text{for} \quad \alpha \le t \le 1. \tag{2.22}$$

Using the reasoning exactly analogous to that which we used to prove Theorem 2.2, it will be shown that due to this assumption

$$u_2'(\alpha) u_1(\alpha) < u_2(\alpha) u_1'(\alpha)$$

we obtain the contradiction. Consequently,

$$u_2'(\alpha)u_1(\alpha) \geq u_2(\alpha)u_1'(\alpha) . \tag{2.23}$$

According to (2.20), (2.22) and (2.23)

$$u_2'(1)u_1(1) - u_1'(1)u_2(1) = u_2'(\alpha)u_1(\alpha) - u_2(\alpha)u_1'(\alpha) +$$

$$+ \int_\alpha^1 [f(t,u_2(t),u_2'(t))u_1(t) - f(t,u_1(t),u_1'(t))u_2(t)]\,dt > 0 .$$

On the other hand, by (2.19) and (2.21),

$$u_2'(1)u_1(1) \leq u_2(1)u_1'(1) .$$

The contradiction thus obtained proves the theorem.

From the above proved theorem as a corollary we obtain some well-known earlier theorems on the existence and uniqueness of the solution of the equation.

$$u'' = f_1(u') + f_2(u)$$

satisfying the boundary conditions (2.2) (see [1] Theorem 1 and [2] Theorem 3.5).

References

1 L. Markus and N.R. Amundson, Nonlinear boundary-value problems arising in chemical reactor theory. J. Differential Equations 4 (1968) 102-113.

2 D.S. Cohen, Multiple stable solutions of nonlinear boundary value problems arising in chemical reactor theory. SIAM J. Appl. Math. 20 (1971) 1-13.

3 M. Hukuhara, Sur une généralisation d'un théorème de Kneser. Proc. Japan Acad. 29 (1953) 154-155.

I.T. Kiguradze and N.R. Lezhava
Institute of Applied Mathematics
Tbilisi University
Tbilisi USSR

M LOWENGRUB

18 Systems of Abel type integral equations

1. INTRODUCTION

In certain mixed boundary value problems arising in the classical theory of elasticity (see Sneddon and Lowengrub [9]), it is possible to reduce the problem to that of finding two unknown functions A_1 and A_2 satisfying the following simultaneous set of dual integral equations

$$\alpha_{11}\,\mathfrak{F}_s[A_1(\xi);x] + \alpha_{12}\,\mathfrak{F}_s[A_2(\xi);x] = f_1(x)\,,\quad x \in (0,1)$$

$$\alpha_{21}\,\mathfrak{F}_c[A_1(\xi);x] + \alpha_{22}\,\mathfrak{F}_c[A_2(\xi);x] = f_2(x)\,,\quad x \in (0,1) \tag{1.1}$$

$$\mathfrak{F}_c[A_1(\xi);x] = 0\,,\quad x \in (1,\infty)$$

$$\mathfrak{F}_s[A_2(\xi);x] = 0\,,\quad x \in (1,\infty)$$

where f_1 and f_2 are continuously differentiable on $[0,1]$ and $_s$ and $_c$ denote the Fourier sine and cosine transforms defined by

$$\mathfrak{F}_s[A(\xi);x] = \sqrt{(\tfrac{2}{\pi})}\int_0^\infty A(\xi)\,\sin(\xi x)\,d\xi$$

and

$$\mathfrak{F}_c[A(\xi);x] = \sqrt{(\tfrac{2}{\pi})}\int_0^\infty A(\xi)\,\cos(\xi x)\,d\xi$$

If in (1.1) we let

$$A_1(\xi) = \int_0^1 p(t)\,J_0(\xi t)\,dt \tag{1.2}$$

and

$$A_2(\xi) = \int_0^1 t\,q(t)\,J_1(\xi t)\,dt\,, \tag{1.3}$$

we see that the problem reduces to determining two functions p and q which satisfy Abel type equations,

$$\alpha_{11} \int_0^x \frac{p(t)\,dt}{(x^2-t^2)^{\frac{1}{2}}} + \alpha_{12}\, x \int_x^1 \frac{q(t)\,dt}{(t^2-x^2)^{\frac{1}{2}}} = f_1(x), \quad 0 < x < 1$$

$$\alpha_{21} \int_x^1 \frac{p(t)\,dt}{(t^2-x^2)^{\frac{1}{2}}} - \alpha_{22}\, x \int_0^x \frac{q(t)\,dt}{(x^2-t^2)^{\frac{1}{2}}} = \tilde{f}_2(x), \quad 0 < x < 1 \tag{1.4}$$

where $\tilde{f}_2(x) = f_2(x) + c$, with c an arbitrary constant.

In this paper, we investigate integral equations which contain (1.3) as a special case. Specifically, we consider the system

$$\alpha(x) \int_0^x \frac{P_1(t)\,dt}{(x^2-t^2)^{\mu}} + \beta(x) \int_x^1 \frac{P_2(t)\,dt}{(t^2-x^2)^{\mu}} = h_1(x), \quad 0 < x < 1 \tag{1.5}$$

$$\gamma(x) \int_x^1 \frac{P_1(t)\,dt}{(t^2-x^2)^{\mu}} + \delta(x) \int_0^x \frac{P_2(t)\,dt}{(x^2-t^2)^{\mu}} = h_2(x), \quad 0 < x < 1 \tag{1.6}$$

where $0 < \mu < 1$ and $\alpha, \beta, \gamma, \delta$ have derivatives satisfying Hölder conditions on $(0,1)$. We also assume that $h_j(x)$, $(j = 1,2)$ satisfies a Hölder condition on $(0,1)$.

By introducing an appropriate function analytic in the entire complex plane except along the cut $(-1,1)$ we show in Section 2 that, in special cases, the problem of determining P_j on $(0,1)$ is equivalent to that of solving a particular Riemann boundary value problem - that is, to determine a function $\Phi(z)$, vanishing at infinity and analytic in the entire complex plane except on $[-1,1]$ where it satisfies a familiar boundary condition. Here we make use of an idea due to Carleman [1]. It is conjectured in that section that such a solution exists only if $\mu = \frac{1}{2}$. Since this is the instance of physical interest, we thoroughly discuss two cases corresponding to that value of μ: (i) $\alpha(x) = \delta(x)$, $\gamma(x) = -\beta(x)$; (ii) $\alpha(x) = \gamma(x)$, $\beta(x) = \delta(x)$. It is also necessary to assume that α and β (hence δ and γ), are real analytic functions.

Section 3 carries out the explicit details of solving the Riemann problem and the corresponding integral equations. The complete solution is obtained by solving two Abel integral equations.

An application to a problem in elasticity is discussed in the fourth section. In particular, our methods are applied to the determination of the stress and displacement fields in the vicinity of a Griffith crack located at the interface of two bonded dissimilar (different elastic constants) half-planes. This problem was originally considered by Erdogan [3] and more recently by Lowengrub and Sneddon [6]. All of the quantities of physical interest are readily obtained. In the final section we briefly discuss some generalizations of our results.

2. REDUCTION TO A RIEMANN BOUNDARY VALUE PROBLEM

Define a function $\Phi_j(z)$ by

$$\Phi_j(z) = \int_0^1 \frac{P_j(t)\,dt}{(z^2 - t^2)^{\mu}}, \quad (j = 1,2) \tag{2.1}$$

where the functions P_j are the functions given in (1.5) and (1.6). The multi-valued function $(z^2 - t^2)^{-\mu}$ is defined in a plane cut along $-1 \leq x \leq 1$ by some branch. In particular, if $\mu = \frac{1}{2}$, then $\sqrt{(z^2 - t^2)} = \xi e^{-\eta}$, $(-\pi \leq \eta \leq \pi)$ where

$$\xi^2 \cos 2\eta = x^2 - y^2 - t^2, \quad \xi^2 \sin(2\eta) = 2xy.$$

The density functions P_j are assumed to satisfy Hölder conditions on $[0,1]$.

Before proceeding further, we summarize some of the properties of $\Phi_j(z)$.

Lemma 2.1: $\Phi_j(z)$ is analytic in the entire complex plane cut along $[-1,1]$. In addition,

(a) for large $|z|$, $\Phi_j(z) = 0\left(\frac{1}{|z|^{2\mu}}\right)$

(G) in a neighbourhood of $+1$ and -1 respectively, we have,

$$\Phi_j(z) = 0\left(\frac{1}{|z-1|^{\mu}}\right) \quad \Phi_j(z) = 0\left(\frac{1}{|z+1|^{\mu}}\right)$$

Proof: This lemma follows directly from the definition of $\Phi_j(z)$.

Lemma 2.2: (Limiting values of $(z^2 - t^2)^{-\mu}$)

$$
\begin{aligned}
&(z^2 - t^2)^{\mu} \to (x^2 - t^2)^{\mu} && (x > t,\ y \to +0) \\
&(z^2 - t^2)^{\mu} \to e^{-i\pi\mu}(t^2 - x^2)^{\mu} && (-t < x < t,\ y \to -0) \\
&(z^2 - t^2)^{\mu} \to e^{i\pi\mu}(t^2 - x^2)^{\mu} && (x < t,\ y \to +0) \qquad (2.2) \\
&(z^2 - t^2)^{\mu} \to e^{2\pi i\mu}(x^2 - t^2)^{\mu} && (x < -t,\ y \to +0) \\
&(z^2 - t^2)^{\mu} \to e^{-2\pi i\mu}(x^2 - t^2)^{\mu} && (x < -t,\ y \to -0)
\end{aligned}
$$

Proof: The proof is straightforward and follows from simple geometrical arguments. (See, for example, Green and Zerna [5].)

Observe that, for $j = 1,2$, we may write

$$\Phi_j(z) = \int_0^{|x|} \frac{P_j(t)dt}{(z^2 - t^2)^{\mu}} + \int_{|x|}^{1} \frac{P_j(t)dt}{(z^2 - t^2)^{\mu}} \qquad (2.3)$$

This leads us to the following very useful lemma,

Lemma 2.3: if $0 < x < 1$, then

$$\text{(a)} \lim_{\substack{z \to x \\ \operatorname{Im} z > 0}} \Phi_j(z) = \Phi_j^+(x) = \int_0^x \frac{P_j(t)dt}{(x^2 - t^2)^{\mu}} + e^{-i\pi\mu} \int_x^1 \frac{P_j(t)dt}{(t^2 - x^2)^{\mu}} \qquad (2.4)$$

$$\text{(b)} \lim_{\substack{z \to x \\ \operatorname{Im} z < 0}} \Phi_j(z) = \Phi_j^-(x) = \int_0^x \frac{P_j(t)dt}{(x^2 - t^2)^{\mu}} + e^{i\pi\mu} \int_x^1 \frac{P_j(t)dt}{(t^2 - x^2)^{\mu}} \qquad (2.5)$$

while if $-1 < x < 0$,

$$\text{(c)}\ \Phi_j^+(x) = e^{-2\pi i\mu} \int_0^{-x} \frac{P_j(t)dt}{(x^2 - t^2)^{\mu}} + e^{-i\pi\mu} \int_{-x}^1 \frac{P_j(t)dt}{(t^2 - x^2)^{\mu}} \qquad (2.6)$$

$$\text{(d)}\ \Phi_j^-(x) = e^{2\pi i\mu} \int_0^{-x} \frac{P_j(t)dt}{(x^2 - t^2)^{\mu}} + e^{i\pi\mu} \int_{-x}^1 \frac{P_j(t)dt}{(t^2 - x^2)^{\mu}} . \qquad (2.7)$$

Proof: Simply apply the results of Lemma 2.2 to the representation given in (2.3).

If we substitute from Lemma 2.3 into equations (1.5) and (1.6) we obtain the condition that on $0 < x < 1$,

$$\alpha(x)\,[e^{+i\pi\mu}\,\Phi_1^+(x) - e^{-i\pi\mu}\,\Phi_1^-(x)] - \beta(x)\,[\Phi_2^+(x) - \Phi_2^-(x)] = 2i\sin\mu\pi\,h_1(x) \tag{2.8}$$

$$-\gamma(x)\,[\Phi_1^+(x) - \Phi_1^-(x)] + \delta(x)\,[e^{i\pi\mu}\,\Phi_2^+(x) - e^{-i\pi\mu}\,\Phi_2^-(x)] = 2i\sin\mu\pi\,h_2(x)\,. \tag{2.9}$$

In order to determine Φ_1 and Φ_2, it is necessary to establish the boundary value problem on $(-1,1)$. Observe from Lemma 2.3 that on $-1 < x < 0$,

$$e^{i\pi\mu}\,\Phi_j^+(x) - e^{-i\pi\mu}\,\Phi_j^-(x) = [-2i\sin\mu\pi]\int_0^{-x}\frac{P_j(t)dt}{(x^2-t^2)^\mu} \tag{2.10}$$

$$\Phi_j^+(x) - \Phi_j^-(x) = [e^{-2\pi i\mu} - e^{2\pi i\mu}]\int_0^{-x}\frac{P_j(t)dt}{(x^2-t^2)^\mu} + [-2i\sin\mu\pi]\int_{-x}^{1}\frac{P_j(t)dt}{(x^2-t^2)^\mu} \tag{2.11}$$

while on $0 < x < 1$,

$$e^{i\pi\mu}\,\Phi_j^+(x) - e^{-i\pi\mu}\,\Phi_j^-(\kappa) = 2i\sin\mu\pi\int_0^{x}\frac{P_j(t)dt}{(x^2-t^2)^\mu} \tag{2.12}$$

and

$$\Phi_j^+(x) - \Phi_j^-(x) = -2i\sin\mu\pi\int_x^{1}\frac{P_j(t)dt}{(t^2-x^2)^\mu} \tag{2.13}$$

Hence, it is clear that only if $\mu = \frac{1}{2}$ do we have the correct symmetry to extend the boundary condition to $(-1,1)$. In order to determine Φ_1 and Φ_2 it is necessary to formulate appropriate conditions on $(-1,1)$. If we denote by $\hat{\alpha},\hat{\beta},\hat{\gamma},\hat{\delta}$, and $\hat{h}_j$ the odd or even extension of $\alpha,\beta,\gamma,\delta$ and h_j on $(-1,1)$, then our boundary conditions take the form

$$i\,\hat{\alpha}(x)\,[\Phi_1^+(x) + \Phi_1^-(x)] - \hat{\beta}(x)\,[\Phi_2^+(x) - \Phi_2^-(x)] = 2i\,\hat{h}_1(x)\,, \tag{2.14}$$

$$-\hat{\gamma}(x)\,[\Phi_1^+(x) - \Phi_1^-(x)] + i\,\hat{\delta}(x)\,[\Phi_2^+(x) + \Phi_2^-(x)] = 2i\,\hat{h}_2(x)\,, \tag{2.15}$$

for $-1 < x < 1$ [where, of course, the $\Phi_j^+ \pm \Phi_j^-$ are extended to $(-1,0)$ in an obvious way].

If we first add (2.14) to (2.15) and then subtract (2.15) from (2.14) we obtain the conditions

$$\{[\hat{\alpha} - \hat{\gamma}]\, \Phi_1^+(x) + i[\hat{\beta} + \hat{\delta}]\, \Phi_2^+(x)\} + \{(\hat{\alpha} + \hat{\gamma})\, \Phi_1^- + i\,(\hat{\delta} - \hat{\beta})\, \Phi_2^-\} = \hat{\omega}(\kappa), \tag{2.16}$$

$$\{[\hat{\alpha} + \hat{\gamma}]\, \Phi_1^+(x) - i[\hat{\delta} - \hat{\beta}]\, \Phi_2^+(x)\} + \{(\hat{\alpha} - \hat{\gamma})\, \Phi_1^- - i\,(\hat{\beta} + \hat{\delta})\, \Phi_2^-\} = \overline{\hat{\omega}(x)} \tag{2.17}$$

where $\hat{\omega}(x) = 2[\hat{h}_1 + i\,\hat{h}_2]$.

Thus if we can construct two auxiliary sectionally analytic functions $\Psi(z)$ and $\Gamma(z)$ satisfying the boundary conditions (2.16) and (2.17) we could possibly have a solution to the integral equations (1.5) and (1.6). The main object would be to obtain an uncoupled set of relations for Ψ and Γ on $(-1,1)$ equivalent to (2.16) and (2.17). It is obvious that this cannot be done for all α, β, γ and δ .

At this point, we consider special cases and state some of the theorems demonstrating the equivalence with the original pair of integral equations. These special cases correspond to the application of physical interest.

Case (i): We set $\hat{\alpha}(x) \equiv \hat{\delta}(x)$, $\hat{\gamma}(x) \equiv -\hat{\beta}(x)$, and $\mu = \frac{1}{2}$ in the original integral equations (1.5) and (1.6). If we substitute into (2.16) and (2.17), we see that our original problem reduces to determining two functions

$$\Psi(z) = \Phi_1(z) + i\,\Phi_2(z), \tag{2.18}$$

$$\chi(z) = \Phi_1(z) - i\,\Phi_2(z) \tag{2.19}$$

analytic everywhere except along the plane cut $-1 \leq x \leq 1$, along which Ψ and χ satisfy the boundary conditions

$$\Psi^+(x) + \frac{\hat{\alpha}(x) - \hat{\beta}(x)}{\hat{\alpha}(x) + \hat{\beta}(x)}\, \Psi^-(x) = \frac{\hat{\omega}(x)}{\alpha(x) + \beta(x)}, \tag{2.20}$$

and

$$\chi^+(x) + \frac{\hat{\alpha}(x) + \hat{\beta}(x)}{\hat{\alpha}(x) - \hat{\beta}(x)}\, \chi^-(x) = \frac{\overline{\hat{\omega}(x)}}{\hat{\alpha}(x) - \hat{\beta}(x)} \tag{2.21}$$

where $\hat{\omega}(x) = 2[\hat{h}_1(x) + i\,\hat{h}_2(x)]$.

It is clear that we must assume, in addition to the criteria stated in Section 1, that along $-1 \leq x \leq 1$, $\hat{\alpha}(x) \neq \pm\hat{\beta}(x)$. If $\hat{\alpha}(x) = \pm\hat{\beta}(x)$, then the case (see (ii) below) must be considered separately.

In the next section, we shall discuss the solution to the above Riemann-Hilbert problem. However, it follows from what has preceded that,

Theorem 2.4: Let Ψ and χ be two functions analytic everywhere except along the plane cut $-1 \leq x \leq 1$ where the conditions (2.20) and (2.21) are satisfied. In addition, assume Ψ and χ are related to Φ_i $(i = 1,2)$ by (2.18) and (2.19). Then if Φ_j is related to P_j $(j = 1,2)$ by (2.6) and (2.7), the solutions to the pair of equations (1.5) and (1.6) with $\mu = \frac{1}{2}$, $\alpha(x) = \delta(x)$, $\gamma(x) = -\beta(x)$ is equivalent to the solution of the Riemann boundary value problems (2.20) and (2.21).

Proof: If $P_j(x)$ $(j = 1,2)$ are solutions to the pair of equations (1.5) and (1.6) and if in addition they satisfy Hölder conditions on $(0,1)$, then Φ_j $(j = 1,2)$ satisfy the conditions of Lemma 2.1 and by Lemma 2.3, Ψ and χ are solutions to the boundary value problems (2.20) and (2.21).

Conversely, if Ψ and χ are solutions of the Riemann boundary value problems (2.20) and (2.21) then we can determine $\Phi_j(z)$, $(j = 1,2)$ uniquely from (2.18) and (2.19). Lemma 2.3 implies that Φ_j satisfy the conditions (2.10) to (2.13) (with $\mu = \frac{1}{2}$). Thus, the functions $P_j(x)$ are uniquely determined by solving the Abel integral equations (2.12) and (2.13). (For a discussion of the uniqueness of Abel integral equations, see [4].) It then follows from Lemma 2.3 that $P_j(x)$ also satisfy the original integral equations (1.5) and (1.6) [with $\mu = \frac{1}{2}$, $\alpha(x) = \delta(x)$, $\gamma(x) = -\beta(x)$]. Thus, the equivalence is established.

Case (ii): We set $\hat{\alpha}(x) = \hat{\gamma}(x)$, $\hat{\beta}(x) = \hat{\delta}(x)$ and $\mu = \frac{1}{2}$ in the original integral equations (1.5) and (1.6). In addition we assume that $\hat{\alpha}$ and $\hat{\beta}$ are real analytic functions and let $\hat{\alpha}(z)$ and $\hat{\beta}(z)$ denote their extensions to the entire complex plane. Substitution into (2.16) and (2.17) reduces our problem to that of determining two functions

$$\Delta(z) = \hat{\alpha}(z)\,\Phi_1(z) + \hat{\beta}(z)\,\Phi_2(z)\ , \tag{2.22}$$

$$\Gamma(z) = \hat{\alpha}(z)\,\Phi_1(z) - \hat{\beta}(z)\,\Phi_2(z)\ , \tag{2.23}$$

analytic everywhere except along the plane cut $-1 \leq x \leq 1$, where Δ and Γ satisfy the boundary conditions

$$\Delta^+(x) + \left(\frac{i+1}{i-1}\right)\Delta^-(x) = \frac{2i[\hat{h}_1(x) + \hat{h}_2(x)]}{(i-1)}, \tag{2.24}$$

$$\Gamma^+(x) + \left(\frac{i-1}{i+1}\right)\Gamma^-(x) = \frac{2i[\hat{h}_1(x) - \hat{h}_2(x)]}{(i+1)}. \tag{2.25}$$

The solution to the Riemann problems for Δ and Γ are straightforward and will be outlined in the next section.

It is clear that a theorem analogous to Theorem 2.4 can be stated for this case. We only require that $\alpha(x)$, $\beta(x)$ do <u>not</u> vanish on $-1 \leq x \leq 1$ This implies that their extension to the entire complex plane cannot vanish. Otherwise, we cannot obtain a unique solution to the system of equations given by (2.22), (2.23) for $\Phi_j(z)$, $(j = 1,2)$ in terms of Δ and Γ. At points where they do vanish, the cases must be considered separately. We shall see this in one of the applications considered below.

3. SOLUTION OF THE PAIR OF INTEGRAL EQUATIONS

We begin by first considering the solution to the pair of equations

$$\alpha(x)\int_0^x \frac{P_1(t)\,dt}{(x^2-t^2)^{\frac{1}{2}}} + \beta(x)\int_x^1 \frac{P_2(t)\,dt}{(t^2-x^2)^{\frac{1}{2}}} = h_1(x), \quad 0 < x < 1 \tag{3.1}$$

$$-\beta(x)\int_x^1 \frac{P_1(t)\,dt}{(t^2-x^2)^{\frac{1}{2}}} + \alpha(x)\int_0^x \frac{P_2(t)\,dt}{(x^2-t^2)^{\frac{1}{2}}} = h_2(x), \quad 0 < x < 1 \tag{3.2}$$

corresponding to Case (ii) above. Recall that α, β, h_1, and h_2 are required to satisfy Hölder conditions on $0 < x < 1$ and additionally, $\alpha(x) \neq \pm\beta(x)$ on $0 < x < 1$. If we denote by $\hat{\alpha}, \hat{\beta}$ and $\hat{h}_j$ $(j = 1,2)$ the odd or even extension of α, β and h_j, we can then determine $\Psi(z)$ and $\chi(z)$ satisfying conditions (2.20) and (2.21). We specifically seek Ψ and χ such that $\Psi(\infty) = \chi(\infty) = 0$.

Let

$$\mathscr{G}_1(x) = \frac{\hat{\alpha}(x) - \hat{\beta}(x)}{\hat{\alpha}(x) + \hat{\beta}(x)}, \quad x \in [-1,1]$$

$$\mathscr{G}_2(x) = \frac{\hat{\alpha}(x) + \hat{\beta}(x)}{\hat{\alpha}(x) - \hat{\beta}(x)}, \quad x \in [-1,1] .$$

It can be shown (Muskhelishvili [7] p.236) that if $\chi_1 = \text{index}\, \mathscr{G}_1(t) \geq 0$, then

$$\Psi(z) = \frac{X_1(z)}{2\pi i} \int_{-1}^{1} \frac{P_1(t)\,dt}{X_1^+(t)\,(t-z)} + X_1(z)\, p_{\chi_1 - 1}(z) , \tag{3.3}$$

where

$$\rho_1(t) = \frac{\hat{\omega}(t)}{\hat{\alpha}(t) + \hat{\beta}(t)} ,$$

$$X_1(z) = [(z^2 - 1)]^{-\chi_1}\, e^{\Gamma_1(z)} ,$$

$$\Gamma_1(z) = \frac{1}{2\pi i} \int_{-1}^{1} \frac{\log \mathscr{G}_1(t)}{t - z}\, dt ,$$

and $P_{\chi_1 - 1}(z)$ is a polynomial of degree not greater than $\chi_1 - 1$. (If $\chi_1 = 0$, then $p_{\chi_1 - 1} \equiv 0$) .

In like manner, if $\chi_2 = \text{index}\, \mathscr{G}_2(t) \geq 0$,

$$\chi(z) = \frac{\chi_2(z)}{2\pi i} \int_{-1}^{1} \frac{\delta(t)\,dt}{\chi_2^+(t)\,(t-z)} + \chi_2(z)\, p_{\chi_2 - 1}(z) , \tag{3.4}$$

where

$$\rho_2(t) = \frac{\bar{\hat{\omega}}(t)}{\hat{\alpha}(t) - \hat{\beta}(t)} ,$$

$$\chi_2(z) = [(z^2 - 1)]^{-\chi_2}\, e^{\Gamma_2(z)}$$

$$\Gamma_2(z) = \frac{1}{2\pi i} \int_{-1}^{1} \frac{\log \mathscr{G}_2(t)}{t - z}\, dt$$

and $p_{\chi_2 - 1}(z)$ is a polynomial of degree less than $\mathcal{X}_2 - 1$ which vanishes identically if $\mathcal{X}_2 = 0$.

If, index $\mathcal{G}_1(t) < 0$ and index $\mathcal{G}_2(t) < 0$ then the solutions of a given class vanishing at ∞ exist if and only if the conditions

$$\int_{-1}^{1} \frac{t^j \rho_i(t)\, dt}{X_i^+(t)} = 0, \quad (j = 1, 2, \ldots, -x_i - 1,) \quad (i = 1, 2). \tag{3.5}$$

Once we have determined $\chi(z)$ and $\Psi(z)$, equations (2.18) and (2.19) uniquely determine $\Phi_j(z)$, $j = 1, 2$. We immediately note from (2.12) that

$$P_j(t) = \frac{1}{\pi} \frac{d}{dt} \int_0^t \frac{x[\Phi_j^+(x) + \Phi_j^-(x)]\, dx}{\sqrt{(t^2 - x^2)}}, \quad (j = 1, 2), \quad 0 \le t \le 1. \tag{3.6}$$

The pair of equations corresponding to Case (ii) lend themselves to a slightly less complicated form of solution. We first write out the solution to the pair of Riemann problems for $\Delta(z)$ and $\Gamma(z)$. It follows (see Muskhelishvili [6] p.450) that the solution $\Delta(z)$ vanishing at infinity is

$$\Delta(z) = \frac{X_1(z)}{2\pi i} \int_{-1}^{1} \frac{\hat{h}^+(t)\, dt}{X_1^+(t)(t - z)} + X_1(z) C_1 \tag{3.7}$$

where C_1 is an arbitrary complex constant,

$$X_1(z) = (z + 1)^{-1/4} (z - 1)^{-3/4} \tag{3.8}$$

and

$$\hat{h}^+(t) = \frac{2i}{i - 1} [\hat{h}_1(t) + \hat{h}_2(t)]. \tag{3.9}$$

In like manner, the solution to the second problem vanishing at infinity is given by

$$\Gamma(z) = \frac{X_2(z)}{2\pi i} \int_{-1}^{1} \frac{\hat{h}^-(t)\, dt}{X_2^+(t)(t - z)} + X_2(z) C_2 \tag{3.10}$$

where C_2 is an arbitrary complex constant,

$$X_2(z) = (z+1)^{-3/4}\,(z-1)^{-1/4} \tag{3.11}$$

and

$$\hat{h}^-(t) = \frac{2i[\hat{h}_1(t) - \hat{h}_2(t)]}{(i+1)} .$$

Once we have determined Γ and Δ, we see that

$$\Phi_1(z) = \frac{\Delta(z) + \Gamma(z)}{2\alpha(z)} \tag{3.12}$$

$$\Phi_2(z) = \frac{\Delta(z) - \Gamma(z)}{2\beta(z)} . \tag{3.13}$$

We again obtain $P_j(t)$ $(j = 1,2)$ from equation (3.6).

4. AN APPLICATION TO A PROBLEM IN ELASTICITY

Lowengrub and Sneddon [6] consider the problem of determining the stress and displacement fields in the vicinity of a Griffith crack $|x| \leq 1$, $y = 0$ located at the interface of two-bonded elastic half-planes (with different elastic constants). In this section we apply our previous results to demonstrate how one can determine the quantities of physical interest without actually determining $P_j(t)$, $(j = 1,2)$. In fact, we first show that the problem reduces to that outlined in Section 1 of this paper and then determine the stresses and displacements in terms of the sectionally analytic functions Δ and Γ given in Section 3.

Following the Lowengrub and Sneddon [6] paper, we assume that if the displacement vector is denoted by $u = (u_y, u_x)$ then

$$u_y(1+, 0+) = u_y(1+, 0-) = 0$$

and (4.1)

$$u_y(1-, 0+) = u_y(1-, 0-) = 0 .$$

In addition, we assume that the normal and shearing stresses σ_{xy} and σ_{yy} satisfy the condition

$$\sigma_{xy}(x,0) = 0(x^{-1}), \quad \sigma_{yy}(x,0) = 0(x^{-1}) \quad \text{as} \quad x \to \infty .$$

The two elastic materials are to occupy the upper half-plane $y > 0$ with elastic constants G_1, κ_1, and the lower half-plane $y < 0$ with elastic constants G_2, κ_2. G_j, $(j = 1,2)$ denote the rigidity moduli of the respective media and $\kappa_j = 3-4\eta_j$, $(j = 1,2)$ where η_j is the Poisson ratio of the elastic material occupying the half-spaces.

Lowengrub and Sneddon [6] show that if the upper and lower surfaces of the crack are subjected to a prescribed even pressure $p(x)$ then inside the crack area the conditions

$$\sigma_{yy}(x,0+) = \sigma_{yy}(x,0-) = -p(x) \quad |x| \leq 1 \tag{4.2}$$

$$\sigma_{xy}(x,0+) = \sigma_{xy}(x,0-) = 0, \quad |x| < 1 \tag{4.3}$$

prevail, while outside the crack area we have the continuity conditions

$$u_x(x,0+) = u_x(x,0-), \quad |x| > 1 \tag{4.4}$$

$$u_y(x,0+) = u_y(x,0-), \quad |x| > 1 \tag{4.5}$$

$$\sigma_{yy}(x,0+) = \sigma_{yy}(x,0-), \quad |x| > 1 \tag{4.6}$$

$$\sigma_{xy}(x,0+) = \sigma_{xy}(x,0-), \quad |x| > 1 . \tag{4.7}$$

In that paper, Lowengrub and Sneddon show that if p is an even function, then the solutions of the equation of elastic equilibrium will be satisfied if we take the displacement field,

$$u_x(x,y) = \begin{cases} \mathfrak{F}_s[\xi^{-1}\{A_1 - \kappa_1^{-1}(A_1 - B_1)y\}e^{-\xi y};\ \xi \to x], & y > 0 \\ \mathfrak{F}_s[\xi^{-1}\{A_2 + \kappa_2^{-1}(A_2 + B_2)y\}e^{\xi y};\ \xi \to x], & y < 0 \end{cases} \tag{4.8}$$

$$u_y(x,y) = \begin{cases} \mathfrak{F}_c[\xi^{-1}\{B_1 + \kappa_1^{-1}(A_1 - B_1)\xi y\}e^{-\xi y};\ \xi \to x], & y > 0 \\ \mathfrak{F}_c[\xi^{-1}\{B_2 - \kappa_2^{-1}(A_2 + B_2)\xi y\}e^{\xi y};\ \xi \to x], & y < 0 . \end{cases} \tag{4.9}$$

$A_i(\xi)$ and $B_i(\xi)$, $(i = 1,2,)$ will be determined from the boundary conditions (4.2)-(4.3). This field yields the following expressions for the stresses on $y = 0$,

$$\sigma_{yy}(x,0) = \frac{G_1\{\alpha \mathfrak{F}_c[\varphi(\xi);x] + \beta \mathfrak{F}_c[\Psi(\xi);x]\}}{\kappa_1(\kappa_2 + \Gamma)(1 + \kappa_1\Gamma)} \tag{4.10}$$

$$\sigma_{xy}(x,0) = \frac{-G_1\{\beta \mathfrak{F}_s[\Psi(\xi);x] + \alpha \mathfrak{F}_s[\varphi(\xi);x]\}}{\kappa_1(\kappa_2 + \Gamma)(1 + \kappa_1\Gamma)}, \tag{4.11}$$

where

$$\alpha = (\kappa_1 - 1)\,\Gamma - (\kappa_2 - 1)$$

$$\beta = (\kappa_1 + 1)\,\Gamma + (\kappa_2 + 1)$$

with $\Gamma = {G_2}/{G_1}$ and φ and Ψ are linear combinations of A_i, B_i $(i = 1,2)$ given in Section 2, equations (2.17) and (2.18) of [6].

The boundary conditions (4.2)-(4.3) are equivalent to the conditions

$$\mathfrak{F}_c[\alpha\varphi + \beta\Psi; x] = f(x), \quad 0 < x < 1 \tag{4.12}$$

$$\mathfrak{F}_s[\beta\varphi + \alpha\Psi; x] = 0, \quad 0 < x < 1 \tag{4.13}$$

$$\mathfrak{F}_s[\xi^{-1}\varphi; x] = 0, \quad x > 1 \tag{4.14}$$

$$\mathfrak{F}_c[\xi^{-1}\Psi; x] = 0, \quad x > 1 \tag{4.15}$$

where

$$f(x) = \kappa_1(\kappa_2 + \Gamma)(1 + \kappa_1\Gamma)\,G_1^{-1}\,p(x). \tag{4.16}$$

One can verify that this system of equations is equivalent to the system,

$$\mathfrak{F}_c[\alpha\varphi + \beta\Psi; x] = f(x), \quad 0 < x < 1 \tag{4.12'}$$

$$\mathfrak{F}_s[\beta\varphi + \alpha\Psi; x] = 0, \quad 0 < x < 1 \tag{4.13'}$$

$$\mathfrak{F}_c[\varphi; x] = 0, \quad x > 1 \tag{4.14'}$$

$$\mathfrak{F}_s[\Psi ; x] = 0 , \qquad x > 1 \tag{4.15'}$$

obtained by differentiating (4.14) and (4.15). This system is precisely (1.1) with $\alpha_{11} = \beta$, $\alpha_{12} = \alpha$, $\alpha_{21} = \alpha$, $\alpha_{22} = \beta$, $f_1(x) = 0$, $f_2(x) = f(x)$. If we let

$$\varphi(\xi) = \int_0^1 P_1(t) J_0(\xi t) dt \tag{4.16'}$$

and

$$\Psi(\xi) = \int_0^1 P_2(t) [\int_0^\xi J_0(ut) du] dt = \int_0^\xi du \int_0^1 P_2(t) J_0(ut) dt \tag{4.17}$$

(see Jones [10] p.473), the conditions (4.14') and (4.15') are automatically satisfied while the remaining conditions are equivalent to the problem of determining $P_j(t)$, $(j = 1,2)$ while satisfying (1.5) and (1.6) with

$$\mu = \tfrac{1}{2}, \quad \alpha(x) = \beta, \quad \beta(x) = \frac{\alpha}{x}, \quad \gamma(x) = \alpha, \quad \delta(x) = \frac{-\beta}{x} \tag{4.18}$$

and $h_1(x) = 0$, $h_2(x) = f(x)$.

If we substitute from (4.18) into the relations (2.16) and (2.17) we find that our problem reduces to that of determining two functions Φ_1 and Φ_2 analytic in the entire complex plane except along the plane cut $-1 \leq x \leq 1$ where Φ_1 and Φ_2 satisfy the boundary conditions

$$\{(\beta-\alpha)\Phi_1^+(x) - i(\beta-\alpha)\frac{1}{x}\Phi_2^+(x)\} + \{(\beta+\alpha)\Phi_1^-(x) - i(\beta+\alpha)\frac{1}{x}\Phi_2^-(x)\} = g(x), \tag{4.19}$$

$$\{(\beta+\alpha)\Phi_1^+(x) + i(\beta+\alpha)\frac{1}{x}\Phi_2^+(x)\} + \{(\beta-\alpha)\Phi_1^-(x) + i(\beta-\alpha)\frac{1}{x}\Phi_2^-(x)\} = \overline{g(x)}, \tag{4.20}$$

where $g(x) = 2i\, f(x)$.

Define:

$$\Delta(z) = \varphi_1(z) - \frac{i}{z}\varphi_2(z) \tag{4.21}$$

and

$$\Sigma(z) = \Phi_1(z) + \frac{i}{z}\varphi_2(z) . \tag{4.22}$$

Observe that both Δ and Σ are analytic in the entire complex plane except along the cut $-1 \leq x \leq 1$, $y = 0$, where they satisfy the conditions

$$\Delta^+(x) + \frac{1}{\kappa}\Delta^-(x) = \frac{2f(x)}{\beta - \alpha}, \tag{4.23}$$

$$\Sigma^+(x) + \kappa\Sigma^-(x) = \frac{-2f(x)}{\beta + \alpha} \tag{4.24}$$

with

$$\kappa = \frac{\beta - \alpha}{\beta + \alpha} = {(\Gamma+1)}/{(\kappa_1\Gamma + \kappa_2)} > 0 .$$

The solutions to both of the above Riemann-Hilbert problems are well known (see [7] p.451) and take the form

$$\Delta(z) = \frac{X_1(z)}{2\pi(\beta-\alpha)}\int_{-1}^{1}\frac{g(t)\,dt}{X_1^+(t)(t-z)} + C_1X_1(z) \tag{4.25}$$

$$\Sigma(z) = \frac{X_2(z)}{\pi(\beta+\alpha)}\int_{-1}^{1}\frac{\overline{g(t)}\,dt}{X_2^+(t)(t-z)} + C_2X_2(z) \tag{4.26}$$

where C_i, $(i = 1,2)$ are arbitrary complex constants, and

$$X_1(z) = (z+1)^{i\rho_1 - \frac{1}{2}}(z-1)^{-i\rho_1 - \frac{1}{2}}, \quad \rho_1 = \frac{1}{2\pi}\log 1/\kappa \tag{4.27}$$

$$X_2(z) = (z+1)^{i\rho_2 - \frac{1}{2}}(z-1)^{-i\rho_2 - \frac{1}{2}}, \quad \rho_2 = \frac{1}{2\pi}\log \kappa \tag{4.28}$$

If f is a polynomial, then (cf. [2] p.18)

$$\Delta(z) = \frac{i}{2\beta}\left[2f(z) - X_1(z)L_1(z) + C_1X_1(z)\right] \tag{4.29}$$

and

$$\Sigma(z) = \frac{i}{2\beta}\left[-2f(z) - X_2(z)L_2(z) + C_2X_2(z)\right] \tag{4.30}$$

where

$$L_1(z) = \frac{1}{\pi} \lim_{R\to\infty} \int_0^{2\pi} \frac{f(Re^{i\Theta}) Re^{i\Theta} d\Theta}{X_1(Re^{i\Theta})(Re^{i\Theta}-z)} \tag{4.31}$$

and

$$L_2(z) = -\frac{1}{\pi} \lim_{R\to\infty} \int_0^{2\pi} \frac{f(Re^{i\Theta}) Re^{i\Theta} d\Theta}{X_2(Re^{i\Theta})(Re^{i\Theta}-z)} . \tag{4.32}$$

We note that, for $x > 1$,

$$\sigma_{yy}(x,0+) = \frac{G_1 \beta}{\kappa_1(\kappa_2 + \Gamma)(1 + \kappa_1 \Gamma)} \cdot \mathfrak{F}_c[\Psi(\xi); x]$$

$$= -\frac{G_1 \beta}{\kappa_1(\kappa_2 + \Gamma)(1 + \kappa_1 \Gamma)} \cdot \frac{1}{x} \int_0^1 \frac{P_2(t)\, dt}{\sqrt{(x^2 - t^2)}} \tag{4.33}$$

and

$$\sigma_{xy}(x,0+) = -\frac{G_1 \beta}{\kappa_1(\kappa_2 + \Gamma)(1 + \kappa_1 \Gamma)} \int_0^1 \frac{P_1(t)\, dt}{\sqrt{(x^2 - t^2)}} \tag{4.34}$$

which from (2.1) and (4.21), (4.22) yields,

$$\sigma_{yy}(x,0+) = \frac{G_1 \beta}{\kappa_1(\kappa_2 + \Gamma)(1 + \kappa_1 \Gamma)} \left[\frac{\Sigma^+(x) - \Delta^+(x)}{2i}\right] \tag{4.35}$$

and

$$\sigma_{xy}(x,0+) = -\frac{G_1 \beta}{\kappa_1(\kappa_2 + \Gamma)(1 + \kappa_1 \Gamma)} \left[\frac{\Sigma^+(x) + \Delta^+(x)}{2}\right] . \tag{4.36}$$

If we assume that the applied pressure is constant, say $p(x) = p_0$, so that $f(x) = f_0 = \kappa_1(\kappa_2 + \Gamma)(1 + \kappa_1 \Gamma) G_1^{-1} p_0$, then we obtain the expressions for $\sigma_{yy}(x,0+)$ and $\sigma_{xy}(x,0+)$ found in Lowengrub and Sneddon ([6] p. 1032). Here we must use the fact that $\sigma_{yy}(x,0+) = 0(\frac{1}{x})$, as $x \to \infty$ and $\sigma_{xy}(x,0+) = 0(\frac{1}{x})$, as $x \to \infty$. This determines the constants C_1 and C_2 in (4.25) and (4.26) which in fact vanish.

It is a simple matter to verify that, for $x > 1$,

$$X_1^+(x) = (x^2 - 1)^{-\frac{1}{2}} e^{-i\rho_2\Theta}$$

$$X_2^+(x) = (x^2 - 1)^{-\frac{1}{2}} e^{i\rho_2\Theta}$$

$$L_1(x) = 2f_0(x + 2i\rho_2) \quad \text{and} \quad L_2(x) = -2f_0(x - 2i\rho_2)$$

so that

$$\sigma_{yy}(x,0+) = p_0\{-1 + (x^2-1)^{-\frac{1}{2}} [x \cos \rho_2\Theta + \rho_2 \sin \rho_2\Theta]\} \tag{4.37}$$

and

$$\sigma_{xy}(x,0+) = \frac{p_0}{\sqrt{(x^2-1)}} \{x \sin \rho_2\Theta - \rho_2 \cos \rho_2\Theta\} \tag{4.38}$$

The components of the displacement vector can be evaluated from the relations

$$u_y(x,0+) = \frac{-1+\kappa_2}{\kappa_1\alpha} \mathfrak{F}_c[\xi^{-1}\varphi(\xi); x], \quad 0 \le x < 1$$

$$u_y(x,0-) = \frac{1+\kappa_2}{\kappa_1\alpha\Gamma} \mathfrak{F}_c[\xi^{-1}\varphi(\xi); x], \quad 0 \le x < 1 .$$

If we use the relation $\mathfrak{F}_c[\xi^{-1}\varphi(\xi); x] = \int_x^1 \mathfrak{F}_s[\varphi(\xi); u]\,du$ (valid since $u_y(1,0+) = u_y(1,0-) = 0$), then

$$u_y(x,0+) = -\frac{1+\kappa_1}{\kappa_1\alpha} \int_x^1 \left[\int_0^u \frac{p_1(t)\,dt}{\sqrt{(x^2-t^2)}}\right] du$$

$$= -\frac{1+\kappa_1}{\kappa_1\alpha} \int_x^1 \left\{\frac{\Phi_1^+(u) + \Phi_1^-(u)}{2}\right\} du$$

$$= -\frac{1+\kappa_1}{\kappa_1\alpha} \int_x^1 \{[\Delta^+(u) + \Delta^-(u)] + [\Sigma^+(u) + \Sigma^-(u)]\}\, du$$

and a similar expression is obtained for $u_y(x,0-)$. Thus, we see that all quantities of physical interest may be obtained in terms of the boundary values of the two sectionally analytic functions $\Delta(z)$ and $\Sigma(z)$.

5. CONCLUSION

The results contained in Section 2 could easily be generalized to a system of equations of the form

$$\alpha(x)\int_a^x \frac{\varphi_1(t)\,dt}{\sqrt{(x^2 - t^2)}} + \beta(x)\int_x^b \frac{\varphi_2(t)\,dt}{\sqrt{(t^2 - x^2)}} = f_1(x), \quad x \in (a,b) \tag{5.1}$$

$$\gamma(x)\int_x^b \frac{\varphi_1(t)\,dt}{\sqrt{(t^2 - x^2)}} + \delta(x)\int_a^x \frac{\varphi_1(t)\,dt}{\sqrt{(t^2 - x^2)}} = f_2(x), \quad x \in (a,b)\ . \tag{5.2}$$

In this case, we introduce the function,

$$\Phi_j(z) = \int_a^b \frac{\varphi_j(t)\,dt}{\sqrt{(z^2 - t^2)}} \tag{5.3}$$

which is analytic everywhere except on the cut $a \leq |x| \leq b$. By $\sqrt{(z^2 - t^2)}$, we mean

$$\sqrt{(z^2 - t^2)} = \xi\, e^{i\eta}, \quad (-\pi \leq \eta \leq \pi)$$

where

$$\xi^2 \cos 2\eta = x^2 - y^2 - t^2, \quad \xi^2 \sin(2\eta) = 2xy\ .$$

The remaining analysis is completely analogous to that considered in Section 3 and will not be redone. Certain obvious modifications are necessary.

We might also note that our methods also apply to general Abel type integral equations of the form

$$\alpha(x)\int_a^x \frac{p_1(t)\,dt}{(x - t)^{\mu}} + \beta(x)\int_x^b \frac{p_2(t)\,dt}{(t - x)^{\mu}} = f_1(x), \quad x \in (a,b) \tag{5.4}$$

$$\gamma(x)\int_x^b \frac{p_1(t)\,dt}{(x - t)^{\mu}} + \delta(x)\int_a^x \frac{p_2(t)\,dt}{(t - x)^{\mu}} = f_2(x), \quad x \in (a,b) \tag{5.5}$$

with $0 < \mu < 1$. It is assumed that $\alpha(x)\,\delta(x) - \beta(x)\,\gamma(x) \neq 0$ on (a,b) and $\alpha, \beta, \gamma, \delta$ have derivatives satisfying Hölder conditions on (a,b). The case of one equation is discussed in a paper by Sakalyuk [8]. If we seek solutions $p_i(t)$ in the class

$$p_i(t) = \frac{p_i^*(t)}{(t-a)^{1-\mu-\epsilon_1}(b-t)^{1-\mu-\epsilon_2}}$$

where $\epsilon_1 > 0$ and $\epsilon_2 > 0$ and p_i^* satisfies Hölder's conditions on (a,b), then the proper sectionally analytic function Φ is

$$\Phi(z) = [(z-a)(b-z)]^{\frac{1}{2}\mu-\frac{1}{2}} \int_a^b \frac{p_j(t)\,dt}{(t-z)^{\mu}} \tag{5.6}$$

where by the expression $[(z-a)(b-z)]^{(\mu-1)/2}(t-z)^{-\mu}$ we mean some branch of this function in the plane cut along $a \le x \le b$.

The details of the analysis will appear as a note elsewhere. We only remark that the techniques are similar to those applied here.

(Research supported in part by National Science Foundation Grants GP 33225X and MPS 75-03894.)

References

1 T. Carleman, Ark. Math. och. Physik 16 No.26 (1932).

2 A.H. England, Complex Variable Methods in Elasticity. John Wiley and Sons, London (1971).

3 F. Erdogan, J.Appl. Mech. 32, 829 (1965).

4 F.D. Ghakov, Boundary Value Problems. Pergamon Press, Oxford (1966). Translation of Krayevye Zadachi, Fizmatgiz, Moscow (1963).

5 A.E. Green and W. Zerna, Theoretical Elasticity. Oxford University Press, Oxford (1954).

6 M. Lowengrub and I.N. Sneddon, Int. J. Eng. Sci. 11 (1973) 1025-1034 .

7 N.I. Muskhelishvili, Some Basic Problems in the Mathematical Theory of Elasticity. Noordhoff (1965).

8 K.D. Sakalyuk, Dokl. Akad. Nauk. SSSR 131 No. 4 (1960) 748-751.

9 I.N. Sneddon and M. Lowengrub, Crack Problems in the Classical Theory of Elasticity. John Wiley and Sons, New York, (1969).

10 D.S. Jones, J. of Math. and Phys. 43, (1964) 263.

M. Lowengrub
Department of Mathematics
Indiana University
Bloomington
Indiana 47401
USA

P RAMANKUTTY

19 The generalized axially symmetric Helmholtz equation

INTRODUCTION

In traditional treatments of Linear Elliptic Partial Differential Equations in bounded and unbounded domains of $\mathbb{R}^n$ the coefficients in the differential equation are usually assumed to be continuous or at least bounded in the closure of the domain. The literature that deals with equations whose coefficients are unbounded on some subset of the domain or of its boundary is somewhat limited in extent and relatively new. A particular equation that has very often been treated in recent times due to its frequent occurrence in mathematical physics is the 'Generalized Axially Symmetric Helmholtz Equation'

$$\Delta u + \frac{s}{\rho}\frac{\partial u}{\partial \rho} + ku = 0 \quad \text{where} \quad \Delta \equiv \sum_{i=1} \frac{\partial^2}{\partial x_i^2} + \frac{\partial^2}{\partial \rho^2}$$

is the Laplacian operator on $\mathbb{R}^{n+1}$ of which the generic point is denoted by $(x_1,\ldots,x_n,\rho)$ and s and k are real numbers. Numerous references to this can be found in [4], [8], [9], [18]. The analytic theory of equations of this type has been extensively treated in [8]. Explicit integral representations for the fundamental solutions of the above equation and related results are to be found in [14], [15].

In this paper a uniqueness theorem modelled on Theorem 3.1 in [2] is obtained for the equation $\Delta u + \frac{s}{\rho}\frac{\partial u}{\partial \rho} + \lambda^2 u = f$ with the restriction that $s > -1$. Also, employing the method of descent in the manner of Diaz and Ludford [6], the 'Identification Principles' of Weinstein [17] and Huber [12, Theorem 2] are extended to the present equation with the value of the parameter s unrestricted, extending previous results of Henrici [10], Weinstein [18] (see also Davis [5, Cor. 2.1]).

1. THE UNIQUENESS THEOREM

1.1 Notation: We will make use of the 'hyperzonal co-ordinates' in a Euclidean space introduced in [1] and further elucidated and used in [4]. Let the open ball

$B_R(x) \subset \mathbb{R}^n$ be defined by

$$B_R(x) = \{y \in \mathbb{R}^n : \|x-y\| < R\}$$

and let

$$\phi : \mathbb{R}^n \times \mathbb{R} \rightarrow [0,\infty) \times \overline{B_1(0)}$$

be defined by $\phi((x,\rho)) = (r,\xi)$, where $r = \sqrt{(\|x\|^2 + \rho^2)}$ and

$$\xi = \begin{cases} \dfrac{x}{r} & \text{if } r > 0 \\ 0 & \text{if } r = 0 . \end{cases}$$

Then r and $\{\xi_i\}_{i=1}^n$ are called the hyperzonal coordinates of the point $(x,\rho) \in \mathbb{R}^n \times \mathbb{R}$. It is obvious that the restriction of ϕ to $\mathbb{R}^n \times [0,\infty)$ is a bijection onto $[0,\infty) \times \overline{B_1(0)}$.

Throughout this paper D will denote a subset of $\mathbb{R}^n \times \mathbb{R}$ of which the generic point is denoted by (x,ρ), $\alpha = \inf_{(x,\rho)\in D} \sqrt{(\|x\|^2+\rho^2)}$, $\beta = \sup_{(x,\rho)\in D} \sqrt{(\|x\|^2+\rho^2)}$ and for $\lambda \in \mathbb{R}$, $s \in \mathbb{R}$, $L_{\lambda,s}$ will denote the partial differential operator defined by

$$L_{\lambda,s}[u] = \sum_{i=1}^{n} \frac{\partial^2 u}{\partial x_i^2} + \frac{\partial^2 u}{\partial \rho^2} + \frac{s}{\rho}\frac{\partial u}{\partial \rho} + \lambda^2 u .$$

Also m will denote a 'multi-index' $(m_1,m_2,\ldots,m_n) \in \mathbb{Z}_+^n$ and $|m| = m_1 + m_2 + \ldots + m_n$, $\mathbb{Z}_+$ being the set of all non-negative integers.

<u>1.2 Definition</u>: A function $u : D \rightarrow \mathbb{R}$ will be called <u>regular</u> if and only if there exists a self-conjugate $\mathbb{C}^n$-neighbourhood N of $\overline{B_1(0)}$ and an $\mathbb{R}$-neighbourhood of M of $[\alpha,\beta]$ and a (complex valued) function $\tilde{u} \in C^2(M\times N)$ such that $\tilde{u}/\phi(D) = u \circ \phi^{-1}$ and for each $r \in [\alpha,\beta]$, $\tilde{u}((r,\xi))$ is analytic on N. (If $\beta = \infty$, then $[\alpha,\beta]$ is to be understood as $[\alpha,\infty)$.)

<u>1.3 Remark</u>: It should be noted that the condition $\tilde{u}/\phi(D) = u \circ \phi^{-1}$ is only another form of the statement that if (x,ρ) and (r,ξ) are corresponding points in D and $[\alpha,\beta] \times \overline{B_1(0)}$, then $u((x,\rho)) = \tilde{u}((r,\xi))$. This implies that if $(x,\rho) \in D$ and $(x,-\rho) \in D$, then $u((x,\rho)) = u((x,-\rho))$ for a regular u and in particular that if D is symmetric about $\rho = 0$, then

$u((x,\rho))$ is even in ρ. If D is symmetric about $\rho = 0$ and $u((x,\rho))$ is analytic in D and even in ρ, then it is clear that u is also regular and indeed we then also have $\tilde{u}((r,\xi))$ analytic in r as well.

In the subset D_+ of D where $\rho > 0$ and also in the subset D_- where $\rho < 0$, the operator $L_{\lambda,s}$ is elliptic with analytic coefficients and hence every solution $u \in C^2(D)$ of $L_{\lambda,s}[u] = 0$ is analytic in these subsets [11]. Also from the equation it is clear that if $s \neq 0$, and $u \in C^2(D)$ is such that $L_{\lambda,s}[u] = 0$ in D_+ and D_-, then $\frac{\partial u}{\partial \rho} = 0$ on the subset D_0 of D where $\rho = 0$. In the case "$s > -1$, $s \neq 0$", it is known [4] that there exist solutions which are analytic also on some portion of D_0.

In view of these, in what follows, hypotheses of analyticity and/or regularity on solutions of the equation $L_{\lambda,s}[u] = 0$ are imposed as found necessary.

1.4 Theorem: Let D be unbounded $s \in (-1,\infty)$ and $u : D \to \mathbb{R}$ a regular solution of $L_{\lambda,s}[u] = 0$ satisfying

(i) $r^{\frac{n+s}{2}} (1 - \|\xi\|^2)^{\gamma\frac{s+1}{2}} \tilde{u}((r,\xi))$ is bounded on $[\alpha,\infty) \times B_1(0)$ for some $\gamma \in [0,1)$

(ii) $\lim_{r\to\infty} r^{\frac{n+s}{2}} \tilde{u}((r,\xi)) = 0$ for each $\xi \in B_1(0)$.

Then u is identically zero.

Proof:

Case I. $s \neq 0$: Since u is regular, we have by [4. Theorem 3],

$$\tilde{u}((r,\xi)) = \sum_{\mu=0}^{\infty} \sum_{|m|=\mu} a_m(r) V_m^s(\xi)$$

where

$$a_m(r) = (h_m^s)^{-1} \int_{B_1(0)} (1 - \|\xi\|^2)^{\frac{s-1}{2}} \tilde{u}((r,\xi)) U_m^s(\xi)\, d\xi \tag{1}$$

$$h_m^s = \int_{B_1(0)} (1 - \|\xi\|^2)^{\frac{s-1}{2}} U_m^s(\xi) V_m^s(\xi)\, d\xi\ , \tag{2}$$

V_m^s being the Appell polynomials in n variables and U_m^s the polynomials related to them, which are defined and studied in [1], [4], [7]. When converted to hyperzonal coordinates $L_{\lambda,s}[u]$ becomes ([8], p.229)

$$\tilde{L}_{\lambda,s}[\tilde{u}] = \frac{\partial^2\tilde{u}}{\partial r^2} + \frac{n+s}{r}\frac{\partial\tilde{u}}{\partial r} + \frac{n(s-1)}{r^2}\tilde{u} + \lambda^2\tilde{u} + \frac{1}{r^2}M[\tilde{u}] \tag{3}$$

where

$$M[\tilde{u}] = \sum_{i=1}^{n} \frac{\partial}{\partial\xi_i}\left(\frac{\partial\tilde{u}}{\partial\xi_i} - \xi_i \sum_{j=1}^{n} \xi_j \frac{\partial\tilde{u}}{\partial\xi_j} - (s-1)\xi_i\tilde{u}\right).$$

Using (3) and the fact that $\tilde{u}(r,\xi)$ is of class C^2 in its variables, it can, as in [3], be deduced from (1) that

$$r^{\delta+\frac{1}{2}} a_m(r) = \sqrt{(r)}\, Z_{|m|+\delta}(\lambda r) \tag{4}$$

where $2\delta = n + s - 1$ and

$$Z_{|m|+\delta}(\lambda r) = A_m H^1_{|m|+\delta}(\lambda r) + B_m H^2_{|m|+\delta}(\lambda r),$$

A_m and B_m being some constants (arbitrary constants) depending on λ, n, s, m, and H^1_μ and H^2_μ being Hankel functions of the first and second kind of order μ.

Since $r^{\frac{n+s}{2}} |\tilde{u}((r,\xi))| (1-\|\xi\|^2)^{\gamma\frac{s+1}{2}}$ is bounded on $[\alpha,\infty) \times B_1(0)$, say by K, the family of functions $\left\{ r^{\frac{n+s}{2}} (1-\|\xi\|^2)^{\frac{s-1}{2}} \tilde{u}((r,\xi)) U_m^s(\xi) \right\}_{r\in[\alpha,\infty)}$ defined on $B_1(0)$ is dominated by the positive function $CK(1-\|\xi\|^2)^{\frac{s-1}{2} - \gamma\frac{s+1}{2}}$ where $C = \max_{\xi\in\overline{B_1(0)}} |U_m^s(\xi)|$. Also since $s > -1$ and $\gamma < 1$, we have $\frac{s-1}{2} - \gamma\frac{s+1}{2} > -1$ whence $\int_{B_1(0)} (1-\|\xi\|^2)^{\frac{s-1}{2} - \gamma\frac{s+1}{2}} d\xi < \infty$. Hence by Lebesgue's dominated convergence theorem, it follows that

$$\lim_{r\to\infty} r^{\frac{n+s}{2}} a_m(r) = \lim_{r\to\infty} r^{\frac{n+s}{2}} (h_m^s)^{-1} \int_{B_1(0)} (1-\|\xi\|^2)^{\frac{s-1}{2}} \tilde{u}((r,\xi))\, U_m^s(\xi)\, d\xi =$$

$$= (h_m^s)^{-1} \int_{B_1(0)} (1-\|\xi\|^2)^{\frac{s-1}{2}} U_m^s(\xi) \lim_{r\to\infty} r^{\frac{n+s}{2}} \tilde{u}((r,\xi))\, d\xi$$

$= 0$ by hypothesis (ii) of the theorem.

By (4), this gives: $\lim_{r\to\infty} \sqrt{(r)}\, Z_{|m|+\delta}(\lambda r) = 0$ and by the asymptotic behaviour ([7], p. 85) of the Hankel functions H_μ^1 and H_μ^2, this implies that $A_m = B_m = 0$. Hence $a_m(r) = 0$ for each m so that $\tilde{u}((r,\xi)) = 0$ in D.

Case II. $s = 0$: If $d\sigma$ denotes the 'surface element' on

$$\Sigma_r(0) \overset{\text{def}}{=} \{(x,\rho) \in \mathbb{R}^n \times \mathbb{R} : \|x\|^2 + \rho^2 = r^2\},$$

we have

$$\int_{\Sigma_r(0)} |u((x,\rho))|^2 d\sigma = \int_{B_1(0)} |\tilde{u}((r,\xi))|^2\, Cr^n d\xi$$

where C is a constant depending only on n.

By hypothesis (i), there is a $K > 0$ and a $\gamma < 1$ such that for every $(r,\xi) \in [\alpha,\infty) \times B_1(0)$, $r^n(1-\|\xi\|^2)^\gamma |\tilde{u}((r,\xi))|^2 < K$. Therefore the family of functions $\{r^n|\tilde{u}((r,\xi))|^2\}_{r\in[\alpha,\infty)}$, defined on $B_1(0)$, is dominated by the positive function $K(1-\|\xi\|^2)^{-\gamma}$; also $\int_{B_1(0)} (1-\|\xi\|^2)^{-\gamma} d\xi < \infty$ since $\gamma < 1$.

Hence by Lebesgue's dominated convergence theorem,

$$\lim_{r\to\infty} \int_{\Sigma_r(0)} |u((x,\rho))|^2 d\sigma = \lim_{r\to\infty} \int_{B_1(0)} Cr^n|\tilde{u}((r,\xi))|^2 d\xi =$$

$$= C \int_{B_1(0)} \lim_{r\to\infty} r^n|\tilde{u}((r,\xi))|^2 d\xi =$$

$$= 0 \text{ by hypothesis (ii)}.$$

But, then by Rellich's well-known growth estimate [13] it follows that $u = 0$ in D.

This completes the proof of the theorem.

In the case $s = -1$, The Appell-series-expansion is not available. However, an analogous result can still be proved with a little extra hypothesis as in Colton [2].

1.5 Theorem: Let D be symmetric about $\rho = 0$, D_1 be its projection on $\rho = 0$ and suppose that $D_1 \subset D$ and D_1 is unbounded. If $u : D \to \mathbb{R}$ is an analytic solution of $L_{\lambda,-1}[u] = 0$ which is even in ρ and satisfies

(i) $r^{\frac{n-1}{2}} |\tilde{u}((r,\xi))|$ is bounded on $[\alpha,\infty) \times B_1(0)$

(ii) $\lim_{r\to\infty} r^{\frac{n-1}{2}} \tilde{u}((r,\xi)) = 0$ for each $\xi \in \overline{B_1(0)}$

then u is identically zero.

Proof: Since u is analytic and even in ρ, $u((x,\rho))$ is analytic in x and ρ^2 and hence

$$u((x,\rho)) = v(x) + \rho^2 w((x,\rho))$$

where $v(x)$ is analytic in x defined on D_1 and $w((x,\rho))$ is an analytic function of x and ρ^2 defined on D. Hence for each $x \in D_1$,

$$\lim_{\rho\to 0}\left(\frac{1}{\rho}\frac{\partial u((x,\rho))}{\partial \rho}\right) = \lim_{\rho\to 0}\left(\frac{\partial^2}{\partial\rho^2} u((x,\rho))\right) = 2w((x,0)) .$$

Therefore, rewriting $L_{\lambda,-1}[u] = 0$ at a point $(x,\rho) \in D$ with $x \in D_1$ and taking limits as $\rho \to 0$, we have

$$\sum_{i=1}^{n} \frac{\partial^2 v(x)}{\partial x_i^2} + \lambda^2 v(x) = 0 . \tag{5}$$

Also, the hypothesis (ii) on u in the special case $\rho = 0$ (i.e., $\|\xi\| = 1$) gives

$$\lim_{\|x\|\to\infty} \|x\|^{\frac{n-1}{2}} v(x) = 0 \tag{6}$$

which, of course, also gives

$$\|x\|^{\frac{n-1}{2}} |v(x)| \text{ is bounded on } D_1 . \tag{7}$$

From (5), (6), (7) it follows by Theorem 1.4 that $v(x) = 0$ for each $x \in D_1$. Hence $u((x,\rho)) = \rho^2 w((x,\rho))$ for each $(x,\rho) \in D$. But this expression for u substituted into $L_{\lambda,-1}[u] = 0$ gives rise to $\rho^2 L_{\lambda,3}[w] = 0$ which, because of the continuity of $L_{\lambda,3}[w]$ over D yields:

$$L_{\lambda,3}[w] = 0 \quad \text{in} \quad D. \tag{8}$$

Since $\rho^2 = r^2(1-\|\xi\|^2)$ and $u((x,\rho)) = \rho^2 w((x,\rho))$ where $w((x,\rho))$ is analytic on D, from the hypotheses (i) and (ii) on u, it follows that

$$r^{\frac{n+3}{2}} |\tilde{w}((r,\xi))| \quad \text{is bounded on} \quad [\alpha,\infty) \times B_1(0) \tag{9}$$

and

$$\lim_{r\to\infty} r^{\frac{n+3}{2}} \tilde{w}((r,\xi)) = 0 \quad \text{for each} \quad \xi \in B_1(0). \tag{10}$$

(8), (9), (10) show that w is a regular solution of $L_{\lambda,3}[w] = 0$ satisfying hypotheses (i) and (ii) of Theorem 1.4 with $\gamma = \frac{1}{2}$. Hence $w((x,\rho)) = 0$ in D and the proof is complete.

Note: The hypothesis 'D_1 is unbounded' may be replaced by 'D is unbounded' provided that it is also assumed that $u = 0$ on ∂D. In this case it can be deduced that $v = 0$ on ∂D_1 which, because of (5), makes v identically zero.

2. THE IDENTIFICATION PRINCIPLE

In this section it is assumed that Int D is connected and non-empty.

With reference to solutions u of $L_{0,s}[u] = 0$ defined on D, the following results are known as Identification Principles.

Weinstein [17]: Let D be symmetric about $\rho = 0$, H a non-empty open subset of the hyperplane $\rho = 0$ contained in Int D, and $s > 0$. If $u(x,\rho)$ is even in ρ, analytic in D and satisfies $L_{0,s}[u] = 0$ in D, then $u = 0$ on H implies $u = 0$ in D (see [12], p.351).

Huber [12]: If $u \in C^2(D) \cap C(\bar{D})$ is a solution of $L_{0,s}[u] = 0$ for $s \geq 1$, where ∂D contains a non-empty open subset H of $\rho = 0$, then $u = 0$ on H implies $u = 0$ in D [Section 4, Theorem 2(a)].

For $\Re e\ s > 0$ Henrici [10] proved the Identification Principle for solutions $L_{\lambda,s}[u] = 0$ assuming analyticity, as in the Weinstein result quoted above. For $s \geq 1$ Weinstein [18], by means of the method of descent of Diaz and Ludford [6], relaxed the assumption of analyticity to that of continuity, thereby obtaining the exact analogue of the Huber result quoted above.

We now proceed to extend these results to all real s, also using the method of descent of Diaz and Ludford [6]. But first we need the following computational result which is essentially Theorem 1.3 in [3].

2.1 Lemma: Suppose that D does not contain the origin and let $u : D \to \mathbb{R}$ be a regular solution of $L_{\lambda,s}[u] = 0$, $m \in \mathbb{Z}_+^n$ and $v : D \to \mathbb{R}$ be defined by

$$\tilde{v}((r,\xi)) = \frac{1}{r^{|m|}} \frac{\partial^{|m|}\tilde{u}((r,\xi))}{\partial\xi_1^{m_1}\ldots\partial\xi_n^{m_n}} .$$

Then

(i) v is a regular solution of $L_{\lambda,\nu}[v] = 0$ where $\nu = s + 2|m|$

(ii) if D is symmetric about $\rho = 0$ and u is analytic in D, then V is also analytic.

Proof:

(i) The regularity of v is obvious from the fact that $\tilde{u}((r,\xi))$ is analytic in ξ and that D does not contain the origin.

By a straightforward calculation it can be verified that

$$\frac{1}{r}\frac{\partial}{\partial\xi_1}(\tilde{L}_{\lambda,s}[\tilde{u}]) \equiv \tilde{L}_{\lambda,s+2}[\tilde{w}] \quad \text{where} \quad \tilde{w}(r,\xi) = \frac{1}{r}\frac{\partial}{\partial\xi_1}\tilde{u}(r,\xi) .$$

Hence, if u is a regular solution of $\tilde{L}_{\lambda,s}[\tilde{u}] = 0$ in D and $w : D \to \mathbb{R}$ is defined by

$$\tilde{w}(r,\xi) = \frac{1}{r}\frac{\partial}{\partial\xi_1}\tilde{u}(r,\xi) ,$$

then w is a regular solution of $L_{\lambda,s+2}[w] = 0$ in D. Repeating this $m_1, m_2, \ldots, m_n$ times for the variables $\xi_1 \ldots \xi_n$, the result (i) follows.

(ii) If D is symmetric about $\rho = 0$ and u is analytic in D, then $u((x,\rho))$ is analytic in x and ρ^2 (see Remark 1.3).

Hence,

$$\frac{\partial \tilde{u}((r,\xi))}{\partial \xi_1} = \sum_{i=1}^{n} \frac{\partial u((x,\rho))}{\partial x_i} \frac{\partial x_i}{\partial \xi_1} + \frac{\partial u((x,\rho))}{\partial \rho^2} \frac{\partial \rho^2}{\partial \xi_1}$$

$$= \frac{\partial u((x,\rho))}{\partial x_1} r + \frac{\partial u((x,\rho))}{\partial \rho^2} r^2 (-2\xi_1) .$$

Therefore,

$$\tilde{w}((r,\xi)) = \frac{\partial u((x,\rho))}{\partial x_1} - 2r\xi_1 \frac{\partial u((x,\rho))}{\partial \rho^2}$$

so that

$$w((x,\rho)) = \frac{\partial u((x,\rho))}{\partial x_1} - 2x_1 \frac{\partial u((x,\rho))}{\partial \rho^2} .$$

This shows that w is analytic in D; the analyticity of v follows.

2.2 Theorem: Let D be symmetric about $\rho = 0$ and contain a non-empty open subset H of the hyperplane $\rho = 0$ in its interior and let $u : D \to \mathbb{R}$ be a regular analytic solution of $L_{\lambda,s}[u] = 0$. If there exists a non-negative integer $p > \frac{-s}{2}$ such that for each $m \in \mathbb{Z}^n_+$,

$$|m| \leq p \Rightarrow \frac{\partial^{|m|}\tilde{u}((r,\xi))}{\partial \xi_1^{m_1} \ldots \partial \xi_n^{m_n}} = 0 \quad \text{on} \quad \phi(H),$$

then $u \equiv 0$.

Proof: Since it is enough to prove that u vanishes at all points other than the origin, it will be assumed that D does not contain the origin. Choose $m \in \mathbb{Z}^n_+$ such that $|m| = p$ and define $v : D \to \mathbb{R}$ by

$$\tilde{v}((r,\xi)) = \frac{1}{r^{|m|}} \frac{\partial^{|m|}\tilde{u}((r,\xi))}{\partial \xi_1^{m_1} \ldots \partial \xi_n^{m_n}} .$$

Then by the foregoing lemma $L_{\lambda,\nu}[v] = 0$, where $\nu = s + 2p$. Let $\Delta = D \times \mathbb{R}$ and define $w : \Delta \to \mathbb{R}$ by the formula: $w((x,\rho,x_{n+1})) = e^{\lambda x_{n+1}} v((x,\rho))$.

Now it follows from the equation $L_{\lambda,\nu}[v] = 0$ that $M_\nu[w] = 0$ where M_ν is the partial differential operator defined by

$$M_\nu [w((x,\dots,x_n,\rho,x_{n+1}))] = \sum_{i=1}^{n+1} \frac{\partial^2 w}{\partial x_i^2} + \frac{\partial^2 w}{\partial \rho^2} + \frac{\nu}{\rho}\frac{\partial w}{\partial \rho} .$$

Also the hypothesis on u implies that $v((x,\rho)) = 0$ for each $(x,\rho) \in H$ and this in turn implies that $w((x,\rho,x_{n+1})) = 0$ on $H \times \mathbb{R} \subset \operatorname{Int}\Delta$. Hence by Weinstein's identification principle, it follows that $w \equiv 0$ in Δ. Hence $v \equiv 0$ in D whence

$$\frac{\partial^{|m|}\tilde{u}((r,\xi))}{\partial\xi_1^{m_1}\dots\partial\xi_n^{m_n}} = 0 \quad \text{in} \quad \phi(D) .$$

This being true for every $m \in \mathbb{Z}_+^n$ such that $|m| = p$, it follows that for each fixed r, $\tilde{u}((r,\xi))$ is a polynomial of degree at most $p-1$ in the variables $\xi_1 \dots \xi_n$. Hence

$$\tilde{u}((r,\xi)) = \sum_{|m|=0}^{p-1} a_m(r)\,\xi^m .$$

Now the hypothesis on u implies that

$$\begin{aligned}&\text{for each } m, \ a_m(r) = 0 \text{ at each } r \text{ such that}\\ &(r,\eta) \in \phi(H) \text{ for some } \eta \in \mathbb{R}^n\end{aligned} \tag{11}$$

Since H is an open subset of the hyperplane $\rho = 0$, we can choose $(a,0) \in H$ and $\delta \in (0,\infty)$ such that $\delta < \|a\|$ and $k \overset{\text{def}}{=} \{(x,0) \in \mathbb{R}^n \times \mathbb{R} : \|x-a\| < \delta\} \subset H$. Now let $N \subset \operatorname{Int} D$ be defined by $N = \{(x,\rho) \in \operatorname{Int} D : \|a\| - \delta < \sqrt{(\|x\|^2 + \rho^2)} < \|a\| + \delta\}$. Then N is an open subset of D containing K and obviously, for each $(r,\xi) \in \phi(N)$ there exists $\eta \in \mathbb{R}^n$ such that $(r,\eta) \in \phi(K) \subset \phi(H)$. Hence it follows from (11) that $a_m(r) = 0$ for each m and for each $(r,\xi) \in \phi(N)$. Therefore, $\tilde{u}((r,\xi)) = 0$ on $\phi(N)$ and consequently, $u((x,\rho)) = 0$ for each $(x,\rho) \in N$. Since N is a non-empty open subset of D, it follows by analytic continuation that $u((x,\rho)) = 0$ for each $(x,\rho) \in D$.

Analogous to the preceding theorem we have also

2.3 Theorem: Let D contain a non-empty open subset H of the hyperplane $\rho = 0$ on its boundary and $u : D \to \mathbb{R}$ be a regular solution of $L_{\lambda, s}[u] = 0$. If there exists a non-negative integer $p \geq \frac{1-s}{2}$ such that for each $m \in Z_+^n$,

$$|m| \leq p \Rightarrow \frac{\partial^{|m|}\tilde{u}((r,\xi))}{\partial\xi_1^{m_1} \ldots \partial\xi_n^{m_n}} = 0 \quad \text{on} \quad \phi(H),$$

then $u \equiv 0$.

(The Proof is the same as that of the preceding theorem if the identification principle due to Huber is invoked.)

Acknowledgement

I wish to thank Professor Robert P Gilbert for his help and guidance in the preparation of this paper.

References

1 P. Appell and J.K. de Fériet, Fonctions Hypergéométriques et Hypersphériques, Polynomes d'Hermite. Gauthier-Villars, Paris (1926).

2 D. Colton, Uniqueness theorems for axially symmetric partial differential equations. J. Math. Mech. 18 (1969) 921-930.

3 D. Colton, Decomposition theorem for the generalized metaharmonic equation in several independent variables. J. Aust. Math. Soc. 13 (1972) 35-46.

4 D. Colton and R.P. Gilbert, A contribution to the Vekua-Rellich theory of metaharmonic functions. Amer. J. Math. 92 (1970) 525-540.

5 J.E. Davis, A Uniqueness Theorem for the Generalized Axially Symmetric Helmholtz Equation. Master's Thesis, University of Delaware, (1970).

6 J.B. Diaz and G.S.S. Ludford, Reflection principles for linear elliptic second order partial differential equations with constant coefficients. Ann. Mat. Pura Appl. (4) 39 (1955) 87-95.

7 Erdélyi, Higher Transcendental Functions, Vol. II. Bateman Manuscript Project, California Institute of Technology, McGraw-Hill (1953).

8 R.P. Gilbert, Function Theoretic Methods in the Theory of Partial Differential Equations. Academic Press, New York (1969).

9 R.P. Gilbert, An investigation of the analytic properties of solutions to the generalized axially symmetric, reduced wave equation in $n+1$ variables, with an application to the theory of potential scattering. SIAM J. Appl. Math 16 (1968) 30-50.

10 P. Henrici, Zur Funktionentheorie der Wellengleichung. Comm. Math. Helv. 27 (1953) 235-293.

11 E. Hopf, Über den funktionalen, insbesondere den analytischen Charakter der Lösungen elliptischer Differentialgleichungen zweiter Ordnung. Math. Z. 34 (2) (1932) 194-233.

12 A. Huber, On the uniqueness of generalized axially symmetric potentials. Ann. of Math 60 (1954) 351-358.

13 F. Rellich, Über das asymptotische Verhalten der Lösungen von $\Delta u + \lambda u = 0$ in unendlichen Gebieten. Jber. Deutsch. Math. Verein. 53 (1943) 57-65.

14 R.J. Weinacht, Fundamental solutions for a class of singular equations. Contr. Diff. Eqns. 3 (1964) 43-55.

15 R.J. Weinacht, Some properties of generalized biaxially symmetric Helmholtz potentials. SIAM J. Math. Anal. 5 (1974) 147-152.

16 A. Weinstein, Generalized axially symmetric potential theory. Bull. Amer. Math. Soc. 59 (1953) 20-38.

17 A. Weinstein, Discontinuous integrals and generalized potential theory. Trans.Amer. Math. Soc. 63 (1948) 342-354.

18 A. Weinstein, Singular partial differential equations and their applications. Proceedings of the Symposium on Fluid Dynamics and Applied Mathematics, University of Maryland. Gordon and Breach, New York (1961) 29-49.

P. Ramankutty
Department of Mathematics
University of Auckland
Auckland
New Zealand

and

Department of Mathematics
University of Utah
Salt Lake City
Utah 84112
USA